1135

Paul L. Meyer, Department of Mathematics, Washington State University

INTRODUCTORY PROBABILITY AND STATISTICAL APPLICATIONS

SECOND EDITION

Addison-Wesley Publishing Company

Reading, Massachusetts · Menlo Park, California · London · Don Mills, Ontario

This book is in the
ADDISON-WESLEY SERIES IN STATISTICS

Second printing, May 1972

Copyright © 1970 by Addison-Wesley Publishing Company, Inc. Philippines copyright 1970
by Addison-Wesley Publishing Company, Inc.

ISBN 0-201-04710-1
BCDEFGHIJK-HA-79876543

To Alan and David

Preface to the First Edition

This text is intended for a one-semester or two-quarter course in introductory probability theory and some of its applications. The prerequisite is one year of differential and integral calculus. No previous knowledge of probability or statistics is presumed. At Washington State University, the course for which this text was developed has been taught for a number of years, chiefly to students majoring in engineering or in the natural sciences. Most of these students can devote only one semester to the study of this subject area. However, since these students are familiar with the calculus, they may begin the study of this subject beyond the strictly elementary level.

Many mathematical topics may be introduced at varying stages of difficulty, and this is certainly true of probability. In this text, an attempt is made to take advantage of the reader's mathematical background without exceeding it. Precise mathematical language is used but care is taken not to become excessively immersed in unnecessary mathematical details. This is most certainly not a "cook book." Although a number of concepts are introduced and discussed in an informal manner, definitions and theorems are carefully stated. If a detailed proof of a theorem is not feasible or desirable at least an outline of the important ideas is provided. A distinctive feature of this text is the "Notes" following most of the theorems and definitions. In these Notes the particular result or concept being presented is discussed from an intuitive point of view.

Because of the self-imposed restriction of writing a relatively brief text on a very extensive subject area, a number of choices had to be made relating to the inclusion or exclusion of certain topics. There seems to be no obvious way of resolving this problem. I certainly do not claim that for some of the topics which are excluded a place might not have been found. Nor do I claim that none of the material could have been omitted. However, for the most part, the emphasis has been on fundamental notions, presented in considerable detail. Only Chapter 11 on reliability could be considered as a "luxury item." But even here I feel that the notions associated with reliability problems are of basic interest to many persons. In addition, reliability concepts represent an excellent vehicle for illustrating many of the ideas introduced earlier in the text.

Even though the coverage is limited by the available time, a fairly wide selection of topics has been achieved. In glancing through the Table of Contents it is evident that about three-quarters of the text deals with probabilistic topics while the last quarter is devoted to a discussion of statistical inference. Although there is nothing magic about this particular division of emphasis between probability and statistics, I do feel that a sound knowledge of the basic principles of probability is imperative for a proper understanding of statistical methods. Ideally, a course in probability should be followed by one in statistical theory and methodology. However, as I indicated earlier, most of the students who take this course do not have time for a two-semester exposure to this subject area and hence I felt compelled to discuss at least some of the more important aspects of the general area of statistical inference.

The potential success of a particular presentation of subject matter should not be judged only in terms of specific ideas learned and specific techniques acquired. The final judgment must also take into account how well the student is prepared to continue his study of the subject either through self-study or through additional formal course work. If this criterion is thought to be important, then it becomes clear that basic concepts and fundamental techniques should be emphasized while highly specialized methods and topics should be relegated to a secondary role. This, too, became an important factor in deciding which topics to include.

The importance of probability theory is difficult to overstate. The appropriate mathematical model for the study of a large number of observational phenomena is a probabilistic one rather than a deterministic one. In addition, the entire subject of statistical inference is based on probabilistic considerations. Statistical techniques are among the most important tools of scientists and engineers. In order to use these techniques intelligently a thorough understanding of probabilistic concepts is required.

It is hoped that in addition to the many specific methods and concepts with which the reader becomes familiar, he also develops a certain point of view: to think probabilistically, replacing questions such as "How long will this component function?" by "How probable is it that this component will function more than 100 hours?" In many situations the second question may be not only the more pertinent one but in fact the only meaningful one to ask.

Traditionally, many of the important concepts of probability are illustrated with the aid of various "games of chance": tossing coins or dice, drawing cards from a deck, spinning a roulette wheel, etc. Although I have not completely avoided referring to such games since they do serve well to illustrate basic notions, an attempt has been made to bring the student into contact with more pertinent illustrations of the applications of probability: the emission of α-particles from a radioactive source, lot sampling, the life length of electronic devices, and the associated problems of component and system reliability, etc.

I am reluctant to mention a most obvious feature of any text in mathematics: the problems. And yet, it might be worthwhile to point out that the working of the problems must be considered as an integral part of the course. Only by becoming personally involved in the setting up and solving of the exercises can the student really develop an understanding and appreciation of the ideas and a familiarity with the pertinent techniques. Hence over 330 problems have been included in the text, over half of which are provided with answers at the end of the book. In addition to the problems for the reader, there are many worked-out examples scattered throughout the text.

This book has been written in a fairly consecutive manner: the understanding of most chapters requires familiarity with the previous ones. However, it is possible to treat Chapters 10 and 11 somewhat lightly, particularly if one is interested in devoting more time to the statistical applications which are discussed in Chapters 13 through 15.

As must be true of anyone writing a text, the debts I owe are to many: To my colleagues for many stimulating and helpful conversations, to my own teachers for the knowledge of and interest in this subject, to the reviewers of early versions of the manuscript for many helpful suggestions and criticisms, to Addison-Wesley Publishing Company for its great help and cooperation from the early stages of this project to the very end, to Miss Carol Sloan for being a most efficient and alert typist, to D. Van Nostrand, Inc., The Free Press, Inc., and Macmillan Publishing Company for their permission to reprint Tables 3, 6, and 1, respectively; to McGraw-Hill Book Co., Inc., Oxford University Press, Inc., Pergamon Press, Ltd., and Prentice-Hall, Inc., for their permission to quote certain examples in the text, and finally to my wife not only for bearing up under the strain but also for "leaving me" and taking our two children with her to visit grandparents for two crucial summer months during which I was able to convert our home into a rather messy but quiet workshop from which emerged, miraculously, the final, final version of this book.

Pullman, Washington
April, 1965

PAUL L. MEYER

Preface to the Second Edition

In view of the considerable number of favorable comments I have received during the past years from both students and instructors who have used the first edition of this book, relatively few changes have been made. During my own repeated use of the book, I found its basic organization of the material and the general level of presentation (e.g. the mixture of rigorous mathematical arguments with more informal expositions and examples) to be just about appropriate for the type of student taking the course.

However, a number of changes and additions were made. First of all, an attempt was made to eliminate the various misprints and errors which were in the first edition. The author is extremely grateful to the many readers who not only discovered some of these but were sufficiently interested to point them out to me.

Secondly, an attempt was made to shed more light on the relationship between various probability distributions so that the student may get a greater understanding of how various probabilistic models may be used to approximate one another.

Finally, some new problems have been added to the already lengthy list included in the first edition.

The author wishes to thank, once again, the Addison-Wesley Publishing Company for its cooperation in all aspects leading to this new edition.

Pullman, Washington   P. L. M.
December, 1969

Contents

Chapter 15   Testing hypotheses

1

Introduction to Probability

1.1 Mathematical Models

In this chapter we shall discuss the type of phenomenon with which we shall be concerned throughout this book. In addition, we shall formulate a mathematical model which will serve us to investigate, quite precisely, this phenomenon.

At the outset it is very important to distinguish between the observable phenomenon itself and the mathematical model for this phenomenon. We have, of course, no influence over what we observe. However, in choosing a model we can use our critical judgment. This has been particularly well expressed by Professor J. Neyman, who wrote:*

> "Whenever we use mathematics in order to study some observational phenomena we must essentially begin by building a mathematical model (deterministic or probabilistic) for these phenomena. Of necessity, the model must simplify matters and certain details must be ignored. The success of the model depends on whether or not the details ignored are really unimportant in the development of the phenomena studied. The solution of the mathematical problem may be correct and yet be in considerable disagreement with the observed data simply because the underlying assumptions made are not warranted. It is usually quite difficult to state with certainty, whether or not a given mathematical model is adequate *before* some observational data are obtained. In order to check the validity of a model, we must *deduce* a number of consequences of our model and then compare these *predicted* results with observations."

We shall keep the above ideas in mind while we consider some observational phenomena and models appropriate for their description. Let us first consider what might suitably be called a *deterministic model*. By this we shall mean a model which stipulates that the conditions under which an experiment is performed *determine* the outcome of the experiment. For example, if we insert a battery into a simple circuit, the mathematical model which would presumably describe the observable flow of current would be $I = E/R$, that is, Ohm's law. The model predicts the value of I as soon as E and R are given. Saying it dif-

---

* *University of California Publications in Statistics*, Vol. I, University of California Press, 1954.

1

ferently, if the above experiment were repeated a number of times, each time using the same circuit (that is, keeping E and R fixed), we would presumably expect to observe the same value for I. Any deviations that might occur would be so small that for most purposes the above description (that is, model) would suffice. The point is that the particular battery, wire, and ammeter used to generate and to observe the current, and our ability to use the measuring instrument, determine the outcome on each repetition. (There are certain factors which may well be different from repetition to repetition that will, however, not affect the outcome in a noticeable way. For instance, the temperature and humidity in the laboratory, or the height of the person reading the ammeter can reasonably be assumed to have no influence on the outcome.)

There are many examples of "experiments" in nature for which deterministic models are appropriate. For example, the gravitational laws describe quite precisely what happens to a falling body under certain conditions. Kepler's laws give us the behavior of the planets. In each situation, the model stipulates that the conditions under which certain phenomena take place determine the value of certain observable variables: the *magnitude* of the velocity, the *area* swept out during a certain time period, etc. These numbers appear in many of the formulas with which we are familiar. For example, we know that under certain conditions the distance traveled (vertically, above the ground) by an object is given by $s = -16t^2 + v_0 t$, where v_0 is the initial velocity and t is the time traveled. The point on which we wish to focus our attention is not the particular form of the above equation (that is, quadratic) but rather on the fact that there is a definite relationship between t and s, determining uniquely the quantity on the left-hand side of the equation if those on the right-hand side are given.

For a large number of situations the deterministic mathematical model described above suffices. However, there are also many phenomena which require a different mathematical model for their investigation. These are what we shall call *nondeterministic* or *probabilistic* models. (Another quite commonly used term is *stochastic* model.) Later in this chapter we shall consider quite precisely how such probabilistic models may be described. For the moment let us consider a few examples.

Suppose that we have a piece of radioactive material which is emitting α-particles. With the aid of a counting device we may be able to record the number of such particles emitted during a specified time interval. It is clear that we cannot predict precisely the number of particles emitted, even if we knew the exact shape, dimension, chemical composition, and mass of the object under consideration. Thus there seems to be no reasonable deterministic model yielding the number of particles emitted, say n, as a function of various pertinent characteristics of the source material. We must consider, instead, a probabilistic model.

For another illustration consider the following meteorological situation. We wish to determine how much precipitation will fall as a result of a particular storm system passing through a specified locality. Instruments are available with which to record the precipitation that occurs. Meteorological observations may give us considerable information concerning the approaching storm system: barometric pressure at various points, changes in pressure, wind velocity, origin and direction of the storm, and various pertinent high-altitude readings. But this information, valuable as it may be for predicting the general nature of the precipitation (light, medium, or heavy, say), simply does not make it possible to state very accurately *how much* precipitation will fall. Again we are dealing with a phenomenon which does not lend itself to a deterministic approach. A probabilistic model describes the situation more accurately.

In principle we might be able to state how much rain fell, if the theory had been worked out (which it has not). Hence we use a probabilistic model. In the example dealing with radioactive disintegration, we must use a probabilistic model *even in principle.*

At the risk of getting ahead of ourselves by discussing a concept which will be defined subsequently, let us simply state that in a deterministic model it is supposed that the actual outcome (whether numerical or otherwise) is determined from the conditions under which the experiment or procedure is carried out. In a non-deterministic model, however, the conditions of experimentation determine only the probabilistic behavior (more specifically, the probabilistic law) of the observable outcome.

Saying it differently, in a deterministic model we use "physical considerations" to predict the outcome, while in a probabilistic model we use the same kind of considerations to specify a probability distribution.

1.2 Introduction to Sets

In order to discuss the basic concepts of the probabilistic model which we wish to develop, it will be very convenient to have available some ideas and concepts of the mathematical theory of sets. This subject is a very extensive one, and much has been written about it. However, we shall need only a few basic notions.

A *set* is a collection of objects. Sets are usually designated by capital letters A, B, etc. In describing which objects are contained in the set A, three methods are available.

(a) We may list the members of A. For example, $A = \{1, 2, 3, 4\}$ describes the set consisting of the positive integers 1, 2, 3, and 4.

(b) We may describe the set A in words. For example, we might say that A consists of all real numbers between 0 and 1, inclusive.

(c) To describe the above set we can simply write $A = \{x \mid 0 \leq x \leq 1\}$; that is, A is the set of all x's, where x is a real number between 0 and 1, inclusive.

The individual objects making up the collection of the set A are called *members* or *elements* of A. When "a" is a member of A we write $a \in A$ and when "a" is not a member of A we write $a \notin A$.

There are two special sets which are often of interest. In most problems we are concerned with the study of a definite set of objects, and no others. For example, we may be concerned with all the real numbers, all items coming off a production line during a 24-hour period, etc. We define the *universal set* as the set of all objects under consideration. This set is usually designated by U.

Another set which must be singled out may rise as follows. Suppose that the set A is described as the set of all *real* numbers x satisfying the equation $x^2 + 1 = 0$. Of course, we know that there are no such numbers. That is, the set A contains no members at all! This situation occurs sufficiently often to warrant the introduction of a special name for such a set. Hence we define the *empty* or *null* set to be the set containing no members. We usually designate this set by $\emptyset$.

It may happen that when two sets A and B are considered, being a member of A implies being a member of B. In that case we say that A is a *subset* of B and we write $A \subset B$. A similar interpretation is given to $B \subset A$. And we say that two sets are the same, $A = B$, if and only if $A \subset B$ and $B \subset A$. That is, two sets are *equal* if and only if they contain the same members.

The following two properties of the empty set and the universal set are immediate.

(a) For every set A, we have $\emptyset \subset A$.

(b) Once the universal set has been agreed upon, then for every set A considered in the context of U, we have $A \subset U$.

EXAMPLE 1.1.   Suppose that U=all real numbers, $A = \{x \mid x^2 + 2x - 3 = 0\}$, $B = \{x \mid (x - 2)(x^2 + 2x - 3) = 0\}$, and $C = \{x \mid x = -3, 1, 2\}$. Then $A \subset B$, and $B = C$.

Next we consider the important idea of *combining* given sets in order to form a new set. Two basic operations are considered. These operations parallel, in certain respects, the operation of addition and multiplication of numbers. Let A and B be two sets. We define C as the *union* of A and B (sometimes called the sum of A and B) as follows:

$$C = \{x \mid x \in A \text{ or } x \in B \text{ (or both)}\}.$$

We write this as $C = A \cup B$. Thus C consists of all elements which are in A, or in B, or in both.

We define D as the *intersection* of A and B (sometimes called the product of A and B) as follows:

$$D = \{x \mid x \in A \text{ and } x \in B\}.$$

We write this as $D = A \cap B$. Thus D consists of all elements which are in A *and* in B.

Finally we introduce the idea of the *complement* of a set A as follows: The set, denoted by $\overline{A}$, consisting of all elements *not* in A (but in the universal set U) is called the complement of A. That is, $\overline{A} = \{x \mid x \notin A\}$.

A graphic device known as a *Venn diagram* can be used to considerable advantage when we are combining sets as indicated above. In each diagram in Fig. 1.1, the *shaded* region represents the set under consideration.

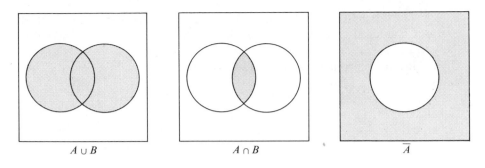

$$A \cup B \qquad\qquad A \cap B \qquad\qquad \overline{A}$$

FIGURE 1.1

EXAMPLE 1.2. Suppose that $U = \{1, 2, 3, 4, 5, 6, 7, 8, 9, 10\}$; $A = \{1, 2, 3, 4\}$, $B = \{3, 4, 5, 6\}$. Then we find that $\overline{A} = \{5, 6, 7, 8, 9, 10\}$, $A \cup B = \{1, 2, 3, 4, 5, 6\}$, and $A \cap B = \{3, 4\}$. Note that in describing a set (such as $A \cup B$) we list an element exactly once.

The above operations of union and intersection defined for just two sets may be extended in an obvious way to any finite number of sets. Thus we define $A \cup B \cup C$ as $A \cup (B \cup C)$ or $(A \cup B) \cup C$, which are the same, as can easily be checked. Similarly, we define $A \cap B \cap C$ as $A \cap (B \cap C)$ or $(A \cap B) \cap C$, which again can be checked to be the same. And it is clear that we may continue these constructions of new sets for *any finite* number of given sets.

We asserted that certain sets were the same, for example $A \cap (B \cap C)$ and $(A \cap B) \cap C$. It turns out that there are a number of such *equivalent* sets, some of which are listed below. If we recall that two sets are the same whenever they contain the same members, it is easy to show that the assertions stated are true. The reader should convince himself of these with the aid of Venn diagrams.

(a) $A \cup B = B \cup A$,   (b) $A \cap B = B \cap A$,

(c) $A \cup (B \cup C) = (A \cup B) \cup C$, (d) $A \cap (B \cap C) = (A \cap B) \cap C$.   (1.1)

We refer to (a) and (b) as the *commutative* laws, and (c) and (d) as the *associative* laws.

There are a number of other such *set identities* involving union, intersection, and complementation. The most important of these are listed below. In each case, their validity may be checked with the aid of a Venn diagram.

(e) $A \cup (B \cap C) = (A \cup B) \cap (A \cup C)$,

(f) $A \cap (B \cup C) = (A \cap B) \cup (A \cap C)$,

(g) $A \cap \emptyset = \emptyset$,   (1.2)

(h) $A \cup \emptyset = A$,   (i) $\overline{(A \cup B)} = \overline{A} \cap \overline{B}$,

(j) $\overline{(A \cap B)} = \overline{A} \cup \overline{B}$,   (k) $\overline{\overline{A}} = A$.

We note that (g) and (h) indicate that ∅ behaves among sets (with respect to the operations ∪ and ∩) very much as does the number zero among numbers (with respect to the operation of addition and multiplication).

One additional set construction given two (or more) sets will be needed in what follows.

> **Definition.** Let A and B be two sets. By the *Cartesian product* of A and B, denoted by $A \times B$, we shall mean the set $\{(a, b), a \in A, b \in B\}$, that is, the set of all ordered pairs where the first element is taken from A and the second from B.

EXAMPLE 1.3. Let $A = \{1, 2, 3\}$; $B = \{1, 2, 3, 4\}$.

Then $A \times B = \{(1, 1), (1, 2), \ldots, (1, 4), (2, 1), \ldots, (2, 4), (3, 1), \ldots, (3, 4)\}$

Note: In general $A \times B \neq B \times A$.

The above notion can be extended as follows: If $A_1, \ldots, A_n$ are sets, then $A_1 \times A_2 \times \cdots \times A_n = \{(a_1, a_2, \ldots, a_n), a_i \in A_i\}$, that is, the set of all ordered n-tuples.

An important special case arises when we take the Cartesian product of a set with itself, that is, $A \times A$ or $A \times A \times A$. Examples of this arise when we deal with the Euclidean plane, $R \times R$, where R is the set of all real numbers and Euclidean 3-space represented as $R \times R \times R$.

The *number of elements* in a set will be of considerable importance to us. If there is a finite number of elements in A, say $a_1, a_2, \ldots, a_n$, we say that A is *finite*. If there is an infinite number of elements in A which may be put into a *one-to-one correspondence* with the positive integers, we say that A is *countably* or *denumerably infinite*. (It can be shown for example, that the set of all rational numbers is countably infinite.) Finally we must consider the case of a nondenumerable infinite set. Such sets contain an infinite number of elements which cannot be enumerated. It can be shown, for instance, that for any two real numbers $b > a$, set $A = \{x \mid a \leq x \leq b\}$ has a nondenumerable number of elements. Since we may associate with each real number a point on the real number line, the above says that any (nondegenerate) interval contains more than a countable number of points.

The concepts introduced above, although representing only a brief glimpse into the theory of sets, are sufficient for our purpose: to describe with considerable rigor and precision, the basic ideas of probability theory.

1.3 Examples of Nondeterministic Experiments

We are now ready to discuss what we mean by a "random" or "nondeterministic" experiment. (More precisely, we shall give examples of phenomena for which nondeterministic models are appropriate. This is a distinction which the reader should keep in mind. Thus we shall repeatedly refer to nondeterministic or random experiments when in fact we are talking about a nondeterministic *model* for an experiment.) We shall not attempt to give a precise dictionary definition of this concept. Instead, we shall cite a large number of examples illustrating what we have in mind.

E_1: Toss a die and observe the number that shows on top.

E_2: Toss a coin four times and observe the total number of heads obtained.

E_3: Toss a coin four times and observe the sequence of heads and tails obtained.

E_4: Manufacture items on a production line and count the number of defective items produced during a 24-hour period.

E_5: An airplane wing is assembled with a large number of rivets. The number of defective rivets is counted.

E_6: A light bulb is manufactured. It is then tested for its life length by inserting it into a socket and the time elapsed (in hours) until it burns out is recorded.

E_7: A lot of 10 items contains 3 defectives. One item is chosen after another (without replacing the chosen item) until the last defective item is obtained. The total number of items removed from the lot is counted.

E_8: Items are manufactured until 10 nondefective items are produced. The total number of manufactured items is counted.

E_9: A missile is launched. At a specified time t, its three velocity components, v_x, v_y, and v_z are observed.

E_{10}: A newly launched missile is observed at times, $t_1, t_2, \ldots, t_n$. At each of these times the missile's height above the ground is recorded.

E_{11}: The tensile strength of a steel beam is measured.

E_{12}: From an urn containing only black balls, a ball is chosen and its color noted.

E_{13}: A thermograph records temperature, continuously, over a 24-hour period. At a specified locality and on a specified date, such a thermograph is "read."

E_{14}: In the situation described in E_{13}, x and y, the minimum and maximum temperatures of the 24-hour period in question are recorded.

What do the above experiments have in common? The following features are pertinent for our characterization of a *random experiment*.

(a) Each experiment is capable of being repeated indefinitely under essentially unchanged conditions.

(b) Although we are in general not able to state what a *particular* outcome will be, we are able to describe the set of *all possible* outcomes of the experiment.

(c) As the experiment is performed repeatedly, the individual outcomes seem to occur in a haphazard manner. However, as the experiment is repeated a *large* number of times, a definite pattern or regularity appears. It is this regularity which makes it possible to construct a precise mathematical model with which to analyze the experiment. We will have much more to say about the nature and importance of this regularity later. For the moment, the reader need only think of the repeated tossings of a fair coin. Although heads and tails will appear, successively, in an almost arbitrary fashion, it is a well-known empirical fact that after a large number of tosses the proportion of heads and tails will be approximately equal.

It should be noted that all the experiments described above satisfy these general characteristics. (Of course, the last mentioned characteristic can only be verified by experimentation; we will leave it to the reader's intuition to believe that if the experiment were repeated a large number of times, the regularity referred to would be evident. For example, if a large number of light bulbs from the same manufacturer were tested, presumably the number of bulbs burning more than 100 hours, say, could be predicted with considerable accuracy.) Note that experiment E_{12} has the peculiar feature that only one outcome is possible. In general such experiments will not be of interest, for the very fact that we do not know which particular outcome will occur when an experiment is performed is what makes it of interest to us.

Note: In describing the various experiments we have specified not only the procedure which is being performed but also what we are interested in observing (see, for example, the difference between E_2 and E_3 mentioned previously). This is a very important point to which we shall refer again later when discussing random variables. For the moment, let us simply note that as a consequence of a single experimental procedure or the occurrence of a single phenomenon, *several* different numerical values could be computed. For instance, if one person is chosen from a large group of persons (and the actual choosing would be the experimental procedure previously referred to), we might be interested in that person's height, weight, annual income, number of children, etc. Of course in most situations we know before beginning our experimentation just what numerical characteristic we are going to be concerned about.

1.4 The Sample Space

Definition. With each experiment $\mathcal{E}$ of the type we are considering we define the *sample space* as the set of all possible outcomes of $\mathcal{E}$. We usually designate this set by S. (In our present context, S represents the universal set described previously.)

Let us consider each of the experiments above and describe a sample space for each. The sample space S_i will refer to the experiment E_i.

S_1: $\{1, 2, 3, 4, 5, 6\}$.

S_2: $\{0, 1, 2, 3, 4\}$.

S_3:   {all possible sequences of the form a_1, a_2, a_3, a_4, where each $a_i = H$ or T depending on whether heads or tails appeared on the ith toss.}

S_4: $\{0, 1, 2, \ldots, N\}$, where N is the maximum number that could be produced in 24 hours.

S_5: $\{0, 1, 2, \ldots, M\}$, where M is the number of rivets installed.

S_6: $\{t \mid t \geq 0\}$.

S_7: $\{3, 4, 5, 6, 7, 8, 9, 10\}$.

S_8: $\{10, 11, 12, \ldots\}$.

S_9: $\{v_x, v_y, v_z \mid v_x, v_y, v_z \text{ real numbers}\}$.

S_{10}: $\{h_1, \ldots, h_n \mid h_i \geq 0, i = 1, 2, \ldots, n\}$.

S_{11}: $\{T \mid T \geq 0\}$.

S_{12}: {black ball}.

S_{13}: This sample space is the most involved of those considered here. We may realistically suppose that the temperature at a specified locality can never get above or below certain values, say M and m. Beyond this restriction, we must allow the possibility of any graph to appear with certain qualifications. Presumably the graph will have no jumps (that is, it will represent a continuous function). In addition, the graph will have certain characteristics of smoothness which can be summarized mathematically by saying that the graph represents a differentiable function. Thus we can finally state that the sample space is

$$\{f \mid f \text{ a differentiable function, satisfying } m \leq f(t) \leq M, \text{ all } t\}.$$

S_{14}: $\{(x, y) \mid m \leq x \leq y \leq M\}$. That is, S_{14} consists of all points in and on a triangle in the two-dimensional x,y-plane.

(In this book we will not concern ourselves with sample spaces of the complexity encountered in S_{13}. However, such sample spaces do arise, but require more advanced mathematics for their study than we are presupposing.)

   In order to describe a sample space associated with an experiment, we must have a very clear idea of what we are measuring or observing. Hence we should speak of "a" sample space associated with an experiment rather than "the" sample space. In this connection note the difference between S_2 and S_3.

   Note also that the outcome of an experiment need not be a number. For example, in E_3 each outcome is a sequence of H's and T's. In E_9 and E_{10} each outcome consists of a vector, while in E_{13} each outcome consists of a function.

It will again be important to discuss the *number* of outcomes in a sample space. Three possibilities arise: the sample space may be finite, countably infinite, or noncountably infinite. Referring to the above examples, we note that S_1, S_2, S_3, S_4, S_5, S_7, and S_{12} are finite, S_8 is countably infinite, and S_6, S_9, S_{10}, S_{11}, S_{13}, and S_{14} are noncountably infinite.

At this point it might be worth while to comment on the difference between a mathematically "idealized" sample space and an experimentally realizable one. For this purpose, let us consider experiment E_6 and its associated sample space S_6. It is clear that when we are actually recording the total time t during which a bulb is functioning, we are "victims" of the accuracy of our measuring instrument. Suppose that we have an instrument which is capable of recording time to two decimal places, for example, 16.43 hours. With this restriction imposed, our sample space becomes *countably infinite:* $\{0.0, 0.01, 0.02, \ldots\}$. Furthermore, it is quite realistic to suppose that no bulb could possibly last more than H hours, where H might be a very large number. Thus it appears that if we are completely realistic about the description of this sample space, we are actually dealing with a *finite* sample space: $\{0.0, 0.01, 0.02, \ldots, H\}$. The total number of outcomes would be $(H/0.01) + 1$, which would be a very large number if H is even moderately large, for example, $H = 100$. It turns out to be far simpler and convenient, mathematically, to assume that *all* values of $t \geq 0$ are possible outcomes and hence to deal with the sample space S_6 as originally defined.

In view of the above comments, a number of the sample spaces described are idealized. In all subsequent situations, the sample space considered will be that one which is mathematically most convenient. In most problems, little question arises as to the proper choice of sample space.

1.5 Events

Another basic notion is the concept of an *event*. An event A (with respect to a particular sample space S associated with an experiment ε) is simply a set of possible outcomes. In set terminology, an event is a *subset* of the sample space S. In view of our previous discussion, this means that S itself is an event and so is the empty set $\emptyset$. Any individual outcome may also be viewed as an event.

The following are some examples of events. Again, we refer to the above-listed experiments: A_i will refer to an event associated with the experiment E_i.

A_1: An even number occurs; that is, $A_1 = \{2, 4, 6\}$.

A_2: $\{2\}$; that is, two heads occur.

A_3: $\{HHHH, HHHT, HHTH, HTHH, THHH\}$; that is, more heads than tails showed.

A_4: $\{0\}$; that is, all items were nondefective.

A_5: $\{3, 4, \ldots, M\}$; that is, more than two rivets were defective.

A_6: $\{t \mid t < 3\}$; that is, the bulb burns less than three hours.

A_{14}: $\{(x, y) \mid y = x + 20\}$; that is, the maximum is 20° greater than the minimum.

When the sample space S is finite or countably infinite, *every* subset may be considered as an event. [It is an easy exercise to show, and we shall do it shortly, that if S has n members, there are exactly 2^n subsets (events).] However, if S is noncountably infinite, a theoretical difficulty arises. It turns out that *not* every conceivable subset may be considered as an event. Certain "nonadmissible" subsets must be excluded for reasons which are beyond the level of this presentation. Fortunately such nonadmissible sets do not really arise in applications and hence will not concern us here. In all that follows it will be tacitly assumed that whenever we speak of an event it will be of the kind we are allowed to consider.

We can now use the various methods of combining sets (that is, events) and obtain the new sets (that is, events) which we introduced earlier.

(a) If A and B are events, $A \cup B$ is the event which occurs if and only if A *or* B (or both) occur.

(b) If A and B are events, $A \cap B$ is the event which occurs if and only if A *and* B occur.

(c) If A is an event, $\overline{A}$ is the event which occurs if and only if A does *not* occur.

(d) If $A_1, \ldots, A_n$ is any finite collection of events, then $\bigcup_{i=1}^n A_i$ is the event which occurs if and only if *at least one* of the events A_i occurs.

(e) If $A_1, \ldots, A_n$ is any finite collection of events, then $\bigcap_{i=1}^n A_i$ is the event which occurs if and only if *all* the events A_i occur.

(f) If $A_1, \ldots, A_n, \ldots$ is any (countably) infinite collection of events, then $\bigcup_{i=1}^\infty A_i$ is the event which occurs if and only if *at least one* of the events A_i occurs.

(g) If $A_1, \ldots A_n, \ldots$ is any (countably) infinite collection of events, then $\bigcap_{i=1}^\infty A_i$ is the event which occurs if and only if *all* the events A_i occur.

(h) Suppose that S represents the sample space associated with some experiment $\mathcal{E}$ and we perform $\mathcal{E}$ twice. Then $S \times S$ may be used to represent all outcomes of these two repetitions. That is, $(s_1, s_2) \in S \times S$ means that s_1 resulted when $\mathcal{E}$ was performed the first time and s_2 when $\mathcal{E}$ was performed the second time.

(i) The example in (h) may obviously be generalized. Consider n repetitions of an experiment $\mathcal{E}$ whose sample space is S. Then $S \times S \times \cdots \times S = \{(s_1, s_2, \ldots, s_n), s_i \in S, i = 1, \ldots, n\}$ represents the set of all possible outcomes when $\mathcal{E}$ is performed n times. In a sense, $S \times S \times \cdots \times S$ is a sample space itself, namely the sample space associated with n repetitions of $\mathcal{E}$.

Definition. Two events, A and B, are said to be *mutually exclusive* if they cannot occur together. We express this by writing $A \cap B = \emptyset$; that is, the intersection of A and B is the empty set.

EXAMPLE 1.4. An electronic device is tested and its total time of service, say t, is recorded. We shall assume the sample space to be $\{t \mid t \geq 0\}$. Let the three events A, B, and C be defined as follows:

$$A = \{t \mid t < 100\}; \qquad B = \{t \mid 50 \leq t \leq 200\}; \qquad C = \{t \mid t > 150\}.$$

Then

$$A \cup B = \{t \mid t \leq 200\}; \qquad A \cap B = \{t \mid 50 \leq t < 100\};$$
$$B \cup C = \{t \mid t \geq 50\}; \qquad B \cap C = \{t \mid 150 < t \leq 200\}; \qquad A \cap C = \emptyset;$$
$$A \cup C = \{t \mid t < 100 \text{ or } t > 150\}; \qquad \overline{A} = \{t \mid t \geq 100\}; \qquad \overline{C} = \{t \mid t \leq 150\}.$$

One of the basic characteristics of the concept of "experiment" as discussed in the previous section is that we do not know which particular outcome will occur when the experiment is performed. Saying it differently, if A is an event associated with the experiment, then we cannot state with certainty that A will or will not occur. Hence it becomes very important to try to associate a number with the event A which will measure, in some sense, how likely it is that the event A occurs. This task leads us to the theory of probability.

1.6 Relative Frequency

In order to motivate the approach adopted for the solution of the above problem, consider the following procedure. Suppose that we repeat the experiment ε n times and let A and B be two events associated with ε. We let n_A and n_B be the number of times that the event A and the event B occurred among the n repetitions, respectively.

Definition. $f_A = n_A/n$ is called the *relative frequency* of the event A in the n repetitions of ε. The relative frequency f_A has the following important properties, which are easily verified.

(1) $0 \leq f_A \leq 1$.
(2) $f_A = 1$ if and only if A occurs *every* time among the n repetitions.
(3) $f_A = 0$ if and only if A *never* occurs among the n repetitions.
(4) If A and B are two mutually exclusive events and if $f_{A \cup B}$ is the relative frequency associated with the event $A \cup B$, then $f_{A \cup B} = f_A + f_B$.
(5) f_A, based on n repetitions of the experiment and considered as a function of n, "converges" in a certain probabilistic sense to $P(A)$ as $n \to \infty$.

Note: Property (5) above is obviously stated somewhat vaguely at this point. Only later (Section 12.2) will we be able to make this idea more precise. For the moment let us simply state that Property (5) involves the fairly intuitive notion that the relative frequency based on an increasing number of observations tends to "stabilize" near some definite value. This is *not* the same as the usual concept of convergence encountered elsewhere in mathematics. In fact, as stated here, this is not a mathematical conclusion at all but simply an empirical fact.

Most of us are intuitively aware of this phenomenon of stabilization although we may never have checked on it. To do so requires a considerable amount of time and patience, since it involves a large number of repetitions of an experiment. However, sometimes we may be innocent observers of this phenomenon as the following example illustrates.

EXAMPLE 1.5. Suppose that we are standing on a sidewalk and fix our attention on two adjacent slabs of concrete. Assume that it begins to rain in such a manner that we are actually able to distinguish individual raindrops and keep track of whether these drops land on one slab or the other. We continue to observe individual drops and note their point of impact. Denoting the ith drop by X_i, where $X_i - 1$ if the drop lands on one slab and 0 if it lands on the other slab, we might observe a sequence such as 1, 1, 0, 1, 0, 0, 0, 1, 0, 0, 1. Now it is clear that we are not able to predict where a particular drop will fall. (Our experiment consists of some sort of meteorological situation causing the release of raindrops.) If we compute the relative frequency of the event $A = \{$the drop lands on slab 1$\}$, then the above sequence of outcomes gives rise to the following relative frequencies (based on the observance of 1, 2, 3, . . . drops): $1, 1, \frac{2}{3}, \frac{3}{4}, \frac{3}{5}, \frac{3}{6}, \frac{3}{7}, \frac{4}{8}, \frac{4}{9}, \frac{4}{10}, \frac{5}{11}, \cdots$ These numbers show a considerable degree of variation, particularly at the beginning. It is intuitively clear that if the above experiment were continued indefinitely, these relative frequencies would stabilize near the value $\frac{1}{2}$. For we have every reason to believe that after some time had elapsed the two slabs would be equally wet.

This stability property of relative frequency is a fairly intuitive notion as yet, and we will be able to make it mathematically precise only later. The essence of this property is that if an experiment is performed a large number of times, the relative frequency of occurrence of some event A tends to vary less and less as the number of repetitions is increased. This characteristic is also referred to as *statistical regularity*.

We have also been somewhat vague in our definition of experiment. Just when is a procedure or mechanism an experiment in our sense, capable of being studied mathematically by means of a nondeterministic model? We have stated previously that an experiment must be capable of being performed repeatedly under essentially unchanged conditions. We can now add another requirement. When the experiment is performed repeatedly it must exhibit the statistical regularity referred to above. Later we shall discuss a theorem (called the Law of Large Numbers) which shows that statistical regularity is in fact a *consequence* of the first requirement: repeatability.

1.7 Basic Notions of Probability

Let us now return to the problem posed above: to assign a number to each event A which will measure how likely it is that A occurs when the experiment is performed. One *possible* approach might be the following one: Repeat the experiment a large number of times, compute the relative frequency f_A, and use this number. When we recall the properties of f_A, it is clear that this number *does* give a very definite indication of how likely it is that A occurs. Furthermore, we know that as the experiment is repeated more and more times, the relative frequency f_A stabilizes near some number, say p. However, there are two serious objections to this approach. (a) It is not clear how large n should be before we know the number. 1000? 2000? 10,000? (b) Once the experiment has been completely described and the event A specified, the number we are seeking should not depend on the experimenter or the particular streak of luck which he ex- periences. (For example, it is possible for a perfectly balanced coin, when tossed 10 times, to come up with 9 heads and 1 tail. The relative frequency of the event A {heads occur} thus equals $\frac{9}{10}$. Yet it is clear that on the next 10 tosses the pattern of heads and tails might be reversed.) What we want is a means of obtaining such a number without resorting to experimentation. Of course, for the number we stipulate to be meaningful, any subsequent experimentation should yield a relative frequency which is "close" to the stipulated value, particularly if the number of repetitions on which the computed relative frequency is based is quite large. We proceed formally as follows.

Definition. Let ε be an experiment. Let S be a sample space associated with ε. With each event A we associate a real number, designated by $P(A)$ and called *the probability of* A satisfying the following properties.

(1) $0 \leq P(A) \leq 1$.
(2) $P(S) = 1$.   (1.3)
(3) If A and B are mutually exclusive events, $P(A \cup B) = P(A) + P(B)$.
(4) If $A_1, A_2, \ldots, A_n, \ldots$ are pairwise mutually exclusive events, then

$$P(\cup_{i=1}^{\infty} A_i) = P(A_1) + P(A_2) + \cdots + P(A_n) + \cdots$$

We note that from Property 3 it immediately *follows* that for any *finite n*,

$$P\left(\bigcup_{i=1}^{n} A_i \right) = \sum_{i=1}^{n} P(A_i).$$

Property 4 does *not* follow; however, when we consider the idealized sample space, this condition will be required and hence is included here.

   The choice of the above-listed properties of probability are obviously motivated by the corresponding characteristics of relative frequency. The property referred to above as statistical regularity will be tied in with this definition of probability later. For the moment we simply state that we shall show that the numbers $P(A)$ and f_A are "close" to each other (in a certain sense), if f_A is based on a large num- ber of repetitions. It is this fact which gives us the justification to use $P(A)$ for measuring how probable it is that A occurs.

   At the moment we do not know *how* to compute $P(A)$. We have simply listed some general properties which $P(A)$ possesses. The reader will have to be patient a little longer (until the next chapter) before he learns how to evaluate $P(A)$. Before we turn to this question, let us state and prove several consequences concerning $P(A)$ which follow from the above conditions, and which do not depend on how we actually compute $P(A)$.

Theorem 1.1. If $\emptyset$ is the empty set, then $P(\emptyset) = 0$.

Proof: We may write, for any event A, $A = A \cup \emptyset$. Since A and $\emptyset$ are mutually exclusive, it follows from Property 3 that $P(A) = P(A \cup \emptyset) = P(A) + P(\emptyset)$. From this the conclusion of the theorem is immediate.

Note: We shall have occasion to see later that the converse of the above theorem is not true. That is, if $P(A) = 0$, we cannot in general conclude that $A = \emptyset$, for there are situations in which we assign probability zero to an event that *can* occur.

Theorem 1.2. If $\overline{A}$ is the complementary event of A, then

$$P(A) = 1 - P(\overline{A}). \tag{1.4}$$

Proof: We may write $S = A \cup \overline{A}$ and, using Properties 2 and 3, we obtain $1 = P(A) + P(\overline{A})$.

Note: This is a particularly useful result, for it means that whenever we wish to evaluate $P(A)$ we may instead compute $P(\overline{A})$ and then obtain the desired result by subtraction. We shall see later that in many problems it is much easier to compute $P(\overline{A})$ than $P(A)$.

Theorem 1.3. If A and B are *any* two events, then

$$P(A \cup B) = P(A) + P(B) - P(A \cap B). \tag{1.5}$$

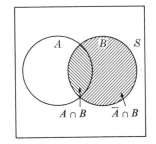

FIGURE 1.2

Proof: The idea of this proof is to decompose $A \cup B$ and B into mutually exclusive events and then to apply Property 3. (See the Venn diagram in Fig. 1.2.)

Thus we write

$$A \cup B = A \cup (B \cap \bar{A}),$$
$$B = (A \cap B) \cup (B \cap \bar{A}).$$

Hence

$$P(A \cup B) = P(A) + P(B \cap \bar{A}),$$
$$P(B) = P(A \cap B) + P(B \cap \bar{A}).$$

Subtracting the second equation from the first yields

$$P(A \cup B) - P(B) = P(A) - P(A \cap B)$$

and hence the result follows.

Note: This theorem represents an obvious extension of Property 3, for if $A \cap B = \emptyset$, we obtain from the above the statement of Property 3.

Theorem 1.4. If A, B, and C are any three events, then

$$P(A \cup B \cup C) = P(A) + P(B) + P(C) - P(A \cap B) - P(A \cap C)$$
$$- P(B \cap C) + P(A \cap B \cap C). \qquad (1.6)$$

Proof: The proof consists of writing $A \cup B \cup C$ as $(A \cup B) \cup C$ and applying the result of the above theorem. The details are left to the reader.

Note: An obvious extension to the above theorem suggests itself. Let $A_1, \ldots, A_k$ be any k events. Then

$$P(A_1 \cup A_2 \cup \cdots \cup A_k) = \sum_{i=1}^{k} P(A_i) - \sum_{i<j=2}^{k} P(A_i \cap A_j)$$

$$+ \sum_{i<j<r=3}^{k} P(A_i \cap A_j \cap A_r) + \cdots + (-1)^{k-1} P(A_1 \cap A_2 \cap \cdots \cap A_k). \qquad (1.7)$$

This result may be established easily by using mathematical induction.

Theorem 1.5. If $A \subset B$, then $P(A) \leq P(B)$.

Proof: We may decompose B into two mutually exclusive events as follows: $B = A \cup (B \cap \bar{A})$. Hence $P(B) = P(A) + P(B \cap \bar{A}) \geq P(A)$, since $P(B \cap \bar{A}) \geq 0$ from Property 1.

Note: This result certainly is intuitively appealing. For it says that if B *must* occur whenever A occurs, then B is at least as probable as A.

1.8 Several Remarks

(a) A word of caution is in order here. From the previous discussion it might be (incorrectly) inferred that when we choose a probabilistic model for the description of some observational phenomenon, we are discarding all deterministic relationships. Nothing could be further from the truth. We still use the fact that, for example, Ohm's law $I = E/R$ holds under certain circumstances. The difference will be one of interpretation. Instead of saying that the above relationship determines I for given E and R, we shall acknowledge that E (and/or) R may vary in some random, unpredictable manner and that hence I will vary in some random manner also. For *given* E and R, I is still determined by the above relationship. The point is that when adopting a probabilistic model for the description of a circuit, we will consider the possibility that E and R may vary in some unpredictable way which can only be described probabilistically. Thus, since it will be meaningful to consider only the *probability* that E and R assume certain values, it becomes meaningful to speak only of the probability that I assumes certain values.

(b) The choice between adopting a deterministic or a probabilistic model may sometimes be difficult to make. It may depend on the intricacy of our measuring technique and its associated accuracy. For example, if accurate measurements are so difficult to obtain that repeated readings of the same quantity will yield varying results, a probabilistic model is undoubtedly best suited to describe the situation.

(c) We shall shortly point out that under certain circumstances we are in a position to make additional assumptions about the probabilistic behavior of our experimental outcomes which will lead to a method of evaluation of the basic probabilities. The choice of these additional assumptions may be based on physical considerations of the experiment (certain symmetry properties, for instance), empirical evidence, or in some cases, simply personal judgment based on previous experience with a similar situation. The relative frequency f_A may play an important role in our deliberation about the numerical assignment of $P(A)$. However, it is important to realize that any assumption we make about $P(A)$ must be such that the basic axioms (1) through (4) of Definition (1.3) are satisfied.

(d) In the course of developing the basic ideas of probability theory, we shall make a number of references to certain analogies to mechanics. The first of these may be appropriate here. In mechanics, we assign to each body B its mass, say $m(B)$. We then make various calculations and reach various conclusions about the behavior of B and its relationship to other bodies, many of which involve its mass $m(B)$. The fact that we may have to resort to some approximation in actually obtaining $m(B)$ for a specified body does not detract from the usefulness of the mass concept. Similarly, we stipulate for each event A, associated with the sample space of an experiment, a number $P(A)$ called the probability of A and satisfying our basic axioms. In actually computing $P(A)$ for a specific event we have to either make additional assumptions or obtain an approximation based on empirical evidence.

(e) It is very important to realize that we have *postulated* the existence of the number $P(A)$ and that we have postulated certain properties which this number possesses. The validity of the various consequences (theorems) derived from these postulates in no way depends on how we go about obtaining a numerical value for $P(A)$. It is vital that this point be clear. For example, we have assumed that $P(A \cup B) = P(A) + P(B)$. In order to use this relationship for the actual *evaluation* of $P(A \cup B)$, we must know the value of $P(A)$ and of $P(B)$. We shall discuss shortly that under certain circumstances we may make additional assumptions leading to a method of evaluation of these probabilities. If these (or other) assumptions are not warranted, we may have to resort to experimentation in order to approximate the value of $P(A)$ from actual data. The relative frequency f_A will play an important role in this and will, in fact, be used to approximate $P(A)$.

However, it is important to keep in mind that f_A and $P(A)$ are not the same, that we simply use f_A to approximate $P(A)$, and that whenever we refer to $P(A)$ we are referring to the postulated value. If we "identify" f_A with $P(A)$ we must realize that we are simply replacing a postulated value with an approximation obtained experimentally. How good or bad this approximation may be in no way influences the logical structure of our model. Although the phenomenon which the model attempts to represent was taken into account in constructing the model, we have separated ourselves (temporarily at least) from the phenomenon itself, when we enter the realm of the model.

PROBLEMS

1.1.   Suppose that the universal set consists of the positive integers from 1 through 10. Let $A = \{2, 3, 4\}$, $B = \{3, 4, 5\}$, and $C = \{5, 6, 7\}$. List the members of the following sets.

(a) $\bar{A} \cap B$ (b) $\bar{A} \cup B$ (c) $\overline{\bar{A} \cap B}$ (d) $\bar{A} \cap (\overline{B \cap C})$ (e) $\bar{A} \cap (B \cup C)$

1.2.   Suppose that the universal set U is given by $U = \{x \mid 0 \le x \le 2\}$. Let the sets A and B be defined as follows: $A = \{x \mid \frac{1}{2} < x \le 1\}$ and $B = \{x \mid \frac{1}{4} \le x < \frac{3}{2}\}$. Describe the following sets.

(a) $\overline{A \cup B}$ (b) $A \cup \bar{B}$ (c) $\overline{A \cap B}$ (d) $\bar{A} \cap B$

1.3.   Which of the following relationships are true?
(a) $(A \cup B) \cap (A \cup C) = A \cup (B \cap C)$ (b) $(A \cup B) = ((A \cap \bar{B}) \cup B$
(c) $\overline{A \cap B} = A \cup B$ (d) $\overline{(A \cup B)} \cap C = \bar{A} \cap \bar{B} \cap \bar{C}$
(e) $(A \cap B) \cap (\bar{B} \cap C) = \emptyset$

1.4.   Suppose that the universal set consists of all points (x, y) both of whose coordinates are integers and which lie inside or on the boundary of the square bounded by the lines $x = 0$, $y = 0$, $x = 6$, and $y = 6$. List the members of the following sets.
(a) $A = \{(x, y) \mid x^2 + y^2 \le 6\}$ (b) $B = \{(x, y) \mid y \le x^2\}$
(c) $C = \{(x, y) \mid x \le y^2\}$ (d) $B \cap C$ (e) $(B \cup A) \cap \bar{C}$

1.5.   Use Venn diagrams to establish the following relationships.

(a) $A \subset B$ and $B \subset C$ imply that $A \subset C$ (b) $A \subset B$ implies that $A = A \cap B$

(c) $A \subset B$ implies that $\bar{B} \subset \bar{A}$ (d) $A \subset B$ implies that $A \cup C \subset B \cup C$

(e) $A \cap B = \emptyset$ and $C \subset A$ imply that $B \cap C = \emptyset$

1.6.   Items coming off a production line are marked defective (D) or nondefective (N). Items are observed and their condition listed. This is continued until two consecutive defectives are produced or four items have been checked, whichever occurs first. Describe a sample space for this experiment.

1.7.   (a) A box of N light bulbs has $r(r < N)$ bulbs with broken filaments. These bulbs are tested, one by one, until a defective bulb is found. Describe a sample space for this experiment.

(b) Suppose that the above bulbs are tested, one by one, until all defectives have been tested. Describe the sample space for this experiment.

1.8.   Consider four objects, say a, b, c, and d. Suppose that the *order* in which these objects are listed represents the outcome of an experiment. Let the events A and B be defined as follows: $A = \{a$ is in the first position$\}$; $B = \{b$ is in the second position$\}$.

(a) List all elements of the sample space.

(b) List all elements of the events $A \cap B$ and $A \cup B$.

1.9.   A lot contains items weighing $5, 10, 15, \ldots, 50$ pounds. Assume that at least two items of each weight are found in the lot. Two items are chosen from the lot. Let X denote the weight of the first item chosen and Y the weight of the second item. Thus the pair of numbers (X, Y) represents a single outcome of the experiment. Using the XY-plane, indicate the sample space and the following events.

(a) $\{X = Y\}$ (b) $\{Y > X\}$

(c) The second item is twice as heavy as the first item.

(d) The first item weighs 10 pounds less than the second item.

(e) The average weight of the two items is less than 30 pounds.

1.10.   During a 24-hour period, at some time X, a switch is put into "ON" position. Subsequently, at some future time Y (still during that same 24-hour period) the switch is put into the "OFF" position. Assume that X and Y are measured in hours on the time axis with the beginning of the time period as the origin. The outcome of the experiment consists of the pair of numbers (X, Y).

(a) Describe the sample space.

(b) Describe and sketch in the XY-plane the following events.

(i) The circuit is on for one hour or less.

(ii) The circuit is on at time z where z is some instant during the given 24-hour period.

(iii) The circuit is turned on before time t_1 and turned off after time t_2 (where again $t_1 < t_2$ are two time instants during the specified period).

(iv) The circuit is on twice as long as it is off.

1.11.   Let A, B, and C be three events associated with an experiment. Express the following verbal statements in set notation.

(a) At least one of the events occurs.

(b) Exactly one of the events occurs.

(c) Exactly two of the events occur.

(d) Not more than two of the events occur simultaneously.

1.12.   Prove Theorem 1.4.

1.13.   (a) Show that for any two events, A_1 and A_2, we have $P(A_1 \cup A_2) \leq P(A_1) + P(A_2)$.

(b) Show that for any n events $A_1, \ldots, A_n$, we have

$$P(A_1 \cup \cdots \cup A_n) \leq P(A_1) + \cdots + P(A_n).$$

[*Hint:* Use mathematical induction. The result stated in (b) is called Boole's inequality.]

1.14.   Theorem 1.3 deals with the probability that *at least one* of the two events A or B occurs. The following statement deals with the probability that *exactly one* of the events A or B occurs.

Show that $[P(A \cap \bar{B}) \cup (B \cap \bar{A})] = P(A) + P(B) - 2P(A \cap B)$.

1.15.   A certain type of electric motor fails either by seizure of the bearings, or by burning out of the electric windings, or by wearing out of the brushes. Suppose that seizure is twice as likely as burning out, which is four times as likely as brush wearout. What is the probability that failure will be by each of these three mechanisms?

1.16.   Suppose that A and B are events for which $P(A) = x$, $P(B) = y$, and $P(A \cap B) = z$. Express each of the following probabilities in terms of x, y, and z.

(a) $P(\bar{A} \cup \bar{B})$ (b) $P(\bar{A} \cap B)$ (c) $P(\bar{A} \cup B)$ (d) $P(\bar{A} \cap \bar{B})$

1.17.   Suppose that A, B, and C are events such that $P(A) = P(B) = P(C) = \frac{1}{4}$, $P(A \cap B) = P(C \cap B) = 0$, and $P(A \cap C) = \frac{1}{8}$. Evaluate the probability that at least one of the events A, B, or C occurs.

1.18.   An installation consists of two boilers and one engine. Let the event A be that the engine is in good condition, while the events B_k ($k = 1, 2$) are the events that the kth boiler is in good condition. The event C is that the installation can operate. If the installation is operative whenever the engine and at least one boiler function, express C and $\bar{C}$ in terms of A and the B_i's.

1.19.   A mechanism has two types of parts, say I and II. Suppose that there are two of type I and three of type II. Define the events A_k, $k = 1, 2$, and B_j, $j = 1, 2, 3$ as follows: A_k: the kth unit of type I is functioning properly; B_j: the jth unit of type II is functioning properly. Finally, let C represent the event: the mechanism functions. Given that the mechanism functions if at least one unit of type I and at least two units of type II function, express the event C in terms of the A_k's and B_j's.

2

Finite Sample Spaces

2.1 Finite Sample Space

In this chapter we shall deal exclusively with experiments for which the sample space S consists of a *finite* number of elements. That is, we suppose that S may be written as $S = \{a_1, a_2, \ldots, a_k\}$. If we refer to the examples of sample spaces in Section 1.4, we note that S_1, S_2, S_3, S_4, S_5, S_7, and S_{12} are all finite.

In order to characterize $P(A)$ for this model we shall first consider the event consisting of a *single outcome*, sometimes called an *elementary* event, say $A = \{a_i\}$. We proceed as follows.

To each elementary event $\{a_i\}$ we assign a number p_i, called the probability of $\{a_i\}$, satisfying the following conditions:

(a) $p_i \geq 0$, $\quad i = 1, 2, \ldots, k$,
(b) $p_1 + p_2 + \cdots + p_k = 1$.

[Since $\{a_i\}$ is an event, these conditions must be consistent with those postulated for probabilities of events in general, as was done in Eq. (1.3). It is an easy matter to check that this is so.]

Next, suppose that an event A consists of r outcomes, $1 \leq r \leq k$, say

$$A = \{a_{j_1}, a_{j_2}, \ldots, a_{j_r}\},$$

where $j_1, j_2, \ldots, j_r$ represent any r indices from $1, 2, \ldots, k$. Hence it follows from Eq. (1.3), Property 4, that

$$P(A) = p_{j_1} + p_{j_2} + \cdots + p_{j_r}. \tag{2.1}$$

To summarize: The assignment of probabilities p_i to each elementary event $\{a_i\}$, subject to the conditions (a) and (b) above, uniquely determines $P(A)$ for each event $A \subset S$, where $P(A)$ is given by Eq. (2.1).

In order to evaluate the individual p_j's, some assumption concerning the individual outcomes must be made.

EXAMPLE 2.1. Suppose that only three outcomes are possible in an experiment, say a_1, a_2, and a_3. Suppose furthermore, that a_1 is twice as probable to occur as a_2, which in turn is twice as probable to occur as a_3.

Hence $p_1 = 2p_2$ and $p_2 = 2p_3$. Since $p_1 + p_2 + p_3 = 1$, we have $4p_3 + 2p_3 + p_3 = 1$, which finally yields

$$p_3 = \tfrac{1}{7}, \qquad p_2 = \tfrac{2}{7}, \qquad \text{and} \quad p_1 = \tfrac{4}{7}.$$

(*Note:* In what follows we shall use the phrase "equally likely" to mean "equally probable.")

2.2 Equally Likely Outcomes

The most commonly made assumption for finite sample spaces is that all outcomes are equally likely. This assumption can by no means be taken for granted, however; it must be carefully justified. There are many experiments for which such an assumption is warranted, but there are also many experimental situations in which it would be quite erroneous to make this assumption. For example, it would be quite unrealistic to suppose that it is as likely for no telephone calls to come into a telephone exchange between 1 a.m. and 2 a.m. as between 5 p.m. and 6 p.m.

If all the k outcomes are equally likely, it follows that each $p_i = 1/k$. For the condition $p_1 + \cdots + p_k = 1$ becomes $kp_i = 1$ for all i. From this it follows that for any event A consisting of r outcomes, we have

$$P(A) = r/k.$$

This method of evaluating $P(A)$ is often stated as follows:

$$P(A) = \frac{\text{number of ways in which } \mathcal{E} \text{ can occur favorable to } A}{\text{total number of ways in which } \mathcal{E} \text{ can occur}}.$$

It is important to realize that the above expression for $P(A)$ is only a consequence of the assumption that all outcomes are equally likely and is only applicable when this assumption is fulfilled. It most certainly does not serve as a general definition of probability.

EXAMPLE 2.2. A die is tossed and all outcomes are assumed to be equally likely. The event A occurs if and only if a number larger than 4 shows. That is, $A = \{5, 6\}$. Hence $P(A) = \tfrac{1}{6} + \tfrac{1}{6} = \tfrac{2}{6}$.

EXAMPLE 2.3. An honest coin is tossed two times. Let A be the event: {one head appears}. In evaluating $P(A)$ one analysis of the problem might be as follows. The sample space is $S = \{0, 1, 2\}$, where each outcome represents the number of heads that occur. Hence $P(A) = \tfrac{1}{3}$! This analysis is obviously incorrect, since for the sample space considered above, all outcomes are *not* equally likely. In order to apply the above methods we should consider, instead, the sample space $S' = \{HH, HT, TH, TT\}$. In this sample space all outcomes are equally likely, and hence we obtain for the correct solution of our problem, $P(A) = \frac{2}{4} = \frac{1}{2}$. We could use the sample space S correctly as follows: The outcomes 0 and 2 are equally likely, while the outcome 1 is twice as likely as either of the others. Hence $P(A) = \frac{1}{2}$, which checks with the above answer.

This example illustrates two points. First, we must be quite sure that all outcomes may be assumed to be equally likely before using the above procedure. Second, we can often, by an appropriate choice of the sample space, reduce the problem to one in which all outcomes *are* equally likely. Whenever possible this should be done, since it usually makes the computation simpler. This point will be referred to again in subsequent examples.

Quite often the manner in which the experiment is executed determines whether or not the possible outcomes are equally likely. For instance, suppose that we choose one bolt from a box containing three bolts of different sizes. If we choose the bolt by simply reaching into the box and picking the one which we happen to touch first, it is obvious that the largest bolt will have a greater probability of being chosen than the other two. However, by carefully labeling each bolt with a number, writing the number on a tag, and choosing a tag, we can try to ensure that each bolt has in fact the same probability of being chosen. Thus we may have to go to considerable trouble in order to ensure that the mathematical assumption of equally likely outcomes is in fact appropriate.

In examples already considered and in many subsequent ones, we are concerned with choosing at random one or more objects from a given collection of objects. Let us define this notion more precisely. Suppose that we have N objects, say $a_1, a_2, \ldots, a_N$.

(a) *To choose one object at random* from the N objects means that each object has the same probability of being chosen. That is,

$$\text{Prob (choosing } a_i) = 1/N, \qquad i = 1, 2, \ldots, N.$$

(b) *To choose two objects at random* from N objects means that *each pair* of objects (disregarding order) has the same probability of being chosen as any other pair. For example, if we must choose two objects at random from (a_1, a_2, a_3, a_4), then obtaining a_1 and a_2 is just as probable as obtaining a_2 and a_3, etc. This formulation immediately raises the question of *how many* different pairs there are. For suppose that there are K such pairs. Then the probability of each pair would be $1/K$. We shall learn shortly how to compute K.

(c) *To choose n objects at random* $(n \leq N)$ from the N objects means that each n-tuple, say $a_{i_1}, a_{i_2}, \ldots, a_{i_n}$, is as likely to be chosen as any other n-tuple.

Note: We have already suggested above that considerable care must be taken during the experimental procedure to ensure that the mathematical assumption of "choosing at random" is fulfilled.

2.3 Methods of Enumeration

We will have to make a digression at this juncture in order to learn how to enumerate. Consider again the above form of $P(A)$, namely $P(A) = r/k$, where k equals the total number of ways in which ε can occur while r equals the number of ways in which A can occur. In the examples exhibited so far, little difficulty was encountered in computing r and k. But we need to consider only slightly more complicated situations to appreciate the need for some systematic counting or enumeration procedures.

EXAMPLE 2.4. A lot of one hundred items consists of 20 defective and 80 non-defective items. Ten of these items are chosen at random, without replacing any item before the next one is chosen. What is the probability that exactly half of the chosen items are defective?

To analyze this problem, let us consider the following sample space S. Each member of S consists of ten possible items from the lot, say $(i_1, i_2, \ldots, i_{10})$. How many such outcomes are there? And among those outcomes, how many have the characteristic that exactly one-half are defective? We clearly need to be able to answer such questions in order to solve the problem being considered. Many similar problems give rise to analogous questions. In the next few sections we shall present some systematic enumeration procedures.

A. Multiplication principle. Suppose that a procedure, designated by 1 can be performed in n_1 ways. Let us assume that a second procedure, designated by 2, can be performed in n_2 ways. Suppose also that each way of doing 1 may be followed by any way of doing 2. Then the procedure consisting of 1 followed by 2 may be performed in $n_1 n_2$ ways.

To indicate the validity of this principle it is simplest to consider the following schematic approach. Consider a point P and two lines, say L_1 and L_2. Let procedure 1 consist of going from P to L_1 while procedure 2 consists of going from L_1 to L_2. Figure 2.1 indicates how the final result is obtained.

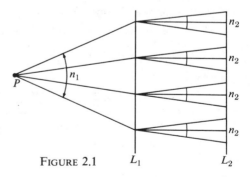

FIGURE 2.1 L_1 L_2

Note: This principle obviously may be extended to any number of procedures. If there are k procedures and the ith procedure may be performed in n_i ways, $i = 1, 2, \ldots, k$, then the procedure consisting of 1, followed by 2, $\ldots$, followed by procedure k may be performed in $n_1 n_2 \cdots n_k$ ways.

EXAMPLE 2.5. A manufactured item must pass through three control stations. At each station the item is inspected for a particular characteristic and marked accordingly. At the first station, three ratings are possible while at the last two stations four ratings are possible. Hence there are $3 \cdot 4 \cdot 4 = 48$ ways in which the item may be marked.

B. Addition principle. Suppose that a procedure, designated by 1, can be performed in n_1 ways. Assume that a second procedure, designated by 2, can be performed in n_2 ways. Suppose furthermore that it is *not* possible that *both* 1 and 2 are performed together. Then the number of ways in which we can perform 1 *or* 2 is $n_1 + n_2$.

We again use the schematic approach to convince ourselves of the validity of the addition principle, as Fig. 2.2 indicates.

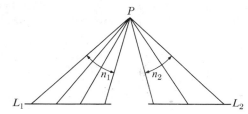

FIGURE 2.2

Note: This principle, too, may be generalized as follows. If there are k procedures and the ith procedure may be performed in n_i ways, $i = 1, 2, \ldots, k$, then the number of ways in which we may perform procedure 1 *or* procedure 2 *or* $\cdots$ *or* procedure k is given by $n_1 + n_2 + \cdots + n_k$, assuming that no two procedures may be performed together.

EXAMPLE 2.6. Suppose that we are planning a trip and are deciding between bus or train transportation. If there are three bus routes and two train routes, then there are $3 + 2 = 5$ different routes available for the trip.

C. Permutations

(a) Suppose that we have n different objects. In how many ways, say $_nP_n$, may these objects be arranged (permuted)? For example, if we have objects a, b, and c, we can consider the following arrangements: abc, acb, bac, bca, cab, and cba. Thus the answer is 6. In general, consider the following scheme. Arranging the n objects is equivalent to putting them into a box with n compartments, in some specified order.

The first slot may be filled in any one of n ways, the second slot in any one of $(n - 1)$ ways, $\ldots$, and the last slot in exactly one way. Hence applying the above multiplication principle, we see that the box may be filled in $n(n - 1)(n - 2) \ldots 1$ ways. This number occurs so often in mathematics that we introduce a special symbol and name for it.

Definition. If n is a positive integer, we define $n! = (n)(n - 1)(n - 2) \cdots 1$ and call it *n-factorial*. We also define $0! = 1$.

Thus the number of permutations of n different objects is given by

$$_nP_n = n!$$

(b) Consider again n *different* objects. This time we wish to choose r of these objects, $0 \leq r \leq n$, *and* permute the chosen r. We denote the number of ways of doing this by $_nP_r$. We again resort to the above scheme of filling a box having n compartments; this time we simply stop after the rth compartment has been filled. Thus the first compartment may be filled in n ways, the second in $(n - 1)$ ways, ..., and the rth compartment in $n - (r - 1)$ ways. Thus the entire procedure may be accomplished, again using the multiplication principle, in

$$n(n - 1)(n - 2) \cdots (n - r + 1)$$

ways. Using the factorial notation introduced above, we may write

$$_nP_r = \frac{n!}{(n - r)!}.$$

D. Combinations. Consider again n different objects. This time we are concerned with counting the number of ways we may choose r out of these n objects *without* regard to order. For example, we have the objects a, b, c, and d, and $r = 2$; we wish to count ab, ac, ad, bc, bd, and cd. In other words, we do *not* count ab and ba since the same objects are involved and only the order differs.

To obtain the general result we recall the formula derived above: The number of ways of choosing r objects out of n *and* permuting the chosen r equals $n!/(n - r)!$ Let C be the number of ways of choosing r out of n, disregarding order. (That is, C is the number sought.) Note that once the r items have been chosen, there are $r!$ ways of permuting them. Hence applying the multiplication principle again, together with the above result, we obtain

$$Cr! = \frac{n!}{(n - r)!}.$$

Thus the number of ways of choosing r out of n different objects, disregarding order, is given by

$$C = \frac{n!}{r!(n - r)!}.$$

This number arises in many contexts in mathematics and hence a special symbol is used for it. We shall write

$$\frac{n!}{r!(n - r)!} = \binom{n}{r}.$$

For the present purpose $\binom{n}{r}$ is defined only if n is a positive integer and if r is an integer $0 \le r \le n$. However, we can define $\binom{n}{r}$ quite generally for any real number n and for any nonnegative integer r as follows:

$$\binom{n}{r} = \frac{n(n-1)(n-2)\cdots(n-r+1)}{r!}.$$

The numbers $\binom{n}{r}$ often are called *binomial coefficients*, for they appear as coefficients in the expansion of the binomial expression $(a+b)^n$. If n is a positive integer, $(a+b)^n = (a+b)(a+b)\cdots(a+b)$. When multiplied out, each term will consist of the product of k a's and $(n-k)$ b's, $k = 0, 1, 2, \ldots, n$. How many terms will there be of the form $a^k b^{n-k}$? We simply count the number of ways in which we can choose k out of n a's, disregarding order. But this is precisely given by $\binom{n}{k}$. Hence we have what is known as the *binomial theorem*.

$$(a+b)^n = \sum_{k=0}^{n} \binom{n}{k} a^k b^{n-k}. \tag{2.2}$$

The numbers $\binom{n}{r}$ have many interesting properties only two of which we mention here. (Unless otherwise stated, we assume n to be a positive integer and r an integer, $0 \le r \le n$.)

$$\text{(a)} \quad \binom{n}{r} = \binom{n}{n-r},$$

$$\text{(b)} \quad \binom{n}{r} = \binom{n-1}{r-1} + \binom{n-1}{r}.$$

It is easy to verify the above two identities algebraically. We simply write out, in each case, the left- and the right-hand sides of the above identities and note that they are the same.

There is, however, another method of verifying these identities which makes use of the interpretation that we have given to $\binom{n}{r}$, namely the number of ways of choosing r out of n things.

(a) When we choose r out of n things we are at the same time "leaving behind" $(n-r)$ things, and hence choosing r out of n is equivalent to choosing $(n-r)$ out of n. But this is exactly the first statement to be verified.

(b) Let us single out any one of the n objects, say the first one, a_1. In choosing r objects, either a_1 is included or it is excluded, but not both. Hence in counting the number of ways we may choose r objects, we may apply the Addition principle referred to earlier.

If a_1 is excluded, then we must choose the desired r objects from the remaining $(n-1)$ objects and there are $\binom{n-1}{r}$ ways of doing this.

If a_1 is to be included, then only $(r-1)$ more objects must be chosen from the remaining $(n-1)$ objects and this can be done in $\binom{n-1}{r-1}$ ways. Thus the required number is the *sum* of these two, which verifies the second identity.

Note: In the above context the binomial coefficients $\binom{n}{k}$ are meaningful only if n and k are nonnegative integers with $0 \le k \le n$. However, if we write

$$\binom{n}{k} = \frac{n!}{k!(n-k)!} = \frac{n(n-1)\cdots(n-k+1)}{k!}$$

we observe that the latter expression is meaningful if n is *any* real number and k is any nonnegative integer. Thus,

$$\binom{-3}{5} = \frac{(-3)(-4)\cdots(-7)}{5!},$$

and so on.

Using this extended version of the binomial coefficients we can state the *generalized form of the binomial theorem:*

$$(1 + x)^n = \sum_{k=0}^{\infty} \binom{n}{k} x^k$$

This series is meaningful for any real n and for all x such that $|x| < 1$. Observe that if n is a positive integer, the infinite series reduces to a finite number of terms since in that case $\binom{n}{k} = 0$ if $k > n$.

EXAMPLE 2.7. (a) From eight persons, how many committees of three members may be chosen? Since two committees are the same if they are made up of the same members (regardless of the order in which they were chosen), we have $\binom{8}{3} = 56$ possible committees.

(b) From eight different flags, how many signals made up of three flags may be obtained? This problem seems very much like the one above. However, here order does make a difference, and hence we obtain $8!/5! = 336$ signals.

(c) A group of eight persons consists of five men and three women. How many committees of three may be formed consisting of exactly two men? Here we must do two things: choose two men (out of five) and choose one woman (out of three). Hence we obtain for the required number $\binom{5}{2} \cdot \binom{3}{1} = 30$ committees.

(d) We can now verify a statement made earlier, namely that the number of subsets of a set having n members is 2^n (counting the empty set and the set itself). Simply label each member with a one or a zero, depending on whether the member is to be included in, or excluded from, the subset. There are two ways of labeling each member, and there are n such members. Hence the multiplication principle tells us that there are $2 \cdot 2 \cdot 2 \cdots 2 = 2^n$ possible labelings. But each particular labeling represents a choice of a subset. For example, $(1, 1, 0, 0, 0, \ldots, 0)$ would consist of the subset made up of just a_1 and a_2. Again, $(1, 1, \ldots, 1)$ would represent S itself, and $(0, 0, \ldots, 0)$ would represent the empty set.

(e) We can obtain the above result by using the Addition principle as follows. In obtaining subsets we must choose the empty set, those subsets consisting of exactly one element, those consisting of exactly 2 elements, ..., and the set itself consisting of all n elements. This may be done in

$$\binom{n}{0} + \binom{n}{1} + \binom{n}{2} + \cdots + \binom{n}{n}$$

ways. However the sum of these binomial coefficients is simply the expansion of $(1 + 1)^n = 2^n$.

Let us now return to Example 2.4. From a lot consisting of 20 defective and 80 nondefective items we chose 10 at random (without replacement). The number of ways of doing this is $\binom{100}{10}$. Hence the probability of finding exactly 5 defective and 5 nondefective items among the chosen 10 is given by

$$\frac{\binom{20}{5}\binom{80}{5}}{\binom{100}{10}}.$$

By means of logarithms of factorials (which are tabulated) the above can be evaluated and equals 0.021.

EXAMPLE 2.8. Let us generalize the above problem. Suppose that we have N items. If we choose n of these at random, without replacement, there are $\binom{N}{n}$ different possible samples, all of which have the same probability of being chosen. If the N items are made up of r_1 A's and r_2 B's (with $r_1 + r_2 = N$), then the probability that the n chosen items contain exactly s_1 A's and $(n - s_1)$ B's is given by

$$\frac{\binom{r_1}{s_1}\binom{r_2}{n - s_1}}{\binom{N}{n}}.$$

(The above is called a *hypergeometric probability* and will be encountered again.)

Note: It is very important to specify, when we speak of choosing items at random, whether we choose *with* or *without replacement*. In most realistic descriptions we intend the latter. For instance, when we inspect a number of manufactured articles in order to discover how many defectives there might be, we usually do not intend to inspect the same item twice. We have noted previously that the number of ways of choosing r things out of n, disregarding order, is given by $\binom{n}{r}$. The number of ways of choosing r things out of n, with replacement, is given by n^r. Here we *are* concerned about the order in which the items were chosen.

EXAMPLE 2.9. Suppose that we choose two objects at random from the four objects labeled a, b, c, and d.

(a) If we choose without replacement, the sample space S may be represented as follows:

$$S = \{(a, b); (a, c); (b, c); (b, d); (c, d); (a, d)\}.$$

There are $\binom{4}{2} = 6$ possible outcomes. Each individual outcome indicates only *which* two objects were chosen and *not* the order in which they were chosen.

(b) If we choose with replacement, the sample space S' may be represented as follows:

$$S' = \begin{cases} (a, a);\ (a, b);\ (a, c);\ (a, d);\ (b, a);\ (b, b);\ (b, c);\ (b, d); \\ (c, a);\ (c, b);\ (c, c);\ (c, d);\ (d, a);\ (d, b);\ (d, c);\ (d, d) \end{cases}.$$

There are $4^2 = 16$ possible outcomes. Here each individual outcome indicates which objects were chosen *and* the order in which they were chosen. Choosing at random implies that if we choose without replacement, all the outcomes in S are equally likely, while if we choose with replacement, then all the outcomes in S' are equally likely. Thus, if A is the event {the object c is chosen}, then we have from S, $P(A) = \frac{3}{6} = \frac{1}{2}$ if we choose without replacement, and from S', $P(A) = \frac{7}{16}$ if we choose with replacement.

E. Permutations when not all objects are different. In all the enumeration methods introduced we have assumed that all the objects under consideration were different (that is, distinguishable). However, this is not always the case.

Suppose, then, that we have n objects such that there are n_1 of one kind, n_2 of a second kind, ..., n_k of a kth kind, where $n_1 + n_2 + \cdots + n_k = n$. Then the number of permutations of these n objects is given by

$$\frac{n!}{n_1! n_2! \cdots n_k!}.$$

The derivation of this formula will be left to the reader. Note that if all the objects *are* different, we have $n_i = 1, i = 1, 2, \ldots, k$, and hence the above formula reduces to $n!$, the previously obtained result.

Note: Let us emphasize once more that the realistic assignment of probabilities to individual outcomes of a sample space (or to a collection of outcomes, i.e., an event) is something which cannot be derived mathematically; it must be obtained from other considerations. For example, we may use certain symmetric features of the experiment to ascertain that all outcomes are equally likely. Again, we may construct a sampling procedure (e.g., choosing one or several individuals from a specified population) in such a manner that it is reasonable to assume that all choices are equally likely. In many other cases, when no apparent underlying assumption is appropriate, we must resort to the relative frequency approach. We repeat the experiment n times and then compute the proportion of times the outcome (or event) under consideration has occurred. In using this as an approximation we know that it is very unlikely that this relative frequency differs from the "true" probability (whose existence has been specified by our theoretical model) by an appreciable amount if n is sufficiently large. When it is impossible to make reasonable assumptions about the probability of an outcome and also impossible to repeat the experiment a large number of times (because of cost or time considerations, for instance), it is really quite meaningless to proceed with a probabilistic study of the experiment except on a purely theoretical basis. (For an additional remark on this same point, see Section 3.5.)

PROBLEMS

2.1.   The following group of persons is in a room: 5 men over 21, 4 men under 21, 6 women over 21, and 3 women under 21. One person is chosen at random. The following events are defined: A = {the person is over 21} ; B = {the person is under 21} ; C = {the person is male} ; D = {the person is female}. Evaluate the following.

(a) $P(B \cup D)$
(b) $P(\bar{A} \cap \bar{C})$

2.2.   Ten persons in a room are wearing badges marked 1 through 10. Three persons are chosen at random, and asked to leave the room simultaneously. Their badge number is noted.

(a) What is the probability that the smallest badge number is 5?
(b) What is the probability that the largest badge number is 5?

2.3.   (a) Suppose that the three digits 1, 2, and 3 are written down in random order. What is the probability that at least one digit will occupy its proper place?

(b) Same as (a) with the digits 1, 2, 3, and 4.
(c) Same as (a) with the digits 1, 2, 3, . . . , n. [Hint: Use (1.7).]
(d) Discuss the answer to (c) if n is large.

2.4.   A shipment of 1500 washers contains 400 defective and 1100 nondefective items. Two-hundred washers are chosen at random (without replacement) and classified.

(a) What is the probability that exactly 90 defective items are found?
(b) What is the probability that at least 2 defective items are found?

2.5.   Ten chips numbered 1 through 10 are mixed in a bowl. Two chips numbered (X, Y) are drawn from the bowl, successively and without replacement. What is the probability that $X + Y = 10$?

2.6.   A lot consists of 10 good articles, 4 with minor defects, and 2 with major defects. One article is chosen at random. Find the probability that:

(a) it has no defects,
(b) it has no major defects,
(c) it is either good or has major defects.

2.7.   If from the lot of articles described in Problem 2.6 two articles are chosen (without replacement), find the probability that:

(a) both are good,   (b) both have major defects,
(c) at least one is good,   (d) at most one is good,
(e) exactly one is good,   (f) neither has major defects,   (g) neither is good.

2.8.   A product is assembled in three stages. At the first stage there are 5 assembly lines, at the second stage there are 4 assembly lines, and at the third stage there are 6 assembly lines. In how many different ways may the product be routed through the assembly process?

2.9.   An inspector visits 6 different machines during the day. In order to prevent operators from knowing when he will inspect he varies the order of his visits. In how many ways may this be done?

2.10.   A complex mechanism may fail at 15 stages. If it fails at 3 stages, in how many ways may this happen?

2.11.   There are 12 ways in which a manufactured item can be a minor defective and 10 ways in which it can be a major defective. In how many ways can 1 minor and 1 major defective occur? 2 minor and 2 major defectives?

2.12.   A mechanism may be set at any one of four positions, say a, b, c, and d. There are 8 such mechanisms which are inserted into a system.

(a) In how many ways may this system be set?

(b) Assume that these mechanisms are installed in some preassigned (linear) order. How many ways of setting the system are available if no two adjacent mechanisms are in the same position?

(c) How many ways are available if only positions a and b are used, and these are used equally often?

(d) How many ways are available if only two different positions are used and one of these positions appears three times as often as the other?

2.13.   Suppose that from N objects we choose n at random, *with* replacement. What is the probability that no object is chosen more than once? (Suppose that $n < N$.)

2.14.   From the letters a, b, c, d, e, and f how many 4-letter code words may be formed if,

(a) no letter may be repeated?

(b) any letter may be repeated any number of times?

2.15.   Suppose that $\binom{99}{5} = a$ and $\binom{99}{4} = b$. Express $\binom{100}{95}$ in terms of a and b. [*Hint:* Do *not* evaluate the above expressions to solve this problem.]

2.16.   A box contains tags marked $1, 2, \ldots, n$. Two tags are chosen at random. Find the probability that the numbers on the tags will be consecutive integers if

(a) the tags are chosen without replacement,

(b) the tags are chosen with replacement.

2.17.   How many subsets can be formed, containing at least one member, from a set of 100 elements?

2.18.   One integer is chosen at random from the numbers $1, 2, \ldots, 50$. What is the probability that the chosen number is divisible by 6 or by 8?

2.19.   From 6 positive and 8 negative numbers, 4 numbers are chosen at random (without replacement) and multiplied. What is the probability that the product is a positive number?

2.20.   A certain chemical substance is made by mixing 5 separate liquids. It is proposed to pour one liquid into a tank, and then to add the other liquids in turn. All possible combinations must be tested to see which gives the best yield. How many tests must be performed?

2.21.   A lot contains n articles. If it is known that r of the articles are defective and the articles are inspected in a random order, what is the probability that the kth article $(k \geq r)$ inspected will be the last defective one in the lot?

2.22. r numbers $(0 < r < 10)$ are chosen at random (without replacement) from the numbers $0, 1, 2, \ldots, 9$. What is the probability that no two are equal?

3

Conditional Probability and Independence

3.1 Conditional Probability

Let us reconsider the difference between choosing an item at random from a lot with or without replacement. In Example 2.4 the lot under consideration had the following makeup: 80 nondefective and 20 defective items. Suppose that we choose two items from this lot, (a) with replacement; (b) without replacement.

We define the following two events.

$A = \{$the first item is defective$\}$, $\qquad B = \{$the second item is defective$\}$.

If we are choosing *with* replacement, $P(A) = P(B) = \frac{20}{100} = \frac{1}{5}$. For each time we choose from the lot there are 20 defective items among the total of 100. However, if we are choosing *without* replacement, the results are not quite immediate. It is still true, of course, that $P(A) = \frac{1}{5}$. But what about $P(B)$? It is clear that in order to compute $P(B)$ we should know the composition of the lot *at the time the second item is chosen*. That is, we should know whether A did or did not occur. This example indicates the need to introduce the following important concept.

Let A and B be two events associated with an experiment $\mathcal{E}$. We denote by $P(B \mid A)$ the *conditional probability* of the event B, *given that* A has occurred.

In the above example, $P(B \mid A) = \frac{19}{99}$. For, if A has occurred, then on the second drawing there are only 99 items left, 19 of which are defective.

Whenever we compute $P(B \mid A)$ we are essentially computing $P(B)$ with respect to the *reduced sample space* A, rather than with respect to the original sample space S. Consider the Venn diagram in Fig. 3.1.

When we evaluate $P(B)$ we are asking ourselves how probable it is that we shall be in B, knowing that we must be in S. And when we compute $P(B \mid A)$ we are asking ourselves how probable it is that we shall be in B, knowing that we must be in A. (That is, the sample space has been *reduced* from S to A.)

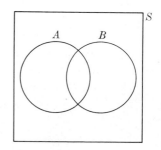

FIGURE 3.1

33

Shortly we shall make a formal definition of $P(B \mid A)$. For the moment, however, we shall proceed with our intuitive notion of conditional probability and consider an example.

EXAMPLE 3.1. Two fair dice are tossed, the outcome being recorded as (x_1, x_2), where x_i is the outcome of the ith die, $i = 1, 2$. Hence the sample space S may be represented by the following array of 36 equally likely outcomes.

$$S = \begin{Bmatrix} (1, 1) & (1, 2) & \cdots & (1, 6) \\ (2, 1) & (2, 2) & \cdots & (2, 6) \\ \vdots & & & \vdots \\ (6, 1) & (6, 2) & \cdots & (6, 6) \end{Bmatrix}.$$

Consider the following two events:

$$A = \{(x_1, x_2) \mid x_1 + x_2 = 10\}, \qquad B = \{(x_1, x_2) \mid x_1 > x_2\}.$$

Thus $A = \{(5, 5), (4, 6), (6, 4)\}$, and $B = \{(2, 1), (3, 1), (3, 2), \ldots, (6, 5)\}$. Hence $P(A) = \frac{3}{36}$ and $P(B) = \frac{15}{36}$. And $P(B \mid A) = \frac{1}{3}$, since the sample space now consists of A (that is three outcomes), and only one of these three outcomes is consistent with the event B. In a similar way we may compute $P(A \mid B) = \frac{1}{15}$.

Finally, let us compute $P(A \cap B)$. The event $A \cap B$ occurs if and only if the sum of the two dice is 10 *and* the first die shows a larger value than the second die. There is only *one* such outcome, and hence $P(A \cap B) = \frac{1}{36}$. If we take a long careful look at the various numbers we have computed above, we note the following:

$$P(A \mid B) = \frac{P(A \cap B)}{P(B)} \qquad \text{and} \qquad P(B \mid A) = \frac{P(A \cap B)}{P(A)}.$$

These relationships did not just happen to arise in the particular example we considered. Rather, they are quite general and give us a means of *formally defining* conditional probability.

To motivate this definition let us return to the concept of relative frequency. Suppose that an experiment ε has been repeated n times. Let n_A, n_B, and $n_{A \cap B}$ be the number of times the events A, B, and $A \cap B$, respectively, have occurred among the n repetitions. What is the meaning of $n_{A \cap B}/n_A$? It represents the relative frequency of B among those outcomes in which A occurred. That is, $n_{A \cap B}/n_A$ is the conditional relative frequency of B, given that A occurred.

We may write $n_{A \cap B}/n_A$ as follows:

$$\frac{n_{A \cap B}}{n_A} = \frac{n_{A \cap B}/n}{n_A/n} = \frac{f_{A \cap B}}{f_A},$$

where $f_{A \cap B}$ and f_A are the relative frequencies of the events $A \cap B$ and A, respectively. As we have already indicated (and as we shall show later), if n, the number of repetitions is large, $f_{A \cap B}$ will be close to $P(A \cap B)$ and f_A will be close to $P(A)$. Hence the above relation suggests that $n_{A \cap B}/n_A$ will be close to $P(B|A)$. Thus we make the following formal definition.

Definition

$$P(B \mid A) = \frac{P(A \cap B)}{P(A)}, \qquad \text{provided that} \quad P(A) > 0. \qquad (3.1)$$

Notes: (a) It is important to realize that the above is not a theorem (we did not prove anything), nor is it an axiom. We have simply introduced the intuitive notion of conditional probability and then made the formal definition of what we mean by this notion. The fact that our formal definition corresponds to our intuitive notion is substantiated by the paragraph preceding the definition.

(b) It is a simple matter to verify that $P(B \mid A)$, for fixed A, satisfies the various postulates of probability Eq. (1.3). (See Problem 3.22.) That is, we have

(1') $\ 0 \leq P(B \mid A) \leq 1$,
(2') $\ P(S \mid A) = 1$,
(3') $\ P(B_1 \cup B_2 \mid A) = P(B_1 \mid A) + P(B_2 \mid A) \qquad$ if $\ B_1 \cap B_2 = \emptyset,$ $\qquad$ (3.2)
(4') $\ P(B_1 \cup B_2 \cup \cdots \mid A) = P(B_1 \mid A) + P(B_2 \mid A) + \cdots \qquad$ if $\ B_i \cap B_j = \emptyset$
$\qquad$ for $i \neq j$.

(c) If $A = S$, $P(B|S) = P(B \cap S)/P(S) = P(B)$.

(d) With every event $B \subset S$ we can associate two numbers, $P(B)$, the (unconditional) probability of B, and $P(B \mid A)$, the conditional probability of B, given that some event A (for which $P(A) > 0$) has occurred. In general, these two probability measures will assign different probabilities to the event B, as the preceding examples indicated. Shortly we shall study an important special case for which $P(B)$ and $P(B \mid A)$ are the same.

(e) Observe that the conditional probability is defined in terms of the unconditional probability measure P. That is, if we know $P(B)$ for every $B \subset S$, we can compute $P(B \mid A)$ for every $B \subset S$.

Thus we have two ways of computing the conditional probability $P(B \mid A)$:

(a) Directly, by considering the probability of B with respect to the reduced sample space A.

(b) Using the above definition, where $P(A \cap B)$ and $P(A)$ are computed with respect to the original sample space S.

Note: If $A = S$, we obtain $P(B \mid S) = P(B \cap S)/P(S) = P(B)$, since $P(S) = 1$ and $B \cap S = B$. This is as it should be, for saying that S has occurred is only saying that the experiment *has* been performed.

EXAMPLE 3.2. Suppose that an office has 100 calculating machines. Some of these machines are electric (E) while others are manual (M). And some of the machines are new (N) while others are used (U). Table 3.1 gives the number of machines in each category. A person enters the office, picks a machine at random,

TABLE 3.1

	E	M	
N	40	30	70
U	20	10	30
	60	40	100

and discovers that it is new. What is the probability that it is electric? In terms of the notation introduced we wish to compute $P(E \mid N)$.

Simply considering the reduced sample space N (e.g., the 70 new machines), we have $P(E \mid N) = \frac{40}{70} = \frac{4}{7}$. Using the definition of conditional probability, we have that

$$P(E \mid N) = \frac{P(E \cap N)}{P(N)} = \frac{40/100}{70/100} = \frac{4}{7}.$$

The most important consequence of the above definition of conditional probability is obtained by writing it in the following form:

$$P(A \cap B) = P(B \mid A)P(A)$$

or, equivalently,   (3.3a)

$$P(A \cap B) = P(A \mid B)P(B).$$

This is sometimes known as the *multiplication theorem* of probability.

We may apply this theorem to compute the probability of the simultaneous occurrence of two events A and B.

EXAMPLE 3.3. Consider again the lot consisting of 20 defective and 80 nondefective items discussed at the beginning of Section 3.1. If we choose two items at random, without replacement, what is the probability that both items are defective? As before, we define the events A and B as follows:

$A = \{$the first item is defective$\}$, $B = \{$the second item is defective$\}$.

Hence we require $P(A \cap B)$, which we may compute, according to the above formula, as $P(B \mid A)P(A)$. But $P(B \mid A) = \frac{19}{99}$, while $P(A) = \frac{1}{5}$. Hence

$$P(A \cap B) = \frac{19}{495}.$$

Note: The above multiplication theorem (3.3a) may be generalized to more than two events in the following way:

$$P[A_1 \cap A_2 \cap \cdots \cap A_n] = P(A_1)P(A_2 \mid A_1)P(A_3 \mid A_1, A_2) \cdots P(A_n \mid A_1, \ldots, A_{n-1}).$$
                                                                                                    (3.3b)

Let us consider for a moment whether we can make a general statement about the relative magnitude of $P(A \mid B)$ and $P(A)$. We shall consider four cases, which are illustrated by the Venn diagrams in Fig. 3.2. We have

(a) $P(A \mid B) = 0 \leq P(A)$, since A cannot occur if B has occurred.

(b) $P(A \mid B) = P(A \cap B)/P(B) = [P(A)/P(B)] \geq P(A)$, since $0 \leq P(B) \leq 1$.

(c) $P(A \mid B) = P(A \cap B)/P(B) = P(B)/P(B) = 1 \geq P(A)$.

(d) In this case we cannot make any statement about the relative magnitude of $P(A \mid B)$ and $P(A)$.

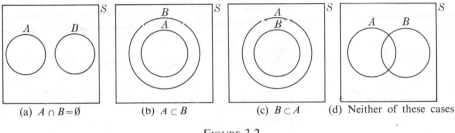

(a) $A \cap B = \emptyset$ (b) $A \subset B$ (c) $B \subset A$ (d) Neither of these cases

FIGURE 3.2

Note that in two of the above cases, $P(A) \leq P(A \mid B)$, in one case, $P(A) \geq P(A \mid B)$, and in the fourth case, we cannot make any comparison at all.

Above, we used the concept of conditional probability in order to evaluate the probability of the simultaneous occurrence of two events. We can apply this concept in another way to compute the probability of a single event A. We need the following definition.

Definition. We say that the events $B_1, B_2, \ldots, B_k$ represent a *partition* of the sample space S if

(a) $B_i \cap B_j = \emptyset$ for all $i \neq j$.

(b) $\bigcup_{i=1}^{k} B_i = S$.

(c) $P(B_i) > 0$ for all i.

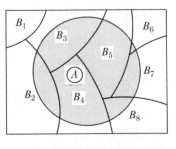

In words: When the experiment ε is performed *one and only* one of the events B_i occurs.

FIGURE 3.3

(For example, for the tossing of a die $B_1 = \{1, 2\}$, $B_2 = \{3, 4, 5\}$, and $B_3 = \{6\}$ would represent a partition of the sample space, while $C_1 = \{1, 2, 3, 4\}$ and $C_2 = \{4, 5, 6\}$ would not.)

Let A be some event with respect to S and let $B_1, B_2, \ldots, B_k$ be a partition of S. The Venn diagram in Fig. 3.3 illustrates this for $k = 8$. Hence we may write

$$A = A \cap B_1 \cup A \cap B_2 \cup \cdots \cup A \cap B_k.$$

Of course, some of the sets $A \cap B_j$ may be empty, but this does not invalidate the above decomposition of A. The important point is that all the events $A \cap B_1, \ldots, A \cap B_k$ are pairwise mutually exclusive. Hence we may apply the addition property for mutually exclusive events (Eq. 1.3) and write

$$P(A) = P(A \cap B_1) + P(A \cap B_2) + \cdots + P(A \cap B_k).$$

However each term $P(A \cap B_j)$ may be expressed as $P(A \mid B_j)P(B_j)$ and hence we obtain what is called the theorem on *total probability:*

$$P(A) = P(A \mid B_1)P(B_1) + P(A \mid B_2)P(B_2) + \cdots + P(A \mid B_k)P(B_k). \qquad (3.4)$$

This result represents an extremely useful relationship. For often when $P(A)$ is required, it may be difficult to compute it directly. However, with the additional information that B_j has occurred, we may be able to evaluate $P(A \mid B_j)$ and then use the above formula.

EXAMPLE 3.4. Consider (for the last time) the lot of 20 defective and 80 non-defective items from which we choose two items *without replacement.* Again defining A and B as

$$A = \{\text{the first chosen item is defective}\},$$
$$B = \{\text{the second chosen item is defective}\},$$

we may now compute $P(B)$ as follows:

$$P(B) = P(B \mid A)P(A) + P(B \mid \overline{A})P(\overline{A}).$$

Using some of the calculations performed in Example 3.3, we find that

$$P(B) = \tfrac{19}{99} \cdot \tfrac{1}{5} + \tfrac{20}{99} \cdot \tfrac{4}{5} = \tfrac{1}{5}.$$

This result may be a bit startling, particularly if the reader recalls that at the beginning of Section 3.1 we found that $P(B) = \tfrac{1}{5}$ when we chose the items *with* replacement.

EXAMPLE 3.5. A certain item is manufactured by three factories, say 1, 2, and 3. It is known that 1 turns out twice as many items as 2, and that 2 and 3 turn out the same number of items (during a specified production period). It is also known that 2 percent of the items produced by 1 and by 2 are defective, while 4 percent of those manufactured by 3 are defective. All the items produced are put into one stockpile, and then one item is chosen at random. What is the probability that this item is defective?

Let us introduce the following events:

$$A = \{\text{the item is defective}\}, \qquad B_1 = \{\text{the item came from 1}\},$$
$$B_2 = \{\text{the item came from 2}\}, \qquad B_3 = \{\text{the item came from 3}\}.$$

We require $P(A)$ and, using the above result, we may write

$$P(A) = P(A \mid B_1)P(B_1) + P(A \mid B_2)P(B_2) + P(A \mid B_3)P(B_3).$$

Now $P(B_1) = \frac{1}{2}$, while $P(B_2) = P(B_3) = \frac{1}{4}$. Also $P(A \mid B_1) = P(A \mid B_2) = 0.02$, while $P(A \mid B_3) = 0.04$. Inserting these values into the above expression, we obtain $P(A) = 0.025$.

Note: The following analogy to the theorem on total probability has been observed in chemistry: Suppose that we have k beakers containing different solutions of the same salt, totaling, say one liter. Let $P(B_i)$ be the volume of the ith beaker and let $P(A \mid B_i)$ be the concentration of the solution in the ith beaker. If we combine all the solutions into one beaker and let $P(A)$ denote the concentration of the resulting solution, we obtain,

$$P(A) = P(A \mid B_1)P(B_1) + \cdots + P(A \mid B_k)P(B_k).$$

3.2 Bayes' Theorem

We may use Example 3.5 to motivate another important result. Suppose that one item is chosen from the stockpile and is found to be defective. What is the probability that it was produced in factory 1?

Using the notation introduced previously, we require $P(B_1 \mid A)$. We can evaluate this probability as a consequence of the following discussion. Let $B_1, \ldots, B_k$ be a partition of the sample space S and let A be an event associated with S. Applying the definition of conditional probability, we may write

$$P(B_i \mid A) = \frac{P(A \mid B_i)P(B_i)}{\sum_{j=1}^{k} P(A \mid B_j)P(B_j)} \qquad i = 1, 2, \ldots, k. \qquad (3.5)$$

This result is known as *Bayes' theorem*. It is also called the formula for the probability of "causes." Since the B_i's are a partition of the sample space, one and only one of the events B_i occurs. (That is, *one* of the events B_i must occur and only one can occur.) Hence the above formula gives us the probability of a particular B_i (that is, a "cause"), given that the event A *has* occurred. In order to apply this theorem we must know the values of the $P(B_i)$'s. Quite often these values are not known, and this limits the applicability of the result. There has been considerable controversy about Bayes' theorem. It is perfectly correct mathematically; only the improper choice for $P(B_i)$ makes the result questionable.

Returning to the question posed above, and now applying Eq. (3.5), we obtain

$$P(B_1 \mid A) = \frac{(0.02)(1/2)}{(0.02)(1/2) + (0.02)(1/4) + (0.04)(1/4)} = 0.40.$$

Note: We can again find an analogy, from chemistry, to Bayes' theorem. In k beakers we have solutions of the same salt, but of different concentrations. Suppose that the total volume of the solutions is one liter. Denoting the volume of the solution in the ith beaker by $P(B_i)$ and denoting the concentration of the salt in the ith beaker by $P(A \mid B_i)$, we find that Eq. (3.5) yields the proportion of the entire amount of salt which is found in the ith beaker.

The following illustration of Bayes' theorem will give us an opportunity to introduce the idea of a *tree diagram*, a rather useful device for analyzing certain problems.

Suppose that a large number of containers of candy are made up of two types, say A and B. Type A contains 70 percent sweet and 30 percent sour ones while for type B these percentages are reversed. Furthermore, suppose that 60 percent of all candy jars are of type A while the remainder are of type B.

You are now confronted with the following decision problem. A jar of unknown type is given to you. You are allowed to sample one piece of candy (an admittedly unrealistic situation but one which will allow us to introduce the relevant ideas without getting too involved), and with this information, you must decide whether to guess that type A or type B has been offered to you. The following "tree diagram" (so called because of the various paths or branches which appear) will help us to analyze the problem. (S_W and S_O stand for choosing a sweet or sour candy, respectively.)

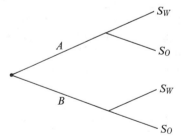

Let us make a few computations:

$$P(A) = 0.6; \quad P(B) = 0.4; \; P(S_W \mid A) = 0.7;$$
$$P(S_O \mid A) = 0.3; \; P(S_W \mid B) = 0.3; \; P(S_O \mid B) = 0.7.$$

What we really wish to know is $P(A \mid S_W)$, $P(A \mid S_O)$, $P(B \mid S_W)$, and $P(B \mid S_O)$. That is, suppose we actually pick a sweet piece of candy. What decision would we be most tempted to make? Let us compare $P(A \mid S_W)$ and $P(B \mid S_W)$. Using

Bayes' formula we have

$$P(A \mid S_W) = \frac{P(S_W \mid A)P(A)}{P(S_W \mid A)P(A) + P(S_W \mid B)P(B)} = \frac{(0.7)(0.6)}{(0.7)(0.6) + (0.3)(0.4)} = \frac{7}{9}.$$

A similar computation yields $P(B \mid S_W) = 2/9$.

Thus, based on the evidence we have (i.e., the obtaining of a sweet candy) it is $2\frac{1}{2}$ times as likely that we are dealing with a container of type A rather than one of type B. Hence we would, presumably, decide that a container of type A was involved. (We could, of course, be wrong. The point of the above analysis is that we are choosing that alternative which appears most likely based on the limited evidence we have.)

In terms of the tree diagram, what was really required (and done) in the preceding calculation was a "backward" analysis. That is, given what we observed, S_W in this case, how probable is it that type A is involved?

A somewhat more interesting situation arises if we are allowed to choose *two* pieces of candy before deciding whether type A or type B is involved. In this case, the tree diagram would appear as follows.

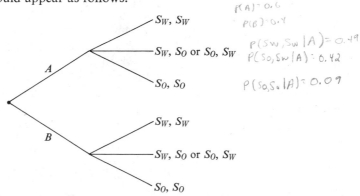

In problem 3.26 you are asked to decide which of the two types, A or B, you are sampling from, depending on which of the three possible experimental outcomes you observe.

3.3 Independent Events

We have considered events A and B which could not occur simultaneously, that is, $A \cap B = \emptyset$. Such events were called mutually exclusive. We have noted previously that if A and B are mutually exclusive, then $P(A \mid B) = 0$, for the given occurrence of B precludes the occurrence of A. At the other extreme we have the situation, also discussed above, in which $B \supset A$ and hence $P(B \mid A) = 1$.

In each of the above situations, knowing that B has occurred gave us some very definite information concerning the probability of the occurrence of A. However, there are many situations in which knowing that some event B did occur has no bearing whatsoever on the occurrence or nonoccurrence of A.

EXAMPLE 3.6. Suppose that a fair die is tossed twice. Define the events A and B as follows:

$$A = \{\text{the first die shows an even number}\},$$
$$B = \{\text{the second die shows a 5 or a 6}\}.$$

It is intuitively clear that events A and B are totally unrelated. Knowing that B did occur yields no information about the occurrence of A. In fact, the following computation bears this out. Taking as our sample space the 36 equally likely outcomes considered in Example 3.1, we find that $P(A) = \frac{18}{36} = \frac{1}{2}$, $P(B) = \frac{12}{36} = \frac{1}{3}$, while $P(A \cap B) = \frac{6}{36} = \frac{1}{6}$. Hence

$$P(A \mid B) = \frac{P(A \cap B)}{P(B)} = \frac{(\frac{1}{6})}{(\frac{1}{3})} = \frac{1}{2}.$$

Thus we find, as we might have expected, that the unconditional probability is equal to the conditional probability $P(A \mid B)$. Similarly,

$$P(B \mid A) = \frac{P(B \cap A)}{P(A)} = \frac{(\frac{1}{6})}{(\frac{1}{2})} = \frac{1}{3} = P(B).$$

Hence we might be tempted to say that A and B are independent if and only if $P(A \mid B) = P(A)$ and $P(B \mid A) = P(B)$. Although this would be essentially appropriate there is another approach which gets around the difficulty encountered here, namely that both $P(A)$ and $P(B)$ must be nonzero before the above equalities are meaningful.

Consider $P(A \cap B)$, assuming that the above conditional probabilities are equal to the corresponding unconditional probabilities. We have

$$P(A \cap B) = P(A \mid B)P(B) = P(A)P(B),$$
$$P(A \cap B) = P(B \mid A)P(A) = P(B)P(A).$$

Thus we find that provided neither $P(A)$ nor $P(B)$ equals zero, the unconditional probabilities are equal to the conditional probabilities if and only if $P(A \cap B) = P(A)P(B)$. Hence we make the following formal definition. [If either $P(A)$ or $P(B)$ equals zero, this definition is still valid.]

Definition. A and B are *independent events* if and only if

$$P(A \cap B) = P(A)P(B). \tag{3.6}$$

Note: This definition is essentially equivalent to the one suggested above, namely, that A and B are independent if $P(B \mid A) = P(B)$ and $P(A \mid B) = P(A)$. This latter form is slightly more intuitive, for it says precisely what we have been trying to say before: that A and B are independent if knowledge of the occurrence of A in no way influences the probability of the occurrence of B.

That the formal definition adopted also has a certain intuitive appeal may be seen by considering the following example.

EXAMPLE 3.7. Let us again consider Example 3.2. We will first consider the table below with only the marginal values given. That is, there are 60 electric and

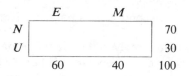

	E	M	
N			70
U			30
	60	40	100

40 manual machines, and of these 70 are new while 30 are used. There are many ways in which the entries of the table could be filled, consistent with the marginal totals given. Below we list a few of these possibilities.

	E	M	
N	60	10	70
U	0	30	30
	60	40	100

(a)

	E	M	
N	30	40	70
U	30	0	30
	60	40	100

(b)

	E	M	
N	42	28	70
U	18	12	30
	60	40	100

(c)

Consider Table (a). Here *all* the electric machines are new, and *all* the used machines are manual. Thus there is an obvious connection (not necessarily causal) between the characteristic of being electric and being new. Similarly in Table (b) *all* the manual machines are new, and *all* the used machines are electric. Again a definite connection seems to exist between these characteristics. However, when we turn to Table (c) the matter is quite different. Here no apparent relationship exists. For example, 60 percent of all machines are electric, and exactly 60 percent of the used machines are electric. Similarly, 70 percent of all the machines are new, and exactly 70 percent of the manual machines are new, etc. Thus no indication is evident that the characteristics of "newness" and "being electric" have any connection with each other. Of course, this table was constructed precisely in a way to exhibit this property. How were the entries of the table obtained? Simply by applying Eq. (3.6); that is, since $P(E) = \frac{60}{100}$ and $P(N) = \frac{70}{100}$, we must have, for independence, $P(E \cap N) = P(E)P(N) = \frac{42}{100}$. Hence the entry in the table indicating the number of new electric machines is given by number 42. The other entries were obtained in a similar fashion.

In most applications we shall *hypothesize* the independence of two events A and B, and then use this assumption to compute $P(A \cap B)$ as $P(A)P(B)$. Usually, physical conditions under which the experiment is performed will make it possible to decide whether such an assumption is justified or at least approximately justified.

EXAMPLE 3.8. Consider a large lot of items, say 10,000. Suppose that 10 percent of these items are defective and 90 percent are nondefective. Two items are chosen. What is the probability that both items are nondefective?

Define the events A and B as follows:

$$A = \{\text{first item is nondefective}\},$$
$$B = \{\text{second item is nondefective}\}.$$

If we assume that the first item is replaced before the second one is chosen, then the events A and B may be assumed independent and hence $P(A \cap B) = (0.9)(0.9) = 0.81$. More realistically, however, the second item is chosen without replacing the first item. In this case,

$$P(A \cap B) = P(B \mid A)P(A) = \tfrac{8999}{9999}\,(0.9)$$

which is approximately 0.81. Thus, although A and B are not independent in the second case, the assumption of independence, which simplifies calculations considerably, causes only a negligible error. (Recall the purpose of a mathematical model as described in Section 1.1.) If there had been only a few items in the lot, say 30, the assumption of independence would have resulted in a larger error. Thus it becomes important to check carefully the conditions under which the experiment is performed in order to establish the validity of an assumption of independence between various events.

EXAMPLE 3.9. Suppose that a mechanism is made up of two components hooked up in series as indicated in Fig. 3.4. Each component has a probability p of not working. What is the probability that the mechanism does work?

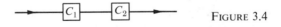

FIGURE 3.4

It is clear that the mechanism will work if and only if *both* components are functioning. Hence

Prob (mechanism works) = Prob (C_1 functions *and* C_2 functions).

The information we are given does not allow us to proceed unless we know (or assume) that the two mechanisms work independently of each other. This may or may not be a realistic assumption, depending on how the two parts are hooked up. If we assume that the two mechanisms work independently, we obtain for the required probability $(1 - p)^2$.

It will be important for us to extend the above notion of independence to more than two events. Let us first consider three events associated with an experiment, say A, B, and C. If A and B, A and C, B and C are each *pairwise* independent (in the above sense), then it does not follow, in general, that there exists no dependence between the three events A, B, and C. The following (somewhat artificial) example illustrates this point.

EXAMPLE 3.10. Suppose that we toss two dice. Define the events A, B, and C as follows:

$$A = \{\text{the first die shows an even number}\},$$
$$B = \{\text{the second die shows an odd number}\},$$
$$C = \{\text{the two dice show both odd or both even numbers}\}.$$

We have $P(A) = P(B) = P(C) = \frac{1}{2}$. Furthermore, $P(A \cap B) = P(A \cap C) = P(B \cap C) = \frac{1}{4}$. Hence the three events are all pairwise independent. However, $P(A \cap B \cap C) = 0 \neq P(A)P(B)P(C)$.

This example motivates the following definition.

Definition. We say that the three events A, B, and C are *mutually independent* if and only if *all* the following conditions hold:

$$P(A \cap B) = P(A)P(B), \qquad P(A \cap C) = P(A)P(C),$$
$$P(B \cap C) = P(B)P(C), \qquad P(A \cap B \cap C) = P(A)P(B)P(C). \tag{3.7}$$

We finally generalize this notion to n events in the following definition.

Definition. The n events $A_1, A_2, \ldots, A_n$ are mutually independent if and only if we have for $k = 2, 3, \ldots, n$,

$$P(A_{i_1} \cap A_{i_2} \cap \cdots \cap A_{i_k}) = P(A_{i_1})P(A_{i_2}) \cdots P(A_{i_k}). \tag{3.8}$$

(There are altogether $2^n - n - 1$ conditions listed; see Problem 3–18.)

Note: In most applications we need not check all these conditions, for we usually *assume* independence (based on what we know about the experiment). We then use this assumption to evaluate, say $P(A_{i_1} \cap A_{i_2} \cap \cdots \cap A_{i_k})$ as $P(A_{i_1})P(A_{i_2}) \cdots P(A_{i_k})$.

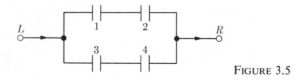

FIGURE 3.5

EXAMPLE 3.11. The probability of the closing of each relay of the circuit shown in Fig. 3.5 is given by p. If all relays function independently, what is the probability that a current exists between the terminals L and R?

Let A_i represent the event $\{$relay i is closed$\}$, $i = 1, 2, 3, 4$. Let E represent the event $\{$current flows from L to $R\}$. Hence $E = (A_1 \cap A_2) \cup (A_3 \cap A_4)$. (Note that $A_1 \cap A_2$ and $A_3 \cap A_4$ are *not* mutually exclusive.) Thus

$$P(E) = P(A_1 \cap A_2) + P(A_3 \cap A_4) - P(A_1 \cap A_2 \cap A_3 \cap A_4)$$
$$= p^2 + p^2 - p^4 = 2p^2 - p^4.$$

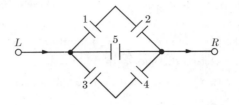

FIGURE 3.6

EXAMPLE 3.12. Assume again that for the circuit in Fig. 3.6, the probability of each relay being closed is p and that all relays function independently. What is the probability that current exists between the terminals L and R?

Using the same notation as in Example 3.11, we have that

$$P(E) = P(A_1 \cap A_2) + P(A_5) + P(A_3 \cap A_4) - P(A_1 \cap A_2 \cap A_5)$$
$$- P(A_1 \cap A_2 \cap A_3 \cap A_4) - P(A_5 \cap A_3 \cap A_4)$$
$$+ P(A_1 \cap A_2 \cap A_3 \cap A_4 \cap A_5)$$
$$= p^2 + p + p^2 - p^3 - p^4 - p^3 + p^5 = p + 2p^2 - 2p^3 - p^4 + p^5.$$

Let us close this chapter by indicating a fairly common, but erroneous, approach to a problem.

EXAMPLE 3.13. Suppose that among six bolts, two are shorter than a specified length. If two bolts are chosen at random, what is the probability that the two short bolts are picked? Let A_i be the event {the ith chosen bolt is short}, $i = 1, 2$.

Hence we want to evaluate $P(A_1 \cap A_2)$. The proper solution is obtained, of course, by writing

$$P(A_1 \cap A_2) = P(A_2 \mid A_1)P(A_1) = \tfrac{1}{5} \cdot \tfrac{2}{6} = \tfrac{1}{15}.$$

The common but *incorrect* approach is to write

$$P(A_1 \cap A_2) = P(A_2)P(A_1) = \tfrac{1}{5} \cdot \tfrac{2}{6} = \tfrac{1}{15}.$$

Of course, the point is that although the answer is correct, the identification of $\tfrac{1}{5}$ with $P(A_2)$ is incorrect; $\tfrac{1}{5}$ represents $P(A_2 \mid A_1)$. To evaluate $P(A_2)$ properly, we write

$$P(A_2) = P(A_2 \mid A_1)P(A_1) + P(A_2 \mid \overline{A}_1)P(\overline{A}_1) = \tfrac{1}{5} \cdot \tfrac{2}{6} + \tfrac{2}{5} \cdot \tfrac{4}{6} = \tfrac{1}{3}.$$

3.4 Schematic Considerations; Conditional Probability and Independence

The following schematic approach may be useful for understanding the concept of conditional probability. Suppose that A and B are two events associated with a sample space for which the various probabilities are indicated in the Venn diagram given in Fig. 3.7.

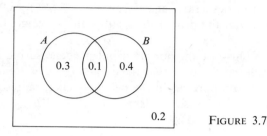

FIGURE 3.7

Hence $P(A \cap B) = 0.1$, $P(A) = 0.1 + 0.3 = 0.4$, and $P(B) = 0.1 + 0.4 = 0.5$.

Next, let us represent the various probabilities by the *areas* of the rectangles as in Fig. 3.8. In each case, the shaded regions indicate the event B: In the left rectangle we are representing $A \cap B$ and in the right one $A' \cap B$.

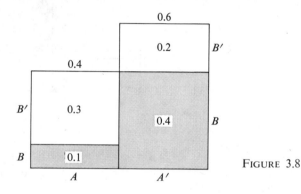

FIGURE 3.8

Now suppose we wish to compute $P(B \mid A)$. Thus we need only to consider A; that is, A' may be ignored in the computation. We note that the proportion of B in A is $1/4$. (We can also check this by applying Eq. (3.1): $P(B \mid A) = P(A \cap B)/P(A) = 0.1/0.4 = 1/4$.) Hence $P(B' \mid A) = 3/4$, and our diagram representing this conditional probability would be given by Fig. 3.9.

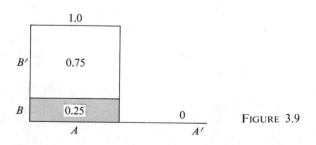

FIGURE 3.9

Observe also that if A is given to have occurred, all of the probability (i.e., 1) must be associated with the event A while none of the probability (i.e., 0) is associated with A'. Furthermore note that in the left rectangle, representing A, only the individual entries have changed from Fig. 3.8 to Fig. 3.9 (adding up to 1 instead of 0.4). However, the proportions within the rectangle have remained the same (i.e., 3 : 1).

Let us also illustrate the notion of independence using the schematic approach introduced above. Suppose that the events A and B are as depicted in Fig. 3.10. In that case the proportions in the two rectangles, representing A and A', are the *same*: 3 : 1 in both cases. Thus we have $P(B) = 0.1 + 0.15 = 0.25$, and $P(B \cap A) = 0.1/0.4 = 0.25$.

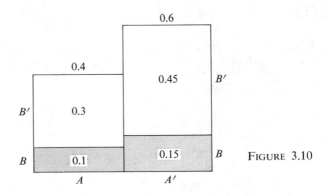

FIGURE 3.10

Finally observe that by simply looking at Fig. 3.8 we can also compute the other conditional probabilities:

$P(A \mid B) = 1/5$ (since 1/5 of the total rectangular area representing B is occupied by A),

$P(A' \mid B) = 4/5.$

PROBLEMS

3.1. Urn 1 contains x white and y red balls. Urn 2 contains z white and v red balls. A ball is chosen at random from urn 1 and put into urn 2. *Then* a ball is chosen at random from urn 2. What is the probability that this ball is white?

3.2. Two defective tubes get mixed up with two good ones. The tubes are tested, one by one, until both defectives are found.

(a) What is the probability that the last defective tube is obtained on the second test?

(b) What is the probability that the last defective tube is obtained on the third test?

(c) What is the probability that the last defective tube is obtained on the fourth test?

(d) Add the numbers obtained for (a), (b), and (c) above. Is the result surprising?

3.3.   A box contains 4 bad and 6 good tubes. Two are drawn out together. One of them is tested and found to be good. What is the probability that the other one is also good?

3.4.   In the above problem the tubes are checked by drawing a tube at random, testing it and repeating the process until all 4 bad tubes are located. What is the probability that the fourth bad tube will be located

(a) on the fifth test?
(b) on the tenth test?

3.5.   Suppose that A and B are independent events associated with an experiment. If the probability that A or B occurs equals 0.6, while the probability that A occurs equals 0.4, determine the probability that B occurs.

3.6.   Twenty items, 12 of which are defective and 8 nondefective, are inspected one after the other. If these items are chosen at random, what is the probability that:

(a) the first two items inspected are defective?
(b) the first two items inspected are nondefective?
(c) among the first two items inspected there is one defective and one nondefective?

3.7.   Suppose that we have two urns, 1 and 2, each with two drawers. Urn 1 has a gold coin in one drawer and a silver coin in the other drawer, while urn 2 has a gold coin in each drawer. One urn is chosen at random; then a drawer is chosen at random from the chosen urn. The coin found in this drawer turns out to be gold. What is the probability that the coin came from urn 2?

3.8.   A bag contains three coins, one of which is coined with two heads while the other two coins are normal and not biased. A coin is chosen at random from the bag and tossed four times in succession. If heads turn up *each* time, what is the probability that this is the two-headed coin?

3.9.   In a bolt factory, machines A, B, and C manufacture 25, 35, and 40 percent of the total output, respectively. Of their outputs, 5, 4, and 2 percent, respectively, are defective bolts. A bolt is chosen at random and found to be defective. What is the probability that the bolt came from machine A? B? C?

3.10.   Let A and B be two events associated with an experiment. Suppose that $P(A) = 0.4$ while $P(A \cup B) = 0.7$. Let $P(B) = p$.

(a) For what choice of p are A and B mutually exclusive?
(b) For what choice of p are A and B independent?

3.11.   Three components of a mechanism, say C_1, C_2, and C_3 are placed in series (in a straight line). Suppose that these mechanisms are arranged in a random order. Let R be the event $\{C_2$ is to the right of $C_1\}$, and let S be the event $\{C_3$ is to the right of $C_1\}$. Are the events R and S independent? Why?

3.12.   A die is tossed, and independently, a card is chosen at random from a regular deck. What is the probability that:

(a) the die shows an even number and the card is from a red suit?
(b) the die shows an even number or the card is from a red suit?

3.13.   A binary number is one composed only of the digits zero and one. (For example, 1011, 1100, etc.) These numbers play an important role in the use of electronic computers. Suppose that a binary number is made up of n digits. Suppose that the probability of an incorrect digit appearing is p and that errors in different digits are independent of one another. What is the probability of forming an incorrect *number*?

3.14.   A die is thrown n times. What is the probability that "6" comes up at least once in the n throws?

3.15.   Each of two persons tosses three fair coins. What is the probability that they obtain the same number of heads?

3.16.   Two dice are rolled. Given that the faces show different numbers, what is the probability that one face is 4?

3.17.   It is found that in manufacturing a certain article, defects of one type occur with probability 0.1 and defects of a second type with probability 0.05. (Assume independence between types of defects.) What is the probability that:

(a) an article does not have both kinds of defects?
(b) an article is defective?
(c) an article has only one type of defect, given that it is defective?

3.18.   Verify that the number of conditions listed in Eq. (3.8) is given by $2^n - n - 1$.

3.19.   Prove that if A and B are independent events, so are A and $\bar{B}$, $\bar{A}$ and B, $\bar{A}$ and $\bar{B}$.

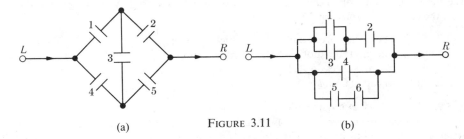

(a)   FIGURE 3.11   (b)

3.20.   In Fig. 3.11(a) and (b), assume that the probability of each relay being closed is p and that each relay is open or closed independently of any other relay. In each case find the probability that current flows from L to R.

TABLE 3.2

Number of breakdowns	0	1	2	3	4	5	6
A	0.1	0.2	0.3	0.2	0.09	0.07	0.04
B	0.3	0.1	0.1	0.1	0.1	0.15	0.15

3.21.   Two machines, A, B, being operated independently, may have a number of breakdowns each day. Table 3.2 gives the probability distribution of breakdowns for each machine. Compute the following probabilities.

(a) A and B have the same number of breakdowns.

(b) The total number of breakdowns is less than 4; less than 5.

(c) A has more breakdowns than B.

(d) B has twice as many breakdowns as A.

(e) B has 4 breakdowns, when it is known that B has at least 2 breakdowns.

(f) The minimum number of breakdowns of the two machines is 3; is less than 3.

(g) The maximum number of breakdowns of the machines is 3; is more than 3.

3.22. By verifying Eq. (3.2), show that for fixed A, $P(B \mid A)$ satisfies the various postulates for probability.

3.23. If each element of a second order determinant is either zero or one, what is the probability that the value of the determinant is positive? (Assume that the individual entries of the determinant are chosen independently, each value being assumed with probability $\frac{1}{2}$.)

3.24. Show that the multiplication theorem $P(A \cap B) = P(A \mid B)P(B)$, established for two events, may be generalized to three events as follows:

$$P(A \cap B \cap C) = P(A \mid B \cap C)P(B \mid C)P(C).$$

3.25. An electronic assembly consists of two subsystems, say A and B. From previous testing procedures, the following probabilities are assumed to be known:

$$P(A \text{ fails}) = 0.20,$$

$$P(B \text{ fails } alone) = 0.15,$$

$$P(A \text{ and } B \text{ fail}) = 0.15,$$

Evaluate the following probabilities.

(a) $P(A \text{ fails} \mid B \text{ has failed})$,

(b) $P(A \text{ fails alone})$.

3.26. Finish the analysis of the example given in Section 3.2 by deciding which of the types of candy jar, A or B, is involved, based on the evidence of two pieces of candy which were sampled.

3.27. Whenever an experiment is performed, the occurrence of a particular event A equals 0.2. The experment is repeated, independently, until A occurs. Compute the probability that it will be necessary to carry out a fourth experiment.

3.28. Suppose that a mechanism has N tubes, all of which are needed for its functioning. To locate a malfunctioning tube one replaces each tube, successively, with a new one. Compute the probability that it will be necessary to check N tubes if the (constant) probability is p that a tube is out of order.

3.29. Prove: If $P(A \mid B) > P(A)$ then $P(B \mid A) > P(B)$.

3.30. A vacuum tube may come from any one of three manufacturers with probabilities $p_1 = 0.25$, $p_2 = 0.50$, and $p_3 = 0.25$. The probabilities that the tube will function properly during a specified period of time equal 0.1, 0.2, and 0.4, respectively, for the three manufacturers. Compute the probability that a randomly chosen tube will function for the specified period of time.

3.31.   An electrical system consists of two switches of type A, one of type B, and four of type C, connected as in Fig. 3.12. Compute the probability that a break in the circuit cannot be eliminated with key K if the switches A, B, and C are open (i.e., out of order) with probabilities 0.3, 0.4, and 0.2, respectively, and if they operate independently.

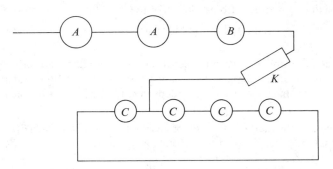

FIGURE 3.12

3.32.   The probability that a system becomes overloaded is 0.4 during each run of an experiment. Compute the probability that the system will cease functioning in three independent trials of the experiment if the probabilities of failure in 1, 2, or 3 trials equal 0.2, 0.5, and 0.8, respectively.

3.33.   Four radio signals are emitted successively. If the reception of any one signal is independent of the reception of another and if these probabilities are 0.1, 0.2, 0.3, and 0.4, respectively, compute the probability that k signals will be received for $k = 0, 1, 2, 3, 4$.

3.34.   The following (somewhat simple-minded) weather forecasting is used by an amateur forecaster. Each day is classified as "dry" or "wet" and the probability that any given day is the same as the preceding one is assumed to be a constant p $(0 < p < 1)$. Based on past records, it is supposed that January 1 has a probability of β of being "dry." Letting β_n = probability (the nth day of the year is "dry"), obtain an expression for β_n in terms of β and p. Also evaluate $\lim_{n \to \infty} \beta_n$ and interpret your result.
(Hint: Express β_n in terms of β_{n-1}.)

3.35.   Three newspapers, A, B, and C, are published in a city and a recent survey of readers indicates the following: 20 percent read A, 16 percent read B, 14 percent read C, 8 percent read A and B, 5 percent read A and C, 2 percent read A, B, and C, and 40 percent read B and C. For one adult chosen at random, compute the probability that (a) he reads none of the papers (b) he reads exactly one of the papers (c) he reads at least A and B if it is known that he reads at least one of the papers published.

3.36.   A fair coin is tossed $2n$ times.
(a) Obtain the probability that there will be an equal number of heads and tails.
(b) Show that the probability computed in (a) is a decreasing function of n.

3.37.   Urn 1, Urn 2, . . . , Urn n each contain α white and β black balls. One ball is taken from Urn 1 into Urn 2 and *then* one is taken from Urn 2 into Urn 3, etc. Finally, one ball is chosen from Urn n. If the first ball transferred was white, what is the probability that the last ball chosen is white? What happens as $n \to \infty$? [*Hint:* Let $p_n = $ Prob (nth ball transferred is white) and express p_n in terms of p_{n-1}.]

3.38   Urn 1 contains α white and β black balls while Urn 2 contains β white and α black balls. One ball is chosen (from one of the urns) and is then returned to that urn. If the chosen ball is white, choose the next ball from Urn 1; if the chosen ball is black, choose the next one from Urn 2. Continue in this manner. Given that the first ball chosen came from Urn 1, obtain Prob (nth ball chosen is white) and also the limit of this probability as $n \to \infty$.

3.39.   A printing machine can print n "letters," say $\alpha_1, \ldots, \alpha_n$. It is operated by electrical impulses, each letter being produced by a *different* impulse. Assume that there exists a constant probability p of printing the correct letter and also assume independence. One of the n impulses, chosen at random, was fed into the machine twice and both times the letter α_1 was printed. Compute the probability that the impulse chosen was meant to print α_1.

4

One-Dimensional Random Variables

4.1 General Notion of a Random Variable

In describing the sample space of an experiment we did not specify that an individual outcome needs to be a number. In fact, we have cited a number of examples in which the result of the experiment was not a numerical quantity. For instance, in classifying a manufactured item we might simply use the categories "defective" and "nondefective." Again, in observing the temperature during a 24-hour period we might simply keep a record of the curve traced by the thermograph. However, in many experimental situations we are going to be concerned with measuring something and recording it as a *number*. Even in the above-cited cases we can assign a number to each (nonnumerical) outcome of the experiment. For example, we could assign the value one to nondefective items and the value zero to defective ones. We could record the maximum temperature of the day, or the minimum temperature, or the average of the maximum and minimum temperatures.

The above illustrations are quite typical of a very general class of problems: In many experimental situations we want to assign a real number x to every element s of the sample space S. That is, $x = X(s)$ is the value of a function X from the sample space to the real numbers. With this in mind, we make the following formal definition.

> **Definition.** Let ε be an experiment and S a sample space associated with the experiment. A *function* X assigning to every element $s \in S$, a real number, $X(s)$, is called a *random variable*.

Notes: (a) The above terminology is a somewhat unfortunate one, but it is so universally accepted that we shall not deviate from it. We have made it as clear as possible that X is a *function* and yet we call it a (random) variable!

(b) It turns out that *not every* conceivable function may be considered as a random variable. One requirement (although not the most general one) is that for every real number x the event $\{X(s) = x\}$ and for every interval I the event $\{X(s) \in I\}$ have well-defined probabilities consistent with the basic axioms. In most of the applications this difficulty does not arise and we shall make no further reference to it.

(c) In some situations the outcome s of the sample space is already the numerical characteristic which we want to record. We simply take $X(s) = s$, the identity function.

(d) In most of our subsequent discussions of random variables we need not indicate the functional nature of X. We are usually interested in the possible *values* of X, rather than where these values came from. For example, suppose that we toss two coins and consider the sample space associated with this experiment. That is,

$$S = \{HH, HT, TH, TT\}.$$

Define the random variable X as follows: X is the number of heads obtained in the two tosses. Hence $X(HH) = 2$, $X(HT) = X(TH) = 1$, and $X(TT) = 0$.

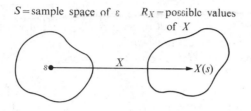

$S = $sample space of ε $R_X - $possible values
                                                   of X

X

$X(s)$

FIGURE 4.1

(e) It is very important to understand a basic requirement of a (single-valued) function: To *every* $s \in S$ there corresponds *exactly one* value $X(s)$. This is shown schematically in Fig. 4.1. Different values of s *may* lead to the same value of X. For example, in the above illustration we found that $X(HT) = X(TH) = 1$.

The space R_X, the set of all possible values of X, is sometimes called the *range space*. In a sense we may consider R_X as another sample space. The (original) sample space S corresponds to the (possibly) nonnumerical outcome of the experiment, while R_X is the sample space associated with the random variable X, representing the numerical characteristic which may be of interest. If $X(s) = s$, we have $S = R_X$.

Although we are aware of the pedagogical danger inherent in giving too many explanations for the same thing, let us nevertheless point out that we may think of a random variable X in two ways:

(a) We perform the experiment ε which results in an outcome $s \in S$. We *then* evaluate the number $X(s)$.

(b) We perform ε, obtaining the outcome s, *and* (immediately) evaluate $X(s)$. The number $X(s)$ is then thought of as the actual outcome of the experiment and R_X becomes *the* sample space of the experiment.

The difference between interpretations (a) and (b) is hardly discernible. It is relatively minor but worthy of attention. In (a) the experiment essentially terminates with the observation of s. The evaluation of $X(s)$ is considered as something that is done subsequently and which is not affected by the randomness of ε. In (b) the experiment is not considered to be terminated until the number $X(s)$ has actually been evaluated, thus resulting in the sample space R_X. Although the first interpretation, (a), is the one usually intended, the second point of view, (b), can be very helpful, and the reader should keep it in mind. What we are saying, and this will become increasingly evident in later sections, is that in studying random variables we are more concerned about the values X assumes than about its functional form. Hence in many cases we shall completely ignore the underlying sample space on which X may be defined.

EXAMPLE 4.1. Suppose that a light bulb is inserted into a socket. The experiment is considered at an end when the bulb ceases to burn. What is a possible outcome, say s? One way of describing s would be by simply recording the date and time of day at which the bulb burns out, for instance May 19, 4:32 p.m. Hence the sample space may be represented as $S = \{(d, t) \mid d = \text{date}, t = \text{time of day}\}$. Presumably the random variable of interest is X, the length of burning time. Note that once $s = (d, t)$ is observed, the *evaluation* of $X(s)$ does not involve any random-ness. When s is specified, $X(s)$ is completely determined.

The two points of view expressed above may be applied to this example as follows. In (a) we consider the experiment to be terminated with the observation $s = (d, t)$, the date and time of day. The computation of $X(s)$ is then performed, involving a simple arithmetic operation. In (b) we consider the experiment to be completed only *after* $X(s)$ is evaluated and the number $X(s) = 107$ hours, say, is then considered to be the outcome of the experiment.

It might be pointed out that a similar analysis could be applied to some other random variables of interest, for instance $Y(s)$ is the temperature in the room at the time the bulb burned out.

EXAMPLE 4.2. Three coins are tossed on a table. As soon as the coins land on the table, the "random" phase of the experiment is over. A single outcome s might consist of a detailed description of how and where the coins landed. Presumably we are only interested in certain numerical characteristics associated with this experiment. For instance, we might evaluate

$X(s) = $ number of heads showing,

$Y(s) = $ maximum distance between any two coins,

$Z(s) = $ minimum distance of coins from any edge of the table.

If the random variable X is of interest, we could, as discussed in the previous example, incorporate the evaluation of $X(s)$ into the description of our experi-ment and hence simply state that the sample space associated with the experiment is $\{0, 1, 2, 3\}$, corresponding to the values of X. Although we shall very often adopt precisely this point of view, it is important to realize that the counting of the number of heads is done *after* the random aspects of the experiment have ended.

Note: In referring to random variables we shall, almost without exception, use capital letters such as X, Y, Z, etc. However, when speaking of the *value* these random variables assume we shall in general use lower case letters such as x, y, z, etc. This is a *very important distinction* to be made and the student might well pause to consider it. For example, when we speak of choosing a person at random from some designated population and measuring his height (in inches, say), we could refer to the *possible* outcomes as a random variable X. We might then ask various questions about X, such as $P(X \geq 60)$. However, once we actually choose a person and measure his height we obtain a specific value of X, say x. Thus it would be meaningless to ask for $P(x \geq 60)$ since x either is or is not ≥ 60. This distinction between a random variable and its value is important, and we shall make subsequent references to it.

As we were concerned about the events associated with the sample space S, so we shall find the need to discuss events with respect to the random variable X, that is, subsets of the range space R_X. Quite often certain events associated with S are "related" (in a sense to be described) to events associated with R_X in the following way.

Definition. Let ε be an experiment and S its sample space. Let X be a random variable defined on S and let R_X be its range space. Let B be an event with respect to R_X; that is, $B \subset R_X$. Suppose that A is defined as

$$A = \{s \in S \mid X(s) \in B\}. \tag{4.1}$$

In words: A consists of all outcomes in S for which $X(s) \in B$ (Fig. 4.2). In this case we say that A and B are *equivalent events*.

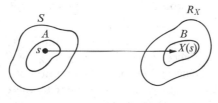

FIGURE 4.2

Notes: (a) Saying the above more informally, A and B are equivalent events whenever they occur together. That is, whenever A occurs, B occurs and conversely. For if A did occur, then an outcome s occurred for which $X(s) \in B$ and hence B occurred. Conversely, if B occurred, a value $X(s)$ was observed for which $s \in A$ and hence A occurred.

(b) It is important to realize that in our definition of equivalent events, A and B are associated with *different* sample spaces.

EXAMPLE 4.3. Consider the tossing of two coins. Hence $S = \{HH, HT, TH, TT\}$. Let X be the number of heads obtained. Hence $R_X = \{0, 1, 2\}$. Let $B = \{1\}$. Since $X(HT) = X(TH) = 1$ if and only if $X(s) = 1$, we have that $A = \{HT, TH\}$ is equivalent to B.

We now make the following important definition.

Definition. Let B be an event in the range space R_X. We then *define* $P(B)$ as follows:

$$P(B) = P(A), \quad \text{where} \quad A = \{s \in S \mid X(s) \in B\}. \tag{4.2}$$

In words: We define $P(B)$ equal to the probability of the event $A \subset S$, which is equivalent to B, in the sense of Eq. (4.1).

Notes: (a) We are assuming that probabilities may be associated with events in S. Hence the above definition makes it possible to assign probabilities to events associated with R_X in terms of probabilities defined over S.

(b) It is actually possible to *prove* that $P(B)$ must be as we defined it. However, this would involve some theoretical difficulties which we want to avoid, and hence we proceed as above.

(c) Since in the formulation of Eq. (4.2) the events A and B refer to different sample spaces, we should really use a different notation when referring to probabilities defined over S and for those defined over R_X, say something like $P(A)$ and $P_X(B)$. However, we shall not do this but continue to write simply $P(A)$ and $P(B)$. The context in which these expressions appear should make the interpretation clear.

(d) The probabilities associated with events in the (original) sample space S are, in a sense, determined by "forces beyond our control" or, as it is sometimes put, "by nature." The makeup of a radioactive source emitting particles, the disposition of a large number of persons who might place a telephone call during a certain hour, and the thermal agitation resulting in a current or the atmospheric conditions giving rise to a storm front illustrate this point. When we introduce a random variable X and its associated range space R_X we are *inducing* probabilities on the events associated with R_X, which are strictly determined if the probabilities associated with events in S are specified.

EXAMPLE 4.4.   If the coins considered in Example 4.3 are "fair," we have $P(HT) = P(TH) = \frac{1}{4}$. Hence $P(HT, TH) = \frac{1}{4} + \frac{1}{4} = \frac{1}{2}$. (The above calculations are a direct consequence of our basic assumption concerning the fairness of the coins.) Since the event $\{X = 1\}$ is equivalent to the event $\{HT, TH\}$, using Eq. (4.1), we have that $P(X = 1) = P(HT, TH) = \frac{1}{2}$. [There was really no choice about the value of $P(X = 1)$ consistent with Eq. (4.2), once $P(HT, TH)$ had been determined. It is in this sense that probabilities associated with events of R_X are *induced*.]

Note: Now that we have established the existence of an induced probability function over the range space of X (Eqs. 4.1 and 4.2) we shall find it convenient to *suppress* the functional nature of X. Hence we shall write (as we did in the above example), $P(X = 1) = \frac{1}{2}$. What is meant is that a certain event in the sample space S, namely $\{HT, TH\} = \{s \mid X(s) = 1\}$ occurs with probability $\frac{1}{2}$. Hence we assign that same probability to the event $\{X = 1\}$ in the range space. We shall continue to write expressions like $P(X = 1)$, $P(X \leq 5)$, etc. It is *very important* for the reader to realize what these expressions really represent.

Once the probabilities associated with various outcomes (or events) in the range space R_X have been determined (more precisely, induced) we shall often ignore the original sample space S which gave rise to these probabilities. Thus in the above example, we shall simply be concerned with $R_X = \{0, 1, 2\}$ and the associated probabilities $(\frac{1}{4}, \frac{1}{2}, \frac{1}{4})$. The fact that these probabilities are *determined* by a probability function defined over the original sample space S need not concern us if we are simply interested in studying the *values* of the random variable X.

In discussing, in detail, many of the important concepts associated with random variables, we shall find it convenient to distinguish between two important cases: the discrete and the continuous random variables.

4.2 Discrete Random Variables

Definition. Let X be a random variable. If the number of possible values of X (that is, R_X, the range space) is finite or countably infinite, we call X a *discrete random variable*. That is, the possible *values* of X may be listed as $x_1, x_2, \ldots, x_n, \ldots$ In the finite case the list terminates and in the countably infinite case the list continues indefinitely.

EXAMPLE 4.5. A radioactive source is emitting α-particles. The emission of these particles is observed on a counting device during a specified period of time. The following random variable is of interest:

$$X = \text{number of particles observed.}$$

What are the possible values of X? We shall assume that these values consist of all nonnegative integers. That is, $R_X = \{0, 1, 2, \ldots, n, \ldots\}$. An objection which we confronted once before may again be raised at this point. It could be argued that during a specified (finite) time interval it is impossible to observe more than, say N particles, where N may be a very large positive integer. Hence the possible values for X should really be: $0, 1, 2, \ldots, N$. However, it turns out to be mathematically simpler to consider the idealized description given above. In fact, whenever we assume that the possible values of a random variable X are countably infinite, we are actually considering an idealized representation of X.

In view of our previous discussions of the probabilistic description of events with a finite or countably infinite number of members, the probabilistic description of a discrete random variable will not cause any difficulty. We proceed as follows.

Definition. Let X be a discrete random variable. Hence R_X, the range space of X, consists of at most a countably infinite number of values, $x_1, x_2, \ldots$ With each possible outcome x_i we associate a number $p(x_i) = P(X = x_i)$, called the probability of x_i. The numbers $p(x_i)$, $i = 1, 2, \ldots$ must satisfy the following conditions:

(a) $p(x_i) \geq 0$ for all i,

(b) $\displaystyle\sum_{i=1}^{\infty} p(x_i) = 1.$ (4.3)

The function p defined above is called the *probability function* (or point probability function) of the random variable X. The collection of pairs $(x_i, p(x_i))$, $i = 1, 2, \ldots$, is sometimes called the *probability distribution* of X.

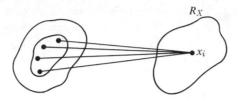

FIGURE 4.3

Notes: (a) The particular choice of the numbers $p(x_i)$ is presumably determined from the probability function associated with events in the sample space S on which X is defined. That is, $p(x_i) = P[s \mid X(s) = x_i]$. (See Eqs. 4.1 and 4.2.) However, since we are interested only in the values of X, that is R_X, and the probabilities associated with these values, we are again suppressing the functional nature of X. (See Fig. 4.3.) Although in most cases the numbers will in fact be determined from the probability distribution in some underlying sample space S, *any* set of numbers $p(x_i)$ satisfying Eq. (4.3) may serve as proper probabilistic description of a discrete random variable.

(b) If X assumes only a finite number of values, say $x_1, \ldots, x_N$, then $p(x_i) = 0$ for $i > N$, and hence the infinite series in Eq. (4.3) becomes a finite sum.

(c) We may again note an analogy to mechanics by considering a total mass of one unit distributed over the real line with the entire mass located at the points $x_1, x_2, \ldots$ The numbers $p(x_i)$ represent the amount of mass located at x_i.

(d) The geometric interpretation (Fig. 4.4) of a probability distribution is often useful.

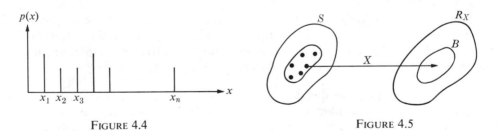

FIGURE 4.4   FIGURE 4.5

Let B be an event associated with the random variable X. That is, $B \subset R_X$ (Fig. 4.5). Specifically, suppose that $B = \{x_{i_1}, x_{i_2}, \ldots\}$. Hence

$$P(B) = P[s \mid X(s) \in B] \quad \text{(since these events are equivalent)}$$

$$= P[s \mid X(s) = x_{i_j}, j = 1, 2, \ldots] = \sum_{j=1}^{\infty} p(x_{i_j}). \tag{4.4}$$

In words: The probability of an event B equals the sum of the probabilities of the individual outcomes associated with B.

Notes: (a) Suppose that the discrete random variable X may assume only a finite number of values, say $x_1, \ldots, x_N$. If each outcome is equally probable, then we obviously have $p(x_1) = \cdots = p(x_N) = 1/N$.

(b) If X assumes a countably infinite number of values, then it is *impossible* to have all outcomes equally probable. For we cannot possibly satisfy the condition $\sum_{i=1}^{\infty} p(x_i) = 1$ if we must have $p(x_i) = c$ for all i.

(c) In every finite interval there will be at most a finite number of the possible values of X. If some such interval contains none of these possible values we assign probability zero to it. That is, if $R_X = \{x_1, x_2, \ldots, x_n\}$ and if no $x_i \in [a, b]$, then $P[a \leq X \leq b] = 0$.

EXAMPLE 4.6. Suppose that a radio tube is inserted into a socket and tested. Assume that the probability that it tests positive equals $\frac{3}{4}$; hence the probability that it tests negative is $\frac{1}{4}$. Assume furthermore that we are testing a large supply of such tubes. The testing continues until the first positive tube appears. Define the random variable X as follows: X is the number of tests required to terminate the experiment. The sample space associated with this experiment is

$$S = \{+, -+, --+, ---+, \ldots\}.$$

To determine the probability distribution of X we reason as follows. The possible values of X are $1, 2, \ldots, n, \ldots$ (we are obviously dealing with the idealized sample space). And $X = n$ if and only if the first $(n - 1)$ tubes are negative and the nth tube is positive. If we suppose that the condition of one tube does not affect the condition of another, we may write

$$p(n) = P(X = n) = (\tfrac{1}{4})^{n-1}(\tfrac{3}{4}), \qquad n = 1, 2, \ldots$$

To check that these values of $p(n)$ satisfy Eq. (4.3) we note that

$$\sum_{n=1}^{\infty} p(n) = \frac{3}{4}\left(1 + \frac{1}{4} + \frac{1}{16} + \cdots\right)$$

$$= \frac{3}{4}\frac{1}{1 - \frac{1}{4}} = 1.$$

Note: We are using here the result that the *geometric series* $1 + r + r^2 + \cdots$ converges to $1/(1 - r)$ whenever $|r| < 1$. This is a result to which we shall refer repeatedly. Suppose that we want to evaluate $P(A)$, where A is defined as {The experiment ends after an even number of repetitions}. Using Eq. (4.4), we have

$$P(A) = \sum_{n=1}^{\infty} p(2n) = \frac{3}{16} + \frac{3}{256} + \cdots$$

$$= \frac{3}{16}(1 + \tfrac{1}{16} + \cdots)$$

$$= \frac{3}{16}\frac{1}{1 - \tfrac{1}{16}} = \frac{1}{5}.$$

4.3 The Binomial Distribution

In later chapters we shall consider, in considerable detail, a number of important discrete random variables. For the moment we shall simply study one of these and then use it to illustrate a number of important concepts.

EXAMPLE 4.7. Suppose that items coming off a production line are classified as defective (D) or nondefective (N). Suppose that three items are chosen at random from a day's production and are classified according to this scheme. The sample space for this experiment, say S, may be described as follows:

$$S = \{DDD, DDN, DND, NDD, NND, NDN, DNN, NNN\}.$$

(Another way of describing S is as $S = S_1 \times S_2 \times S_3$, the Cartesian product of S_1, S_2, and S_3, where each $S_i = \{D, N\}$.)

Let us suppose that with probability 0.2 an item is defective and hence with probability 0.8 an item is nondefective. Let us assume that these probabilities are the *same* for each item at least throughout the duration of our study. Finally let us suppose that the classification of any particular item is independent of the classification of any other item. Using these assumptions, it follows that the probabilities associated with the various outcomes of the sample space S as described above are

$$(0.2)^3, (0.8)(0.2)^2, (0.8)(0.2)^2, (0.8)(0.2)^2, (0.2)(0.8)^2, (0.2)(0.8)^2, (0.2)(0.8)^2, (0.8)^3.$$

Our interest usually is not focused on the individual outcomes of S. Rather, we simply wish to know *how many* defectives were found (irrespective of the order in which they occurred). That is, we wish to consider the random variable X which assigns to each outcome $s \in S$ the number of defectives found in s. Hence the set of possible values of X is $\{0, 1, 2, 3\}$.

We can obtain the probability distribution for $X, p(x_i) = P(X = x_i)$ as follows:

$$
\begin{aligned}
X &= 0 &&\text{if and only if} && NNN \text{ occurs};\\
X &= 1 &&\text{if and only if} && DNN, NDN, \text{ or } NND \text{ occurs};\\
X &= 2 &&\text{if and only if} && DDN, DND, \text{ or } NDD \text{ occurs};\\
X &= 3 &&\text{if and only if} && DDD \text{ occurs}.
\end{aligned}
$$

(Note that $\{NNN\}$ is equivalent to $\{X = 0\}$, etc.) Hence

$$p(0) = P(X = 0) = (0.8)^3, \qquad p(1) = P(X = 1) = 3(0.2)(0.8)^2,$$
$$p(2) = P(X = 2) = 3(0.2)^2(0.8), \qquad p(3) = P(X = 3) = (0.2)^3.$$

Observe that the sum of these probabilities equals 1, for the sum may be written as $(0.8 + 0.2)^3$.

Note: The above discussion illustrates how the probabilities in the range space R_X (in this case $\{0, 1, 2, 3\}$) are *induced* by the probabilities defined over the sample space S. For the assumption that the eight outcomes of

$$S = \{DDD, DDN, DND, NDD, NND, NDN, DNN, NNN\}$$

have the probabilities given in Example 4.7, *determined* the value of $p(x)$ for all $x \in R_X$.

Let us now generalize the notions introduced in the above example.

Definition. Consider an experiment ε and let A be some event associated with ε. Suppose that $P(A) = p$ and hence $P(\overline{A}) = 1 - p$. Consider n independent repetitions of ε. Hence the sample space consists of all possible sequences $\{a_1, a_2, \ldots, a_n\}$, where each a_i is either A or $\overline{A}$, depending on whether A or $\overline{A}$ occurred on the ith repetition of ε. (There are 2^n such sequences.) Furthermore, assume that $P(A) = p$ remains the same for all repetitions. Let the random variable X be defined as follows: $X =$ number of times the event A occurred. We call X a *binomial* random variable with parameters n and p. Its possible values are obviously $0, 1, 2, \ldots, n$. (Equivalently we say that X has a *binomial distribution.*) The individual repetitions of ε will be called *Bernoulli trials.*

Theorem 4.1. Let X be a binomial variable based on n repetitions. Then

$$P(X = k) = \binom{n}{k} p^k (1 - p)^{n-k}, \qquad k = 0, 1, \ldots, n. \qquad (4.5)$$

Proof: Consider a particular element of the sample space of ε satisfying the condition that $X = k$. One such outcome would arise, for instance, if the first k repetitions of ε resulted in the occurrence of A, while the last $n - k$ repetitions resulted in the occurrence of $\overline{A}$, that is

$$\underbrace{AAA \cdots A}_{k}\ \underbrace{\overline{A}\,\overline{A}\,\overline{A} \cdots \overline{A}}_{n - k}.$$

Since all repetitions are independent, the probability of this particular sequence would be $p^k (1 - p)^{n-k}$. But exactly the same probability would be associated with any other outcome for which $X = k$. The total number of such outcomes equals $\binom{n}{k}$, for we must choose exactly k positions (out of n) for the A's. But this yields the above result, since these $\binom{n}{k}$ outcomes are all mutually exclusive.

Notes: (a) To verify our calculation we note that, using the binomial theorem, we have $\sum_{k=0}^{n} P(X = k) = \sum_{k=0}^{n} \binom{n}{k} p^k (1 - p)^{n-k} = [p + (1 - p)]^n = 1^n = 1$, as it should be. Since the probabilities $\binom{n}{k} p^k (1 - p)^{n-k}$ are obtained by expanding the binomial expression $[p + (1 - p)]^n$, we call this the binomial distribution.

(b) Whenever we perform independent repetitions of an experiment and are interested only in a dichotomy—defective *or* nondefective, hardness above *or* below a certain

standard, noise level in a communication system above *or* below a preassigned threshold—
we are potentially dealing with a sample space on which we may define a binomial random
variable. So long as the conditions of experimentation stay sufficiently uniform so that
the probability of some attribute, say A, stays constant, we may use the above model.

(c) If n is small, the individual terms of the binomial distribution are relatively easy to
compute. However, if n is reasonably large, these computations become rather cumbersome. Fortunately, the binomial probabilities have been tabulated. There are many
such tabulations. (See Appendix.)

EXAMPLE 4.8. Suppose that a radio tube inserted into a certain type of set has
a probability of 0.2 of functioning more than 500 hours. If we test 20 tubes, what
is the probability that exactly k of these function more than 500 hours, $k =$
0, 1, 2, . . . , 20?

If X is the number of tubes functioning more than 500 hours, we shall assume
that X has a binomial distribution. Thus $P(X = k) = \binom{20}{k}(0.2)^k(0.8)^{20-k}$.

The following values may be read from Table 4.1.

TABLE 4.1.

$P(X = 0) = 0.012$	$P(X = 4) = 0.218$	$P(X = 8) = 0.022$
$P(X = 1) = 0.058$	$P(X = 5) = 0.175$	$P(X = 9) = 0.007$
$P(X = 2) = 0.137$	$P(X = 6) = 0.109$	$P(X = 10) = 0.002$
$P(X = 3) = 0.205$	$P(X = 7) = 0.055$	$P(X = k) = 0^+$ for $k \geq 11$
(The remaining probabilities are less than 0.001.)		

If we plot this probability distribution, we obtain the graph shown in Fig. 4.6.
The pattern which we observe here is quite general: The binomial probabilities
increase monotonically until they reach a maximum value and then decrease
monotonically. (See Problem 4.8.)

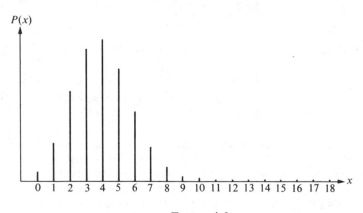

FIGURE 4.6

EXAMPLE 4.9. In operating a certain machine, there is a certain probability that the machine operator makes an error. It may be realistically assumed that the operator learns in the sense that the probability of his making an error decreases as he uses the machine repeatedly. Suppose that the operator makes n attempts and that the n trials are statistically independent. Suppose specifically that P(an error is made on the ith repetition) $= 1/(i + 1)$, $i = 1, 2, \ldots, n$. Assume that 4 attempts are contemplated (that is, $n = 4$) and we define the random variable X as the number of machine operations made without error. Note that X is not binomially distributed because the probability of "success" is not constant.

To compute the probability that $X = 3$, for instance, we proceed as follows: $X = 3$ if and only if there is exactly one unsuccessful attempt. This can happen on the first, second, third, or fourth trial. Hence

$$P(X = 3) = \tfrac{1}{2}\tfrac{2}{3}\tfrac{3}{4}\tfrac{4}{5} + \tfrac{1}{2}\tfrac{1}{3}\tfrac{3}{4}\tfrac{4}{5} + \tfrac{1}{2}\tfrac{2}{3}\tfrac{1}{4}\tfrac{4}{5} + \tfrac{1}{2}\tfrac{2}{3}\tfrac{3}{4}\tfrac{1}{5} = \tfrac{5}{12}.$$

EXAMPLE 4.10. Consider a situation similar to the one described in Example 4.9. This time we shall assume that there is a constant probability p_1 of making no error on the machine during each of the first n_1 attempts and a constant probability $p_2 \le p_1$ of making no error on each of the next n_2 repetitions. Let X be the number of successful operations of the machine during the $n = n_1 + n_2$ independent attempts. Let us find a general expression for $P(X = k)$. For the same reason as given in the preceding example, X is not binomially distributed. To obtain $P(X = k)$ we proceed as follows.

Let Y_1 be the number of correct operations during the first n_1 attempts and let Y_2 be the number of correct operations during the second n_2 attempts. Hence Y_1 and Y_2 are independent random variables and $X = Y_1 + Y_2$. Thus $X = k$ if and only if $Y_1 = r$ and $Y_2 = k - r$, for any integer r satisfying $0 \le r \le n_1$ and $0 \le k - r \le n_2$.

The above restrictions on r are equivalent to $0 \le r \le n_1$ and $k - n_2 \le r \le k$. Combining these we may write

$$\max (0, k - n_2) \le r \le \min (k, n_1).$$

Hence we have

$$P(X = k) = \sum_{r=\max(0,k-n_2)}^{\min(k,n_1)} \binom{n_1}{r} p_1^r (1 - p_1)^{n_1-r} \binom{n_2}{k - r} p_2^{k-r}(1 - p_2)^{n_2-(k-r)}.$$

With our usual convention that $\binom{a}{b} = 0$ whenever $b > a$ or $b < 0$, we may write the above probability as

$$P(X = k) = \sum_{r=0}^{n_1} \binom{n_1}{r} p_1^r (1 - p_1)^{n_1-r} \binom{n_2}{k - r} p_2^{k-r}(1 - p_2)^{n_2-k+r}. \tag{4.6}$$

For instance, if $p_1 = 0.2$, $p_2 = 0.1$, $n_1 = n_2 = 10$, and $k = 2$, the above probability becomes

$$P(X = 2) = \sum_{r=0}^{2} \binom{10}{r} (0.2)^r(0.8)^{10-r} \binom{10}{2-r} (0.1)^{2-r}(0.9)^{8+r} = 0.27,$$

after a straightforward calculation.

Note: Suppose that $p_1 = p_2$. In this case, Eq. (4.6) should reduce to $\binom{n}{k} p_1^k(1 - p_1)^{n-k}$, since now the random variable X *does* have a binomial distribution. To see that this is so, note that we may write (since $n_1 + n_2 = n$)

$$P(X = k) = p_1^k(1 - p_1)^{n-k} \sum_{r=0}^{n_1} \binom{n_1}{r} \binom{n_2}{k-r}.$$

To show that the above sum equals $\binom{n}{k}$ simply compare coefficients for the powers of x^k on both sides of the identity $(1 + x)^{n_1}(1 + x)^{n_2} = (1 + x)^{n_1+n_2}$.

4.4. Continuous Random Variables

Suppose that the range space of X is made up of a very large finite number of values, say all values x in the interval $0 \le x \le 1$ of the form 0, 0.01, 0.02, ... , 0.98, 0.99, 1.00. With each of these values is associated a nonnegative number $p(x_i) = P(X = x_i)$, $i = 1, 2, \ldots$, whose sum equals 1. This situation is represented geometrically in Fig. 4.7.

We have pointed out before that it might be mathematically easier to idealize the above probabilistic description of X by supposing that X can assume *all* possible values, $0 \le x \le 1$. If we do this, what happens to the point probabilities $p(x_i)$? Since the possible values of X are noncountable, we cannot really speak of the ith value of X, and hence $p(x_i)$ becomes meaningless. What we shall do is to replace the function p, defined only for $x_1, x_2, \ldots$, by a function f defined (in the present context) for *all* values of x, $0 \le x \le 1$. The properties of Eq. (4.3) will be replaced by $f(x) \ge 0$ and $\int_0^1 f(x) \, dx = 1$. Let us proceed formally as follows.

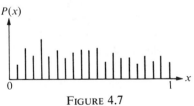

FIGURE 4.7

Definition. X is said to be a *continuous random variable* if there exists a function f, called the probability density function (pdf) of X, satisfying the following conditions:

(a) $f(x) \ge 0$ for all x,

(b) $\displaystyle\int_{-\infty}^{+\infty} f(x) \, dx = 1.$ (4.7)

(c) For any a, b, with $-\infty < a < b + \infty$,

$$\text{we have } P(a \le X \le b) = \int_a^b f(x)\, dx. \tag{4.8}$$

Notes: (a) We are essentially saying that X is a continuous random variable if X may assume all values in some interval (c, d) where c and d may be $-\infty$ and $+\infty$, respectively. The stipulated existence of a pdf is a mathematical device which has considerable intuitive appeal and makes our computations simpler. In this connection it should again be pointed out that when we suppose that X is a continuous random variable, we are dealing with the *idealized* description of X.

(b) $P(c < X < d)$ represents the area under the graph in Fig. 4.8 of the pdf f between $x = c$ and $x = d$.

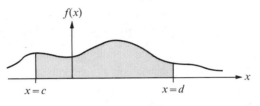

FIGURE 4.8

(c) It is a consequence of the above probabilistic description of X that for *any* specified value of X, say x_0, we have $P(X = x_0) = 0$, since $P(X = x_0) = \int_{x_0}^{x_0} f(x)\, dx = 0$. This result may seem quite contrary to our intuition. We must realize, however, that if we allow X to assume *all* values in some interval, then probability zero is not equivalent with impossibility. Hence in the continuous case, $P(A) = 0$ does not imply $A = \emptyset$, the empty set. (See Theorem 1.1.) Saying this more informally, consider choosing a point at random on the line segment $\{x \mid 0 \le x \le 2\}$. Although we might be willing to agree (for mathematical purposes) that *every* conceivable point on the segment could be the outcome of our experiment, we would be extremely surprised if in fact we chose precisely the midpoint of the segment, or any other *specified* point, for that matter. When we state this in precise mathematical language we say the event has "probability zero." In view of these remarks, the following probabilities are all the *same* if X is a continuous random variable:

$$P(c \le X \le d), \qquad P(c \le X < d), \qquad P(c < X \le d), \qquad \text{and} \qquad P(c < X < d).$$

(d) Although we shall not verify the details here, it may be shown that the above assignment of probabilities to events in R_X satisfies the basic axioms of probability (Eq. 1.3), where we may take $\{x \mid -\infty < x < +\infty\}$ as our sample space.

(e) If a function f^* satisfies the conditions, $f^*(x) \ge 0$, for all x, and $\int_{-\infty}^{+\infty} f^*(x)\, dx = K$, where K is a positive real number (not necessarily equal to 1), then f^* does *not* satisfy all the conditions for being a pdf. However, we may easily define a new function, say f, in terms of f^* as follows:

$$f(x) = \frac{f^*(x)}{K} \qquad \text{for all } x.$$

Hence f satisfies all the conditions for a pdf.

(f) If X assumes values only in some finite interval $[a, b]$, we may simply set $f(x) = 0$ for all $x \notin [a, b]$. Hence the pdf is defined for *all* real values of x, and we may require that $\int_{-\infty}^{+\infty} f(x)\, dx = 1$. Whenever the pdf is specified only for certain values of x, we shall suppose that it is zero elsewhere.

(g) $f(x)$ does not represent the probability of anything! We have noted before that $P(X = 2) = 0$, for example, and hence $f(2)$ certainly does not represent this probability. Only when the function is integrated between two limits does it yield a probability. We can, however, give an interpretation of $f(x)\,\Delta x$ as follows. From the mean-value theorem of the calculus it follows that

$$P(x \leq X \leq x + \Delta x) = \int_{x}^{x+\Delta x} f(s)\, ds = \Delta x f(\xi), \qquad x \leq \xi \leq x + \Delta x.$$

If Δx is small, $f(x)\,\Delta x$ equals *approximately* $P(x \leq X \leq x + \Delta x)$. (If f is continuous from the right, this approximation becomes more accurate as $\Delta x \to 0$.)

(h) We should again point out that the probability distribution (in this case the pdf) is induced on R_X by the underlying probability associated with events in S. Thus, when we write $P(c < X < d)$, we mean, as always, $P[c < X(s) < d]$, which in turn equals $P[s \mid c < X(s) < d]$, since these events are equivalent. The above definition, Eq. (4.8), essentially stipulates the existence of a pdf f defined over R_X such that

$$P[s \mid c < X(s) < d] = \int_{c}^{d} f(x)\, dx.$$

We shall again suppress the functional nature of X and hence we shall be concerned only with R_X and the pdf f.

(i) In the continuous case we can again consider the following *analogy to mechanics:* Suppose that we have a total mass of one unit, continuously distributed over the interval $a \leq x \leq b$. Then $f(x)$ represents the mass density at the point x and $\int_{c}^{d} f(x)\, dx$ represents the total mass contained in the interval $c \leq x \leq d$.

EXAMPLE 4.11. The existence of a pdf was assumed in the above discussion of a continuous random variable. Let us consider a simple example in which we can easily determine the pdf by making an appropriate assumption about the probabilistic behavior of the random variable. Suppose that a point is chosen in the interval (0, 1). Let X represent the random variable whose value is the x-coordinate of the chosen point.

Assume: If I is any interval in (0, 1), then $\text{Prob}[X \in I]$ is directly proportional to the length of I, say $L(I)$. That is, $\text{Prob}[X \in I] = kL(I)$, where k is the constant of proportionality. (It is easy to see, by taking $I = (0, 1)$ and observing that $L((0, 1)) = 1$ and $\text{Prob }[X \in (0, 1)] = 1$, that $k = 1$.)

Obviously X assumes all values in (0, 1). What is its pdf? That is, can we find a function f such that

$$P(a < X < b) = \int_{a}^{b} f(x)\, dx?$$

Note that if $a < b < 0$ or $1 < a < b$, $P(a < X < b) = 0$ and hence $f(x) = 0$. If $0 < a < b < 1$, $P(a < X < b) = b - a$ and hence $f(x) = 1$. Thus we find,

$$f(x) = \begin{cases} 1, & 0 < x < 1, \\ 0, & \text{elsewhere.} \end{cases}$$

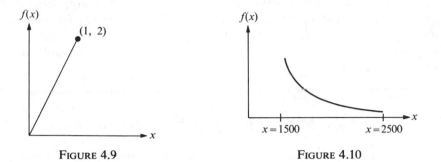

FIGURE 4.9   FIGURE 4.10

EXAMPLE 4.12. Suppose that the random variable X is continuous. (See Fig. 4.9.) Let the pdf f be given by

$$f(x) = 2x, \quad 0 < x < 1,$$
$$= 0, \quad \text{elsewhere.}$$

Clearly, $f(x) \geq 0$ and $\int_{-\infty}^{+\infty} f(x)\, dx = \int_0^1 2x\, dx = 1$. To compute $P(X \leq \frac{1}{2})$, we must simply evaluate the integral $\int_0^{1/2} (2x)\, dx = \frac{1}{4}$.

The concept of conditional probability discussed in Chapter 3 can be meaningfully applied to random variables. For instance, in the above example we may evaluate $P(X \leq \frac{1}{2} \mid \frac{1}{3} \leq X \leq \frac{2}{3})$. Directly applying the definition of conditional probability, we have

$$P(X \leq \tfrac{1}{2} \mid \tfrac{1}{3} \leq X \leq \tfrac{2}{3}) = \frac{P(\tfrac{1}{3} \leq X \leq \tfrac{1}{2})}{P(\tfrac{1}{3} \leq X \leq \tfrac{2}{3})}$$
$$= \frac{\int_{1/3}^{1/2} 2x\, dx}{\int_{1/3}^{2/3} 2x\, dx} = \frac{5/36}{1/3} = \frac{5}{12}.$$

EXAMPLE 4.13. Let X be the life length of a certain type of light bulb (in hours). Assuming X to be a continuous random variable, we suppose that the pdf f of X is given by

$$f(x) = a/x^3, \quad 1500 \leq x \leq 2500,$$
$$= 0, \quad \text{elsewhere.}$$

(That is, we are assigning probability zero to the events $\{X < 1500\}$ and $\{X > 2500\}$.) To evaluate the constant a, we invoke the condition $\int_{-\infty}^{+\infty} f(x)\, dx =$

1 which in this case becomes $\int_{1500}^{2500} a/x^3 \, dx = 1$. From this we obtain $a = 7$, 031, 250. The graph of f is shown in Fig. 4.10.

In a later chapter we shall study, in considerable detail, a number of important random variables, both discrete and continuous. We know from our use of deterministic models that certain functions play a far more important role than others. For example, the linear, quadratic, exponential, and trigonometric functions play a vital role in describing deterministic models. We shall find in developing non-deterministic (that is, probabilistic) models that certain random variables are of particular importance.

4.5 Cumulative Distribution Function

We want to introduce another important, general concept in this chapter.

Definition. Let X be a random variable, discrete or continuous. We define F to be the *cumulative distribution function* of the random variable X (abbreviated as cdf) where $F(x) = P(X \leq x)$.

Theorem 4.2. (a) If X is a discrete random variable,

$$F(x) = \sum_j p(x_j),\tag{4.9}$$

where the sum is taken over all indices j satisfying $x_j \leq x$.

(b) If X is a continuous random variable with pdf f,

$$F(x) = \int_{-\infty}^{x} f(s) \, ds.\tag{4.10}$$

Proof: Both of these results follow immediately from the definition.

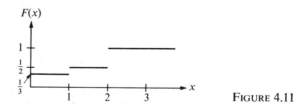

FIGURE 4.11

EXAMPLE 4.14. Suppose that the random variable X assumes the three values 0, 1, and 2 with probabilities $\frac{1}{3}$, $\frac{1}{6}$, and $\frac{1}{2}$, respectively. Then

$$
\begin{aligned}
F(x) &= 0 && \text{if } x < 0,\\
&= \tfrac{1}{3} && \text{if } 0 \leq x < 1,\\
&= \tfrac{1}{2} && \text{if } 1 \leq x < 2,\\
&= 1 && \text{if } x \geq 2.
\end{aligned}
$$

(Note that it is very important to indicate the inclusion or exclusion of the endpoints in describing the various intervals.) The graph of F is given in Fig. 4.11.

EXAMPLE 4.15. Suppose that X is a continuous random variable with pdf

$$f(x) = 2x, \quad 0 < x < 1,$$
$$= 0, \quad \text{elsewhere.}$$

Hence the cdf F is given by

$$F(x) = 0 \quad \text{if} \quad x \le 0,$$
$$= \int_0^x 2s \, ds = x^2$$
$$\qquad \text{if} \quad 0 < x \le 1,$$
$$= 1 \quad \text{if} \quad x > 1.$$

The graph is shown in Fig. 4.12.

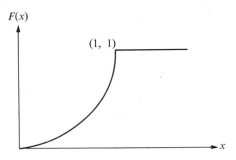

FIGURE 4.12

The graphs obtained in Figs. 4.11 and 4.12 for the cdf's are (in each case) quite typical in the following sense.

(a) If X is a discrete random variable with a finite number of possible values, the graph of the cdf F will be made up of horizontal line segments (it is called a step function). The function F is continuous except at the possible values of X, namely, $x_1, \ldots, x_n$. At the value x_j the graph will have a "jump" of magnitude $p(x_j) = P(X = x_j)$.

(b) If X is a continuous random variable, F will be a continuous function for all x.

(c) The cdf F is defined for *all* values of x, which is an important reason for considering it.

There are two important properties of the cdf which we shall summarize in the following theorem.

Theorem 4.3. (a) The function F is nondecreasing. That is, if $x_1 \le x_2$, we have $F(x_1) \le F(x_2)$.

(b) $\lim_{x \to -\infty} F(x) = 0$ and $\lim_{x \to \infty} F(x) = 1$.   [We often write this as $F(-\infty) = 0, F(\infty) = 1.$]

Proof: (a) Define the events A and B as follows: $A = \{X \leq x_1\}$, $B = \{X \leq x_2\}$. Then, since $x_1 \leq x_2$, we have $A \subset B$ and by Theorem 1.5, $P(A) \leq P(B)$, which is the required result.

(b) In the continuous case we have

$$F(-\infty) = \lim_{x \to -\infty} \int_{-\infty}^{x} f(s)\, ds = 0,$$

$$F(\infty) = \lim_{x \to \infty} \int_{-\infty}^{x} f(s)\, ds = 1.$$

In the discrete case the argument is analogous.

The cumulative distribution function is important for a number of reasons. This is true particularly when we deal with a continuous random variable, for in that case we cannot study the probabilistic behavior of X by computing $P(X = x)$. That probability always equals zero in the continuous case. However, we can ask about $P(X \leq x)$ and, as the next theorem demonstrates, obtain the pdf of X.

Theorem 4.4. (a) Let F be the cdf of a continuous random variable with pdf f. Then

$$f(x) = \frac{d}{dx} F(x),$$

for all x at which F is differentiable.

(b) Let X be a discrete random variable with possible values $x_1, x_2, \ldots,$ and suppose that it is possible to label these values so that $x_1 < x_2 < \cdots$ Let F be the cdf of X. Then

$$p(x_j) = P(X = x_j) = F(x_j) - F(x_{j-1}). \tag{4.12}$$

Proof: (a) $F(x) = P(X \leq x) = \int_{-\infty}^{x} f(s)\, ds$. Thus applying the fundamental theorem of the calculus we obtain, $F'(x) = f(x)$.

(b) Since we assumed $x_1 < x_2 < \ldots,$ we have

$$F(x_j) = P(X = x_j \cup X = x_{j-1} \cup \cdots \cup X = x_1)$$
$$= p(j) + p(j - 1) + \cdots + p(1).$$

And

$$F(x_{j-1}) = P(X = x_{j-1} \cup X = x_{j-2} \cup \cdots \cup X = x_1)$$
$$= p(j - 1) + p(j - 2) + \cdots + p(1).$$

Hence $F(x_j) - F(x_{j-1}) = P(X = x_j) = p(x_j).$

Note: Let us briefly reconsider (a) of the above theorem. Recall the definition of the derivative of the function F:

$$F'(x) = \lim_{h \to 0} \frac{F(x + h) - F(x)}{h}$$

$$= \lim_{h \to 0^+} \frac{P(X \leq x + h) - P(X \leq x)}{h}$$

$$= \lim_{h \to 0^+} \frac{1}{h}[P(x < X \leq x + h)].$$

Thus if h is small and positive,

$$F'(x) = f(x) \cong \frac{P(x < X \leq x + h)}{h}.$$

That is, $f(x)$ is approximately equal to the "amount of probability in the interval $(x, x + h]$ per length h." Hence the name *probability density function*.

EXAMPLE 4.16. Suppose that a continuous random variable has cdf F given by

$$F(x) = 0, \qquad x \leq 0,$$
$$= 1 - e^{-x}, \qquad x > 0.$$

Then $F'(x) = e^{-x}$ for $x > 0$, and thus the pdf f is given by

$$f(x) = e^{-x}, \qquad x \geq 0,$$
$$= 0, \qquad \text{elsewhere.}$$

Note: A final word on terminology may be in order. This terminology, although not quite uniform, has become rather standardized. When we speak of the *probability distribution* of a random variable X we mean its pdf f if X is continuous, or its point probability function p defined for $x_1, x_2, \ldots$ if X is discrete. When we speak of the cumulative distribution function, or sometimes just the *distribution function*, we always mean F, where $F(x) = P(X \leq x)$.

4.6 Mixed Distributions

We have restricted our discussion entirely to random variables which are either discrete or continuous. Such random variables are certainly the most important in applications. However, there are situations in which we may encounter the *mixed* type: the random variable X may assume certain distinct values, say $x_1, \ldots, x_n$, with positive probability and also assume all values in some interval, say $a \leq x \leq b$. The probability distribution of such a random variable would be obtained by combining the ideas considered above for the description of discrete and continuous random variables as follows. To each value x_i assign a number $p(x_i)$ such that $p(x_i) \geq 0$, all i, and such that $\sum_{i=1}^{n} p(x_i) = p < 1$. Then define a function f satisfying $f(x) \geq 0$, $\int_a^b f(x)\,dx = 1 - p$. For all a, b, with $-\infty < a < b < +\infty$,

$$P(a \leq X \leq b) = \int_a^b f(x)\,dx + \sum_{\{i:a\leq x_i\leq b\}} p(x_i).$$

In this way we satisfy the condition

$$P(S) = P(-\infty < X < \infty) = 1.$$

A random variable of mixed type might arise as follows. Suppose that we are testing some equipment and we let X be the time of functioning. In most problems we would describe X as a continuous random variable with possible values $x \geq 0$. However, situations may arise in which there is a positive probability that the item does not function at all, that is, it fails at time $X = 0$. In such a case we would want to modify our model and assign a positive probability, say p, to the outcome $X = 0$. Hence we would have $P(X = 0) = p$ and $P(X > 0) = 1 - p$. Thus the number p would describe the distribution of X at 0, while the pdf f would describe the distribution for values of $X > 0$ (Fig. 4.13).

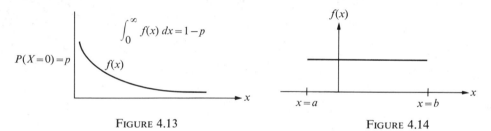

FIGURE 4.13   FIGURE 4.14

4.7 Uniformly Distributed Random Variables

In Chapters 8 and 9 we shall study in considerable detail a number of important discrete and continuous random variables. We have already introduced the important binomial random variable. Let us now consider briefly an important continuous random variable.

Definition. Suppose that X is a continuous random variable assuming all values in the interval $[a, b]$, where both a and b are finite. If the pdf of X is given by

$$f(x) = \frac{1}{b - a}, \qquad a \leq x \leq b,$$
$$= 0, \qquad \text{elsewhere,} \tag{4.13}$$

we say that X is *uniformly distributed* over the interval $[a, b]$. (See Fig. 4.14.)

Notes: (a) A uniformly distributed random variable has a pdf which is *constant* over the interval of definition. In order to satisfy the condition $\int_{-\infty}^{+\infty} f(x)\,dx = 1$, this constant must be equal to the reciprocal of the length of the interval.

(b) A uniformly distributed random variable represents the continuous analog to equally likely outcomes in the following sense. For any subinterval $[c, d]$, where $a \leq c < d \leq b$, $P(c \leq X \leq d)$ is the *same* for all subintervals having the same length. That is,

$$P(c \leq X \leq d) = \int_c^d f(x)\,dx = \frac{d - c}{b - a}$$

and thus depends only on the length of the interval and not on the location of that interval.

(c) We can now make precise the intuitive notion of *choosing a point P at random* on an interval, say $[a, b]$. By this we shall simply mean that the x-coordinate of the chosen point, say X, is uniformly distributed over $[a, b]$.

EXAMPLE 4.17. A point is chosen at random on the line segment $[0, 2]$. What is the probability that the chosen point lies between 1 and $\frac{3}{2}$?

Letting X represent the coordinate of the chosen point, we have that the pdf of X is given by $f(x) = \frac{1}{2}$, $0 < x < 2$, and hence $P(1 \leq X \leq \frac{3}{2}) = \frac{1}{4}$.

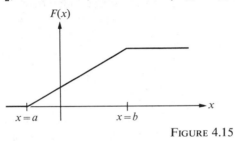

FIGURE 4.15

EXAMPLE 4.18. The hardness, say H, of a specimen of steel (measured on the Rockwell scale) may be assumed to be a continuous random variable uniformly distributed over $[50, 70]$ on the B scale. Hence

$$f(h) = \frac{1}{20}, \qquad 50 < h < 70,$$
$$= 0, \qquad \text{elsewhere.}$$

EXAMPLE 4.19. Let us obtain an expression for the cdf of a uniformly distributed random variable.

$$F(x) = P(X \leq x) = \int_{-\infty}^{x} f(s)\,ds$$

$$= 0 \qquad \text{if } x < a,$$
$$= \frac{x - a}{b - a} \qquad \text{if } a \leq x < b,$$
$$= 1 \qquad \text{if } x \geq b.$$

The graph is shown in Fig. 4.15.

4.8 A Remark

We have pointed out repeatedly that at some stage in our development of a probabilistic model, some probabilities must be assigned to outcomes on the basis of either experimental evidence (such as relative frequencies, for example) or some other considerations, such as past experience with the phenomena being studied. The following question might occur to the student: Why could we not obtain all probabilities we are interested in by some such nondeductive means? The answer is that many events whose probabilities we wish to know are so involved that our intuitive knowledge is insufficient. For instance, suppose that 1000 items are coming off a production line every day, some of which are defective. We wish to know the probability of having 50 or fewer defective items on a given day. Even if we are familiar with the general behavior of the production process, it might be difficult for us to associate a quantitative measure with the event: 50 or fewer items are defective. However, we might be able to make the statement that any individual item had probability 0.10 of being defective. (That is, past experience gives us the information that about 10 percent of the items are defective.) Furthermore, we might be willing to assume that individual items are defective or nondefective independently of one another. *Now* we can proceed deductively and *derive* the probability of the event under consideration. That is, if X = number of defectives,

$$P(X \leq 50) = \sum_{k=0}^{50} \binom{1000}{k}(0.10)^k(0.90)^{1000-k}.$$

The point being made here is that the various methods for computing probabilities which we have derived (and others which we shall study subsequently) are of great importance since with them we can evaluate probabilities associated with rather involved events which would be difficult to obtain by intuitive or empirical means.

PROBLEMS

4.1.   A coin is known to come up heads three times as often as tails. This coin is tossed three times. Let X be the number of heads that appear. Write out the probability distribution of X and also the cdf. Make a sketch of both.

4.2.   From a lot containing 25 items, 5 of which are defective, 4 are chosen at random. Let X be the number of defectives found. Obtain the probability distribution of X if

(a) the items are chosen with replacement,
(b) the items are chosen without replacement.

4.3.   Suppose that the random variable X has possible values $1, 2, 3, \ldots$, and $P(X = j) = 1/2^j, j = 1, 2, \ldots$

(a) Compute $P(X$ is even).

(b) Compute $P(X \geq 5)$.

(c) Compute $P(X$ is divisible by 3).

4.4. Consider a random variable X with possible outcomes: $0, 1, 2, \ldots$ Suppose that $P(X = j) = (1 - a)a^j, j = 0, 1, 2, \ldots$

(a) For what values of a is the above model meaningful?

(b) Verify that the above does represent a legitimate probability distribution.

(c) Show that for any two positive integers s and t,

$$P(X > s + t \mid X > s) = P(X \geq t).$$

4.5. Suppose that twice as many items are produced (per day) by machine 1 as by machine 2. However, about 4 percent of the items from machine 1 tend to be defective while machine 2 produces only about 2 percent defectives. Suppose that the daily output of the two machines is combined. A random sample of 10 is taken from the combined output. What is the probability that this sample contains 2 defectives?

4.6. Rockets are launched until the first successful launching has taken place. If this does not occur within 5 attempts, the experiment is halted and the equipment inspected. Suppose that there is a constant probability of 0.8 of having a successful launching and that successive attempts are independent. Assume that the cost of the first launching is K dollars while subsequent launchings cost $K/3$ dollars. Whenever a successful launching takes place, a certain amount of information is obtained which may be expressed as financial gain of, say C dollars. If T is the net cost of this experiment, find the probability distribution of T.

4.7. Evaluate $P(X = 5)$, where X is the random variable defined in Example 4.10. Suppose that $n_1 = 10$, $n_2 = 15$, $p_1 = 0.3$ and $p_2 = 0.2$.

4.8. (*Properties of the binomial probabilities.*) In the discussion of Example 4.8 a general pattern for the binomial probabilities $\binom{n}{k} p^k (1 - p)^{n-k}$ was suggested. Let us denote these probabilities by $p_n(k)$.

(a) Show that for $0 \leq k < n$ we have

$$p_n(k + 1)/p_n(k) = [(n - k)/(k + 1)] [p/(1 - p)].$$

(b) Using (a) show that

   (i) $p_n(k + 1) > p_n(k)$ if $k < np - (1 - p)$,

   (ii) $p_n(k + 1) = p_n(k)$ if $k = np - (1 - p)$,

   (iii) $p_n(k + 1) < p_n(k)$ if $k > np - (1 - p)$.

(c) Show that if $np - (1 - p)$ is an integer, $p_n(k)$ assumes its maximum value for two values of k, namely $k_0 = np - (1 - p)$ and $k_0' = np - (1 - p) + 1$.

(d) Show that if $np - (1 - p)$ is not an integer then $p_n(k)$ assumes its maximum value when k is equal to the smallest integer greater than k_0.

(e) Show that if $np - (1 - p) < 0$, $p_n(0) > p_n(1) > \cdots > p_n(n)$ while if $np - (1 - p) = 0$, $p_n(0) = p_n(1) > p_n(2) > \cdots > p_n(n)$.

4.9.   The continuous random variable X has pdf $f(x) = x/2, 0 \leq x \leq 2$.   Two independent determinations of X are made.   What is the probability that both these determinations will be greater than one?   If three independent determinations had been made, what is the probability that exactly two of these are larger than one?

4.10.   Let X be the life length of an electron tube and suppose that X may be represented as a continuous random variable with pdf $f(x) = be^{-bx}, x \geq 0$.   Let $p_j = P(j \leq X < j + 1)$.   Show that p_j is of the form $(1 - a)a^j$ and determine a.

4.11.   The continuous random variable X has pdf $f(x) = 3x^2, -1 \leq x \leq 0$.   If b is a number satisfying $-1 < b < 0$, compute $P(X > b \mid X < b/2)$.

4.12.   Suppose that f and g are pdf's on the same interval, say $a \leq x \leq b$.

(a) Show that $f + g$ is *not* a pdf on that interval.

(b) Show that for every number $\beta, 0 < \beta < 1, \beta f(x) + (1 - \beta)g(x)$ is a pdf on that interval.

4.13.   Suppose that the graph in Fig. 4.16 represents the pdf of a random variable X.

(a) What is the relationship between a and b?

(b) If $a > 0$ and $b > 0$, what can you say about the largest value which b may assume?   (See Fig. 4.16.)

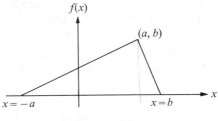

FIGURE 4.16

4.14.   The percentage of alcohol $(100X)$ in a certain compound may be considered as a random variable, where $X, 0 < X < 1$, has the following pdf:

$$f(x) = 20x^3(1 - x), \quad 0 < x < 1.$$

(a) Obtain an expression for the cdf F and sketch its graph.

(b) Evaluate $P(X \leq \frac{2}{3})$.

(c) Suppose that the selling price of the above compound depends on the alcohol content.   Specifically, if $\frac{1}{3} < X < \frac{2}{3}$, the compound sells for C_1 dollars/gallon; otherwise it sells for C_2 dollars/gallon.   If the cost is C_3 dollars/gallon, find the probability distribution of the net profit per gallon.

4.15.   Let X be a continuous random variable with pdf f given by:

$$\begin{aligned} f(x) &= ax, & 0 \leq x \leq 1, \\ &= a, & 1 \leq x \leq 2, \\ &= -ax + 3a, & 2 \leq x \leq 3, \\ &= 0, & \text{elsewhere.} \end{aligned}$$

(a) Determine the constant a.   (b) Determine F, the cdf, and sketch its graph.

(c) If X_1, X_2, and X_3 are three independent observations from X, what is the probability that exactly one of these three numbers is larger than 1.5?

4.16.   The diameter on an electric cable, say X, is assumed to be a continuous random variable with pdf $f(x) = 6x(1 - x), 0 \leq x \leq 1$.

(a) Check that the above is a pdf and sketch it.

(b) Obtain an expression for the cdf of X and sketch it.

(c) Determine a number b such that $P(X < b) = 2P(X > b)$.

(d) Compute $P(X \leq \frac{1}{2} | \frac{1}{3} < X < \frac{2}{3})$.

4.17. Each of the following functions represents the cdf of a continuous random variable. In each case $F(x) = 0$ for $x < a$ and $F(x) = 1$ for $x > b$, where $[a, b]$ is the indicated interval. In each case, sketch the function F, determine the pdf f and sketch it. Also verify that f is a pdf.

(a) $F(x) = x/5, 0 \leq x \leq 5$ (b) $F(x) = (2/\pi) \sin^{-1} (\sqrt{x}), 0 \leq x \leq 1$

(c) $F(x) = e^{3x}, -\infty < x \leq 0$ (d) $F(x) = x^3/2 + \frac{1}{2}, -1 \leq x \leq 1$.

4.18. Let X be the life length of an electronic device (measured in hours). Suppose that X is a continuous random variable with pdf $f(x) = k/x^n$, $2000 \leq x \leq 10,000$.

(a) For $n = 2$, determine k.

(b) For $n = 3$, determine k.

(c) For general n, determine k.

(d) What is the probability that the device will fail before 5000 hours have elapsed?

(e) Sketch the cdf $F(t)$ for (c) and determine its algebraic form.

4.19. Let X be a binomially distributed random variable based on 10 repetitions of an experiment. If $p = 0.3$, evaluate the following probabilities using the table of the binomial distribution in the Appendix.

(a) $P(X \leq 8)$ (b) $P(X = 7)$ (c) $P(X > 6)$.

4.20. Suppose that X is uniformly distributed over $[-\alpha, +\alpha]$, where $\alpha > 0$. Whenever possible, determine α so that the following are satisfied.

(a) $P(X > 1) = \frac{1}{3}$ (b) $P(X > 1) = \frac{1}{2}$ (c) $P(X < \frac{1}{2}) = 0.7$

(d) $P(X < \frac{1}{2}) = 0.3$ (e) $P(|X| < 1) = P(|X| > 1)$.

4.21. Suppose that X is uniformly distributed over $[0, \alpha]$, $\alpha > 0$. Answer the questions of Problem 4.20.

4.22. A point is chosen at random on a line of length L. What is the probability that the ratio of the shorter to the longer segment is less than $\frac{1}{4}$?

4.23. A factory produces 10 glass containers daily. It may be assumed that there is a constant probability $p = 0.1$ of producing a defective container. Before these containers are stored they are inspected and the defective ones are set aside. Suppose that there is a constant probability $r = 0.1$ that a defective container is misclassified. Let X equal the number of containers classified as defective at the end of a production day. (Suppose that all containers which are manufactured on a particular day are also inspected on that day.)

(a) Compute $P(X = 3)$ and $P(X > 3)$.   (b) Obtain an expression for $P(X = k)$.

4.24. Suppose that 5 percent of all items coming off a production line are defective. If 10 such items are chosen and inspected, what is the probability that at most 2 defectives are found?

4.25. Suppose that the life length (in hours) of a certain radio tube is a continuous random variable X with pdf $f(x) = 100/x^2$, $x > 100$, and 0 elsewhere.

(a) What is the probability that a tube will last less than 200 hours if it is known that the tube is still functioning after 150 hours of service?

(b) What is the probability that if 3 such tubes are installed in a set, exactly one will have to be replaced after 150 hours of service?

(c) What is the maximum number of tubes that may be inserted into a set so that there is a probability of 0.5 that after 150 hours of service all of them are still functioning?

4.26.   An experiment consists of n independent trials. It may be supposed that because of "learning," the probability of obtaining a successful outcome increases with the number of trials performed. Specifically, suppose that P(success on the ith repetition) = $(i + 1)/(i + 2), i = 1, 2, \ldots, n$.

(a) What is the probability of having at least 3 successful outcomes in 8 repetitions?

(b) What is the probability that the first successful outcome occurs on the eighth repetition?

4.27.   Referring to Example 4.10,

(a) evaluate $P(X = 2)$ if $n = 4$,

(b) for arbitrary n, show that $P(X = n - 1) = P$ (exactly one unsuccessful attempt) is equal to $[1/(n + 1)] \sum_{i=1}^{n}(1/i)$.

4.28.   If the random variable K is uniformly distributed over $(0, 5)$, what is the probability that the roots of the equation $4x^2 + 4xK + K + 2 = 0$ are real?

4.29.   Suppose that the random variable X has possible values $1, 2, 3, \ldots$ and that $P(X = r) = k(1 - \beta)^{r-1}, 0 < \beta < 1$.

(a) Determine the constant k.

(b) Find the *mode* of this distribution (i.e., that value of r which makes $P(X = r)$ largest).

4.30.   A random variable X may assume four values with probabilities $(1 + 3x)/4$, $(1 - x)/4$, $(1 + 2x)/4$, and $(1 - 4x)/4$. For what values of x is this a probability distribution?

5

Functions of Random Variables

5.1 An Example

Suppose that the radius X of the opening of a finely calibrated tube is considered as a continuous random variable with pdf f. Let $A = \pi X^2$ be the cross-sectional area of the opening. It is intuitively clear that since the value of X is the outcome of a random experiment, the value of A also is. That is, A is a (continuous) random variable, and we might wish to obtain its pdf, say g. We would expect that since A is a function of X, the pdf g is in some way derivable from a knowledge of the pdf f. We shall be concerned with problems of this general nature in this chapter. Before we familiarize ourselves with some of the specific techniques needed, let us formulate the above concepts more precisely.

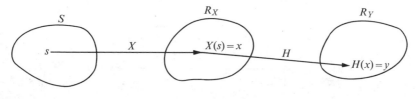

FIGURE 5.1

5.2 Equivalent Events

Let $\mathcal{E}$ be an experiment and let S be a sample space associated with $\mathcal{E}$. Let X be a random variable defined on S. Suppose that $y = H(x)$ is a real-valued function of x. Then $Y = H(X)$ is a random variable since for every $s \in S$, a value of Y is determined, say $y = H[X(s)]$. Schematically we have Fig. 5.1. As before, we call R_X the range space of X, the set of all possible values of the function X. Similarly we define R_Y as the *range space of the random variable Y*, the set of all possible values of Y. We have previously (Eq. 4.1) defined the notion of equivalent events in S and in R_X. We now extend this concept in the following natural way.

Definition. Let C be an event (subset) associated with the range space of Y, R_Y, as described above. Let $B \subset R_X$ be defined as follows:

$$B = \{x \in R_X : H(x) \in C\}. \tag{5.1}$$

In words: B is the set of all values of X such that $H(x) \in C$. If B and C are related in this way we call them *equivalent events*.

Notes: (a) As before, the informal interpretation of the above is that B and C are equivalent events if and only if B and C occur together. That is, when B occurs, C occurs and conversely.

(b) Suppose that A is an event associated with S which is equivalent to an event B associated with R_X. Then, if C is an event associated with R_Y which is equivalent to B, we have that A is equivalent to C.

(c) It is again important to realize that when we speak of equivalent events (in the above sense), these events are associated with different sample spaces.

EXAMPLE 5.1. Suppose that $H(x) = \pi x^2$ as in Section 5.1. Then the events B: $\{X > 2\}$ and C: $\{Y > 4\pi\}$ are equivalent. For if $Y = \pi X^2$, then $\{X > 2\}$ occurs if and only if $\{Y > 4\pi\}$ occurs, since X cannot assume negative values in the present context. (See Fig. 5.2.)

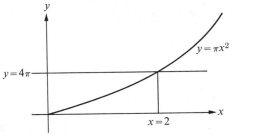

FIGURE 5.2

Note: It is again important to point out that a notational shorthand is being used when we write expressions such as $\{X > 2\}$ and $\{Y > 4\pi\}$. What we are referring to, of course, are the values of X and the values of Y, that is, $\{s \mid X(s) > 2\}$ and $\{x \mid Y(x) > 4\pi\}$.

As we did in Chapter 4, (Eq. 4.2), we shall again make the following definition.

Definition. Let X be a random variable defined on the sample space S. Let R_X be the range space of X. Let H be a real-valued function and consider the random variable $Y = H(X)$ with range space R_Y. For any event $C \subset R_Y$, we *define* $P(C)$ as follows:

$$P(C) = P[\{x \in R_X : H(x) \in C\}]. \tag{5.2}$$

In words: The probability of an event associated with the range space of Y is defined as the probability of the equivalent event (in terms of X) as given by Eq. (5.2).

Notes: (a) The above definition will make it possible to evaluate probabilities involving events associated with Y if we know the probability distribution of X and if we can determine the equivalent event in question.

(b) Since we discussed previously (Eq. 4.1 and 4.2) how to relate probabilities associated with R_X to probabilities associated with S, we may write Eq. (5.2) as follows:

$$P(C) = P[\{x \in R_X : H(x) \in C\}] = P[\{s \in S : H(X(s)) \in C\}].$$

EXAMPLE 5.2. Let X be a continuous random variable with pdf

$$f(x) = e^{-x}, \qquad x > 0.$$

(A simple integration reveals that $\int_0^\infty e^{-x}\,dx = 1$.)

Suppose that $H(x) = 2x + 1$. Hence $R_X = \{x \mid x > 0\}$, while $R_Y = \{y \mid y > 1\}$. Suppose that the event C is defined as follows: $C = \{Y \geq 5\}$. Now $y \geq 5$ if and only if $2x + 1 \geq 5$ which in turn yields $x \geq 2$. Hence C is equivalent to $B = \{X \geq 2\}$. (See Fig. 5.3.) Now $P(X \geq 2) = \int_2^\infty e^{-x}\,dx = 1/e^2$. Hence applying Eq. (5.2) we find that

$$P(Y \geq 5) = 1/e^2.$$

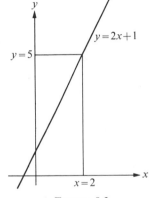

FIGURE 5.3

Notes: (a) It is again worthwhile to point out that we may consider incorporating both the evaluation of $x = X(s)$ and the evaluation of $y = H(x)$ into our experiment and hence simply consider R_Y, the range space of Y, as the sample space of our experiment.

Strictly speaking, of course, the sample space of our experiment is S and the outcome of the experiment is s. Everything we do subsequently is not influenced by the random nature of the experiment. The determination of $x = X(s)$ and the evaluation of $y = H(x)$ are strictly deterministic processes once s has been observed. However, as we discussed previously, we can incorporate these computations into the description of our experiment and thus deal directly with the range space R_Y.

(b) Just as the probability distribution was induced in R_X by the probability distribution over the original sample space S, so the probability distribution of Y is determined if the probability distribution of X is known. For instance, in Example 5.2 above, the specified distribution of X determined completely the value of $P(Y \geq 5)$.

(c) In considering a function of a random variable X, say $Y = H(X)$, we must comment that not every conceivable function H may be permitted. However, the functions that arise in applications are inevitably among those we may consider and hence we shall not refer to this minor difficulty subsequently.

5.3 Discrete Random Variables

Case 1. X is a discrete random variable. If X is a discrete random variable and $Y = H(X)$, then it follows immediately that Y is also a discrete random variable.

For suppose that the possible values of X may be enumerated as $x_1, x_2, \ldots, x_n, \ldots$ Then certainly the possible values of Y may be enumerated as $y_1 = H(x_1), y_2 = H(x_2), \ldots$ (Some of the above Y-values may be the same, but this certainly does not distract from the fact that these values may be enumerated.)

EXAMPLE 5.3. Suppose that the random variable X assumes the three values $-1, 0$, and 1 with probabilities $\frac{1}{3}, \frac{1}{2}$, and $\frac{1}{6}$, respectively. Let $Y = 3X + 1$. Then the possible values of Y are $-2, 1$, and 4, assumed with probabilities $\frac{1}{3}, \frac{1}{2}$, and $\frac{1}{6}$.

This example suggests the following *general procedure:* If $x_1, \ldots, x_n, \ldots$ are the possible values of X, $p(x_i) = P(X = x_i)$, and H is a function such that to each value y there corresponds exactly one value x, then the probability distribution of Y is obtained as follows.

Possible values of Y: $\qquad y_i = H(x_i), \qquad i = 1, 2, \ldots, n, \ldots ;$

Probabilities of Y: $\qquad q(y_i) = P(Y = y_i) = p(x_i).$

Quite often the function H does not possess the above characteristic, and it may happen that several values of X lead to the same value of Y, as the following example illustrates.

EXAMPLE 5.4. Suppose that we consider the same random variable X as in Example 5.3 above. However, we introduce $Y = X^2$. Hence the possible values of Y are zero and one, assumed with probabilities $\frac{1}{2}, \frac{1}{2}$. For $Y = 1$ if and only if $X = -1$ or $X = 1$ and the probability of this latter event is $\frac{1}{3} + \frac{1}{6} = \frac{1}{2}$. In terms of our previous terminology the events $B: \{X = \pm 1\}$ and $C: \{Y = 1\}$ are equivalent events and hence by Eq. (5.2) have equal probabilities.

The *general procedure* for situations as described in the above example is as follows: Let $x_{i_1}, x_{i_2}, \ldots, x_{i_k}, \ldots,$ represent the X-values having the property $H(x_{i_j}) = y_i$ for all j. Then

$$q(y_i) = P(Y = y_i) = p(x_{i_1}) + p(x_{i_2}) + \cdots$$

In words: To evaluate the probability of the event $\{Y = y_i\}$, find the equivalent event in terms of X (in the range space R_X) and then add all the corresponding probabilities. (See Fig. 5.4.)

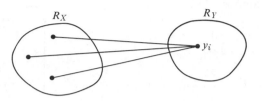

R_X R_Y

$\bullet\, y_i$

FIGURE 5.4

EXAMPLE 5.5. Let X have possible values $1, 2, \ldots, n, \ldots$ and suppose that $P(X = n) = \frac{1}{2}^n$. Let

$$Y = 1 \qquad \text{if} \quad X \text{ is even,}$$
$$= -1 \qquad \text{if} \quad X \text{ is odd.}$$

Hence Y assumes the two values -1 and $+1$. Since $Y = 1$ if and only if $X = 2$, or $X = 4$, or $X = 6$, or $\ldots$, applying Eq. (5.2) yields

$$P(Y = 1) = \tfrac{1}{4} + \tfrac{1}{16} + \tfrac{1}{64} + \cdots = \tfrac{1}{3}.$$

Hence

$$P(Y = -1) = 1 - P(Y = 1) = \tfrac{2}{3}.$$

Case 2. X is a continuous random variable. It may happen that X is a continuous random variable while Y is discrete. For example, suppose that X may assume all real values while Y is defined to be $+1$ if $X \geq 0$ while $Y = -1$ if $X < 0$. In order to obtain the probability distribution of Y, simply determine the equivalent event (in the range space R_X) corresponding to the different values of Y. In the above case, $Y = 1$ if and only if $X \geq 0$, while $Y = -1$ if and only if $X < 0$. Hence $P(Y = 1) = P(X \geq 0)$ while $P(Y = -1) = P(X < 0)$. If the pdf of X is known, these probabilities may be evaluated. In the general case, if $\{Y = y_i\}$ is equivalent to an event, say A, in the range space of X, then

$$q(y_i) = P(Y = y_i) = \int_A f(x)\, dx.$$

5.4 Continuous Random Variables

The most important (and most frequently encountered) case arises when X is a continuous random variable with pdf f and H is a continuous function. Hence $Y = H(X)$ is a continuous random variable, and it will be our task to obtain its pdf, say g.

The *general procedure* will be as follows:

(a) Obtain G, the cdf of Y, where $G(y) = P(Y \leq y)$, by finding the event A (in the range space of X) which is equivalent to the event $\{Y \leq y\}$.

(b) Differentiate $G(y)$ with respect to y in order to obtain $g(y)$.

(c) Determine those values of y in the range space of Y for which $g(y) > 0$.

EXAMPLE 5.6. Suppose that X has pdf

$$f(x) = 2x, \quad 0 < x < 1,$$
$$= 0, \quad \text{elsewhere.}$$

Let $H(x) = 3x + 1$. Hence to find the pdf of $Y = H(X)$ we have (see Fig. 5.5)

$$G(y) = P(Y \le y) = P(3X + 1 \le y)$$
$$= P(X \le (y - 1)/3)$$
$$= \int_0^{(y-1)/3} 2x \, dx = [(y - 1)/3]^2.$$

Thus

$$g(y) = G'(y) = \tfrac{2}{9}(y - 1).$$

Since $f(x) > 0$ for $0 < x < 1$, we find that $g(y) > 0$ for $1 < y < 4$.

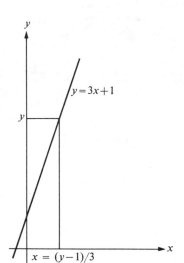

FIGURE 5.5

Note: The event A, referred to above, equivalent to the event $\{Y \le y\}$ is simply $\{X \le (y - 1)/3\}$.

There is another, slightly different way of obtaining the same result which will be useful later on. Consider again

$$G(y) = P(Y \le y) = P\left(X \le \frac{y - 1}{3}\right) = F\left(\frac{y - 1}{3}\right),$$

where F is the cdf of X; that is,

$$F(x) = P(X \le x).$$

In order to evaluate the derivative of G, $G'(y)$, we use the chain rule for differentiation as follows:

$$\frac{dG(y)}{dy} = \frac{dG(y)}{du} \cdot \frac{du}{dy}, \quad \text{where} \quad u = \frac{y - 1}{3}$$

Hence

$$G'(y) = F'(u) \cdot \frac{1}{3} = f(u) \cdot \frac{1}{3} = 2\left(\frac{y - 1}{3}\right) \cdot \frac{1}{3},$$

FIGURE 5.6

as before. The pdf of Y has the graph shown in Fig. 5.6. (To check the computation note that $\int_1^4 g(y) \, dy = 1$.)

EXAMPLE 5.7. Suppose that a continuous random variable has pdf as given in Example 5.6. Let $H(x) = e^{-x}$. To find the pdf of $Y = H(X)$ we proceed as follows (see Fig. 5.7):

$$G(y) = P(Y \leq y) = P(e^{-X} \leq y)$$
$$= P(X \geq -\ln y) = \int_{-\ln y}^{1} 2x \, dx$$
$$= 1 - (-\ln y)^2.$$

Hence $g(y) = G'(y) = -2 \ln y/y$. Since $f(x) > 0$ for $0 < x < 1$, we find that $g(y) > 0$ for $1/e < y < 1$. (Note that the algebraic sign for $g(y)$ is correct since $\ln y < 0$ for $1/e < y < 1$.) The graph of $g(y)$ is sketched in Fig. 5.8.

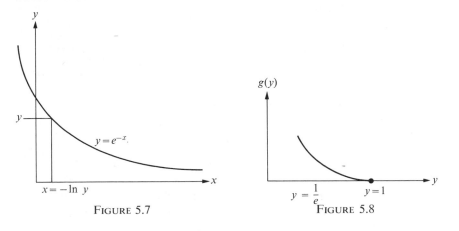

FIGURE 5.7   FIGURE 5.8

Again we can obtain the above result by a slightly different approach which we briefly outline. As before,

$$G(y) = P(Y \leq y) = P(X \geq -\ln y)$$
$$= 1 - P(X \leq -\ln y) = 1 - F(-\ln y),$$

where as before F is the cdf of X. In order to obtain the derivative of G we again use the chain rule of differentiation as follows:

$$\frac{dG(y)}{dy} = \frac{dG}{du}\frac{du}{dy}, \quad \text{where} \quad u = -\ln y.$$

Thus

$$G'(y) = -F'(u)\left(-\frac{1}{y}\right) = +2 \ln y \cdot \left(-\frac{1}{y}\right),$$

as before.

Let us now generalize the approach suggested by the above examples. The crucial step in each of the examples was taken when we replaced the event $\{Y \leq y\}$ by the equivalent event in terms of the random variable X. In the above problems this was relatively easy since in each case the function was a strictly increasing or strictly decreasing function of X.

In Fig. 5.9, y is a strictly increasing function of x. Hence we may solve $y = H(x)$ for x in terms of y, say $x = H^{-1}(y)$, where H^{-1} is called the inverse function of H. Thus if H is strictly increasing, $\{H(X) \leq y\}$ is equivalent to $\{X \leq H^{-1}(y)\}$ while if H is strictly decreasing, $\{H(X) \leq y\}$ is equivalent to $\{X \geq H^{-1}(y)\}$.

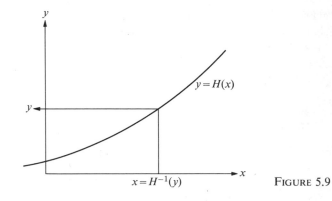

FIGURE 5.9

The method used in the above examples may now be generalized as follows.

Theorem 5.1. Let X be a continuous random variable with pdf f, where $f(x) > 0$ for $a < x < b$. Suppose that $y = H(x)$ is a strictly monotone (increasing or decreasing) function of x. Assume that this function is differentiable (and hence continuous) for all x. Then the random variable Y defined as $Y = H(X)$ has a pdf g given by,

$$g(y) = f(x) \left| \frac{dx}{dy} \right| ,\tag{5.3}$$

where x is expressed in terms of y. If H is increasing, then g is nonzero for those values of y satisfying $H(a) < y < H(b)$. If H is decreasing, then g is nonzero for those values of y satisfying $H(b) < y < H(a)$.

Proof: (a) Assume that H is a strictly increasing function. Hence

$$G(y) = P(Y \leq y) = P(H(X) \leq y)$$
$$= P(X \leq H^{-1}(y)) = F(H^{-1}(y)).$$

Differentiating $G(y)$ with respect to y, we obtain, using the chain rule for derivatives,

$$\frac{dG(y)}{dy} = \frac{dG(y)}{dx} \frac{dx}{dy}, \qquad \text{where} \quad x = H^{-1}(y).$$

Thus

$$G'(y) = \frac{dF(x)}{dx} \frac{dx}{dy} = f(x) \frac{dx}{dy}.$$

(b) Assume that H is a decreasing function. Therefore

$$G(y) = P(Y \le y) = P(H(X) \le y) = P(X \ge H^{-1}(y))$$
$$= 1 - P(X \le H^{-1}(y)) = 1 - F(H^{-1}(y)).$$

Proceeding as above, we may write

$$\frac{dG(y)}{dy} = \frac{dG(y)}{dx}\frac{dx}{dy} = \frac{d}{dx}[1 - F(x)]\frac{dx}{dy} = -f(x)\frac{dx}{dy}.$$

Note: The algebraic sign obtained in (b) is correct since, if y is a decreasing function of x, x is a decreasing function of y and hence $dx/dy < 0$. Thus, by using the absolute-value sign around dx/dy, we may combine the result of (a) and (b) and obtain the final form of the theorem.

EXAMPLE 5.8. Let us reconsider Examples 5.6 and 5.7, by applying Theorem 5.1.
(a) For Example 5.6 we had $f(x) = 2x$, $0 < x < 1$, and $y = 3x + 1$. Hence $x = (y - 1)/3$ and $dx/dy = \frac{1}{3}$. Thus $g(y) = 2[(y - 1)/3]\frac{1}{3} = \frac{2}{9}(y - 1)$, $1 < y < 4$, which checks with the result obtained previously.
(b) In Example 5.7 we had $f(x) = 2x$, $0 < x < 1$, and $y = e^{-x}$. Hence $x = -\ln y$ and $dx/dy = -1/y$. Thus $g(y) = -2(\ln y)/y$, $1/e < y < 1$ which again checks with the above result.

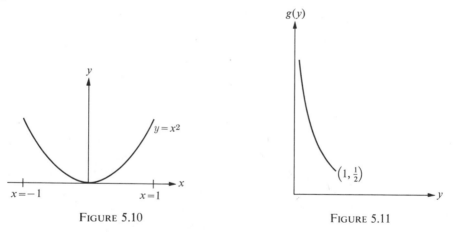

FIGURE 5.10   FIGURE 5.11

If $y = H(x)$ is not a monotone function of x we cannot apply the above method directly. Instead, we shall return to the general method outlined above. The following example illustrates this procedure.

EXAMPLE 5.9. Suppose that

$$f(x) = \tfrac{1}{2}, \qquad -1 < x < 1,$$
$$= 0, \qquad \text{elsewhere.}$$

Let $H(x) = x^2$. This is obviously *not* a monotone function over the interval $[-1, 1]$ (Fig. 5.10). Hence we obtain the pdf of $Y = X^2$ as follows:

$$G(y) = P(Y \leq y) = P(X^2 \leq y)$$
$$= P(-\sqrt{y} \leq X \leq \sqrt{y})$$
$$= F(\sqrt{y}) - F(-\sqrt{y}),$$

where F is the cdf of the random variable X. Therefore

$$g(y) = G'(y) = \frac{f(\sqrt{y})}{2\sqrt{y}} - \frac{f(-\sqrt{y})}{-2\sqrt{y}}$$

$$= \frac{1}{2\sqrt{y}} [f(\sqrt{y}) + f(-\sqrt{y})].$$

Thus $g(y) = (1/2\sqrt{y}) (\frac{1}{2} + \frac{1}{2}) = 1/2\sqrt{y}, 0 < y < 1$. (See Fig. 5.11.)
    The method used in the above example yields the following general result.

Theorem 5.2. Let X be a continuous random variable with pdf f. Let $Y = X^2$.
Then the random variable Y has pdf given by

$$g(y) = \frac{1}{2\sqrt{y}} [f(\sqrt{y}) + f(-\sqrt{y})].$$

Proof: See Example 5.9.

PROBLEMS

5.1.   Suppose that X is uniformly distributed over $(-1, 1)$. Let $Y = 4 - X^2$. Find the pdf of Y, say $g(y)$, and sketch it. Also verify that $g(y)$ *is* a pdf.

5.2.   Suppose that X is uniformly distributed over $(1, 3)$. Obtain the pdf of the following random variables:
    (a) $Y = 3X + 4$ (b) $Z = e^X$.
Verify in each case that the function obtained is a pdf. Sketch the pdf.

5.3.   Suppose that the continuous random variable X has pdf $f(x) = e^{-x}, x > 0$. Find the pdf of the following random variables:
    (a) $Y = X^3$ (b) $Z = 3/(X + 1)^2$.

5.4.   Suppose that the discrete random variable X assumes the values 1, 2, and 3 with equal probability. Find the probability distribution of $Y = 2X + 3$.

5.5.   Suppose that X is uniformly distributed over the interval $(0, 1)$. Find the pdf of the following random variables:
    (a) $Y = X^2 + 1$ (b) $Z = 1/(X + 1)$.

5.6. Suppose that X is uniformly distributed over $(-1, 1)$. Find the pdf of the following random variables:

(a) $Y = \sin(\pi/2)X$ (b) $Z = \cos(\pi/2)X$ (c) $W = |X|$.

5.7. Suppose that the radius of a sphere is a continuous random variable. (Due to inaccuracies of the manufacturing process, the radii of different spheres may be different.) Suppose that the radius R has pdf $f(r) = 6r(1 - r), 0 < r < 1$. Find the pdf of the volume V and the surface area S of the sphere.

5.8. A fluctuating electric current I may be considered as a uniformly distributed random variable over the interval $(9, 11)$. If this current flows through a 2-ohm resistor, find the pdf of the power $P = 2I^2$.

5.9. The speed of a molecule in a uniform gas at equilibrium is a random variable V whose pdf is given by

$$f(v) = av^2e^{-bv^2}, \qquad v > 0,$$

where $b = m/2kT$ and k, T, and m denote Boltzman's constant, the absolute temperature, and the mass of the molecule, respectively.

(a) Evaluate the constant a (in terms of b). [*Hint:* Use the fact that $\int_0^\infty e^{-x^2}\,dx = \sqrt{\pi}/2$ and integrate by parts.]

(b) Derive the distribution of the random variable $W = mV^2/2$, which represents the kinetic energy of the molecule.

5.10. A random voltage X is uniformly distributed over the interval $(-k, k)$. If X is the input of a nonlinear device with the characteristics shown in Fig. 5.12, find the probability distribution of Y in the following three cases:

(a) $k < a$ (b) $a < k < x_0$ (c) $k > x_0$.

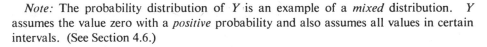

FIGURE 5.12

Note: The probability distribution of Y is an example of a *mixed* distribution. Y assumes the value zero with a *positive* probability and also assumes all values in certain intervals. (See Section 4.6.)

5.11. The radiant energy (in Btu/hr/ft^2) is given as the following function of temperature T (in degree fahrenheit): $E = 0.173(T/100)^4$. Suppose that the temperature T is considered to be a continuous random variable with pdf

$$f(t) = 200t^{-2}, \qquad 40 \le t \le 50,$$
$$= 0, \qquad \text{elsewhere.}$$

Find the pdf of the radiant energy E.

5.12. To measure air velocities, a tube (known as Pitot static tube) is used which enables one to measure differential pressure. This differential pressure is given by $P = (1/2) dV^2$, where d is the density of the air and V is the wind speed (mph). If V is a random variable uniformly distributed over (10, 20), find the pdf of P.

5.13. Suppose that $P(X \leq 0.29) = 0.75$, where X is a continuous random variable with some distribution defined over (0, 1). If $Y = 1 - X$, determine k so that $P(Y \leq k) = 0.25$.

9

Two- and Higher-Dimensional Random Variables

6.1 Two-Dimensional Random Variables

In our study of random variables we have, so far, considered only the one-dimensional case. That is, the outcome of the experiment could be recorded as a single number x.

In many situations, however, we are interested in observing two or more numerical characteristics simultaneously. For example, the hardness H and the tensile strength T of a manufactured piece of steel may be of interest, and we would consider (h, t) as a single experimental outcome. We might study the height H and the weight W of some chosen person, giving rise to the outcome (h, w). Finally we might observe the total rainfall R and the average temperature T at a certain locality during a specified month, giving rise to the outcome (r, t).

We shall make the following formal definition.

Definition. Let ε be an experiment and S a sample space associated with ε. Let $X = X(s)$ and $Y = Y(s)$ be two functions each assigning a real number to each outcome $s \in S$ (Fig. 6.1). We call (X, Y) a *two-dimensional random variable* (sometimes called a *random vector*).

If $X_1 = X_1(s), X_2 = X_2(s), \ldots, X_n = X_n(s)$ are n functions each assigning a real number to every outcome $s \in S$, we call $(X_1, \ldots, X_n)$ an *n-dimensional random variable* (or an n-dimensional random vector.)

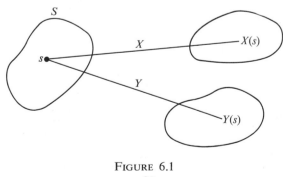

FIGURE 6.1

93

Note: As in the one-dimensional case, our concern will be not with the functional nature of $X(s)$ and $Y(s)$, but rather with the values which X and Y assume. We shall again speak of the *range space* of (X, Y), say $R_{X \times Y}$, as the set of all possible values of (X, Y). In the two-dimensional case, for instance, the range space of (X, Y) will be a subset of the Euclidean plane. Each outcome $X(s)$, $Y(s)$ may be represented as a point (x, y) in the plane. We will again suppress the functional nature of X and Y by writing, for example, $P[X \leq a, Y \leq b]$ instead of $P[X(s) \leq a, Y(s) \leq b]$.

As in the one-dimensional case, we shall distinguish between two basic types of random variables: the discrete and the continuous random variables.

Definition. (X, Y) is a *two-dimensional discrete* random variable if the possible values of (X, Y) are finite or countably infinite. That is, the possible values of (X, Y) may be represented as (x_i, y_j), $i = 1, 2, \ldots, n, \ldots$; $j = 1, 2, \ldots$, $m, \ldots$

(X, Y) is a *two-dimensional continuous* random variable if (X, Y) can assume all values in some noncountable set of the Euclidean plane. [For example, if (X, Y) assumes all values in the rectangle $\{(x, y) \mid a \leq x \leq b, c \leq y \leq d\}$ or all values in the circle $\{(x, y) \mid x^2 + y^2 \leq 1\}$, we would say that (X, Y) is a continuous two-dimensional random variable.]

Notes: (a) Speaking informally, (X, Y) is a two-dimensional random variable if it represents the outcome of a random experiment in which we have measured the *two* numerical characteristics X and Y.

(b) It may happen that one of the components of (X, Y), say X, is discrete, while the other is continuous. However, in most applications we deal only with the cases discussed above, in which either both components are discrete or both are continuous.

(c) In many situations the two random variables X and Y, when considered jointly, are in a very natural way the outcome of a single experiment, as illustrated in the above examples. For instance, X and Y may represent the height and weight of the same individual, etc. However this sort of connection need not exist. For example, X might be the current flowing in a circuit at a specified moment, while Y might be the temperature in the room at that moment, and we could then consider the two-dimensional random variable (X, Y). In most applications there is a very definite reason for considering X and Y jointly.

We proceed in analogy with the one-dimensional case in describing the probability distribution of (X, Y).

Definition. (a) Let (X, Y) be a two-dimensional discrete random variable. With each possible outcome (x_i, y_j) we associate a number $p(x_i, y_j)$ representing $P(X = x_i, Y = y_j)$ and satisfying the following conditions:

$$(1) \quad p(x_i, y_j) \geq 0 \qquad \text{for all} \quad (x, y),$$

$$(2) \quad \sum_{j=1}^{\infty} \sum_{i=1}^{\infty} p(x_i, y_j) = 1. \tag{6.1}$$

The function p defined for all (x_i, y_j) in the range space of (X, Y) is called the *probability function* of (X, Y). The set of triples $(x_i, y_j, p(x_i, y_j))$, $i, j = 1, 2,$ $\ldots$, is sometimes called the *probability distribution* of (X, Y).

(b) Let (X, Y) be a continuous random variable assuming all values in some region R of the Euclidean plane. The *joint probability density function* f is a function satisfying the following conditions:

$$(3) \quad f(x, y) \geq 0 \qquad \text{for all} \quad (x, y) \in R,$$
$$(4) \quad \iint_R f(x, y) \, dx \, dy = 1. \tag{6.2}$$

Notes: (a) The analogy to a mass distribution is again clear. We have a unit mass distributed over a region in the plane. In the discrete case, all the mass is concentrated at a finite or countably infinite number of places with mass $p(x_i, y_j)$ located at (x_i, y_j). In the continuous case, mass is found at all points of some noncountable set in the plane.

(b) Condition 4 states that the total *volume* under the surface given by the equation $z = f(x, y)$ equals 1.

(c) As in the one-dimensional case, $f(x, y)$ does not represent the probability of anything. However, for positive Δx and Δy sufficiently small, $f(x, y) \Delta x \Delta y$ is approximately equal to $P(x \leq X \leq x + \Delta x, y \leq Y \leq y + \Delta y)$.

(d) As in the one-dimensional case we shall adopt the convention that $f(x, y) = 0$ if $(x, y) \notin R$. Hence we may consider f defined for all (x, y) in the plane and the requirement 4 above becomes $\int_{-\infty}^{+\infty} \int_{-\infty}^{+\infty} f(x, y) \, dx \, dy = 1$.

(e) We shall again *suppress* the functional nature of the two-dimensional random variable (X, Y). We *should* always be writing statements of the form $P[X(s) = x_i, Y(s) = y_j]$, etc. However, if our shortcut notation is understood, no difficulty should arise.

(f) Again, as in the one-dimensional case, the probability distribution of (X, Y) is actually *induced* by the probability of events associated with the original sample space S. However, we shall be concerned mainly with the values of (X, Y) and hence deal directly with the range space of (X, Y). Nevertheless, the reader should not lose sight of the fact that if $P(A)$ is specified for all events $A \subset S$, then the probability associated with events in the range space of (X, Y) is determined. That is, if B is in the range space of (X, Y), we have

$$P(B) = P\big[(X(s), Y(s)) \in B\big] = P\big[s \mid (X(s), Y(s)) \in B\big].$$

This latter probability refers to an event in S and hence *determines* the probability of B. In terms of our previous terminology, B and $\{s \mid (X(s), Y(s)) \in B\}$ are *equivalent* events (Fig. 6.2).

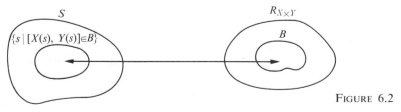

FIGURE 6.2

If B is in the range space of (X, Y) we have

$$P(B) = \sum\sum_{B} p(x_i, y_j), \tag{6.3}$$

if (X, Y) is discrete, where the sum is taken over all indices (i, j) for which $(x_i, y_j) \in B$. And

$$P(B) = \iint_{B} f(x, y) \, dx \, dy, \tag{6.4}$$

if (X, Y) is continuous.

EXAMPLE 6.1. Two production lines manufacture a certain type of item. Suppose that the capacity (on any given day) is 5 items for line I and 3 items for line II. Assume that the number of items actually produced by either production line is a random variable. Let (X, Y) represent the two-dimensional random variable yielding the number of items produced by line I and line II, respectively. Table 6.1 gives the joint probability distribution of (X, Y). Each entry represents

$$p(x_i, y_j) = P(X = x_i, Y = y_j).$$

Thus $p(2, 3) = P(X = 2, Y = 3) = 0.04$, etc. Hence if B is defined as

$$B = \{\text{More items are produced by line I than by line II}\}$$

we find that

$$
\begin{aligned}
P(B) &= 0.01 + 0.03 + 0.05 + 0.07 + 0.09 + 0.04 + 0.05 + 0.06 \\
&\quad + 0.08 + 0.05 + 0.05 + 0.06 + 0.06 + 0.05 \\
&= 0.75.
\end{aligned}
$$

EXAMPLE 6.2. Suppose that a manufacturer of light bulbs is concerned about the number of bulbs ordered from him during the months of January and February. Let X and Y denote the number of bulbs ordered during these two months, respectively. We shall assume that (X, Y) is a two-dimensional continuous random

TABLE 6.1

Y \ X	0	1	2	3	4	5
0	0	0.01	0.03	0.05	0.07	0.09
1	0.01	0.02	0.04	0.05	0.06	0.08
2	0.01	0.03	0.05	0.05	0.05	0.06
3	0.01	0.02	0.04	0.06	0.06	0.05

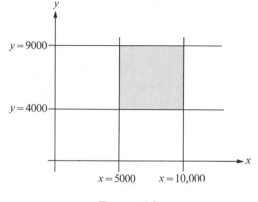

FIGURE 6.3

variable with the following joint pdf (see Fig. 6.3):

$$f(x, y) = c \qquad \text{if} \quad 5000 \le x \le 10{,}000 \quad \text{and} \quad 4000 \le y \le 9000,$$
$$= 0, \qquad \text{elsewhere.}$$

To determine c we use the fact that $\int_{-\infty}^{+\infty} \int_{-\infty}^{+\infty} f(x, y) \, dx \, dy = 1$. Therefore

$$\int_{-\infty}^{+\infty} \int_{-\infty}^{+\infty} f(x, y) \, dx \, dy = \int_{4000}^{9000} \int_{5000}^{10{,}000} f(x, y) \, dx \, dy = c[5000]^2.$$

Thus $c = (5000)^{-2}$. Hence if $B = \{X \ge Y\}$, we have

$$P(B) = 1 - \frac{1}{(5000)^2} \int_{5000}^{9000} \int_{5000}^{y} dx \, dy$$

$$= 1 - \frac{1}{(5000)^2} \int_{5000}^{9000} [y - 5000] \, dy = \frac{17}{25}.$$

Note: In the above example, X and Y should obviously be integer-valued since we cannot order a fractional number of bulbs! However, we are again dealing with the idealized situation in which we allow X to assume *all* values between 5000 and 10,000 (inclusive).

EXAMPLE 6.3. Suppose that the two-dimensional continuous random variable (X, Y) has joint pdf given by

$$f(x, y) = x^2 + \frac{xy}{3}, \qquad 0 \le x \le 1, \quad 0 \le y \le 2,$$
$$= 0, \qquad \text{elsewhere.}$$

To check that $\int_{-\infty}^{+\infty} \int_{-\infty}^{+\infty} f(x, y)\, dx\, dy = 1$:

$$\int_{-\infty}^{+\infty} \int_{-\infty}^{+\infty} f(x, y)\, dx\, dy = \int_0^2 \int_0^1 \left(x^2 + \frac{xy}{3} \right) dx\, dy$$

$$= \int_0^2 \left. \frac{x^3}{3} + \frac{x^2 y}{6} \right|_{x=0}^{x=1} dy$$

$$= \int_0^2 \left(\frac{1}{3} + \frac{y}{6} \right) dy = \left. \frac{1}{3} y + \frac{y^2}{12} \right|_0^2$$

$$= \tfrac{2}{3} + \tfrac{4}{12} = 1.$$

Let $B = \{X + Y \geq 1\}$. (See Fig. 6.4.) We shall compute $P(B)$ by evaluating $1 - P(\bar{B})$, where $\bar{B} = \{X + Y < 1\}$. Hence

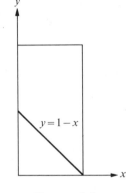

$$P(B) = 1 - \int_0^1 \int_0^{1-x} \left(x^2 + \frac{xy}{3} \right) dy\, dx$$

$$= 1 - \int_0^1 \left[x^2(1 - x) + \frac{x(1 - x)^2}{6} \right] dx$$

$$= 1 - \tfrac{7}{72} = \tfrac{65}{72}.$$

In studying one-dimensional random variables, we found that F, the cumulative distribution function, played an important role. In the two-dimensional case we can again define a cumulative function as follows.

$y = 1 - x$

FIGURE 6.4

Definition. Let (X, Y) be a two-dimensional random variable. The *cumulative distribution function* (cdf) F of the two-dimensional random variable (X, Y) is defined by

$$F(x, y) = P(X \leq x, Y \leq y).$$

Note: F is a function of *two* variables and has a number of properties analogous to those discussed for the one-dimensional cdf. (See Section 4.5.) We shall mention only the following important property.

If F is the cdf of a two-dimensional random variable with joint pdf f, then

$$\partial^2 F(x, y)/\partial x\, \partial y = f(x, y)$$

wherever F is differentiable. This result is analogous to Theorem 4.4 in which we proved that $(d/dx)F(x) = f(x)$, where f is the pdf of the one-dimensional random variable X.

6.2 Marginal and Conditional Probability Distributions

With each two-dimensional random variable (X, Y) we associate two one-dimensional random variables, namely X and Y, individually. That is, we may be interested in the probability distribution of X or the probability distribution of Y.

EXAMPLE 6.4. Let us again consider Example 6.1. In addition to the entries of Table 6.1, let us also compute the "marginal" totals, that is, the sum of the 6 columns and 4 rows of the table. (See Table 6.2.)

The probabilities appearing in the row and column margins represent the probability distribution of Y and X, respectively. For instance, $P(Y = 1) = 0.26$, $P(X = 3) = 0.21$, etc. Because of the appearance of Table 6.2 we refer, quite generally, to the *marginal* distribution of X or the *marginal* distribution of Y, whenever we have a two-dimensional random variable (X, Y), whether discrete or continuous.

TABLE 6.2

Y \ X	0	1	2	3	4	5	Sum
0	0	0.01	0.03	0.05	0.07	0.09	0.25
1	0.01	0.02	0.04	0.05	0.06	0.08	0.26
2	0.01	0.03	0.05	0.05	0.05	0.06	0.25
3	0.01	0.02	0.04	0.06	0.06	0.05	0.24
Sum	0.03	0.08	0.16	0.21	0.24	0.28	1.00

In the *discrete* case we proceed as follows: Since $X = x_i$ must occur with $Y = y_j$ for some j and can occur with $Y = y_j$ for only one j, we have

$$p(x_i) = P(X = x_i) = P(X = x_i, Y = y_1 \text{ or } X = x_i, Y = y_2 \text{ or } \cdots)$$

$$= \sum_{j=1}^{\infty} p(x_i, y_j).$$

The function p defined for $x_1, x_2, \ldots$, represents the *marginal probability distribution* of X. Analogously we define $q(y_j) = P(Y = Y_j) = \sum_{i=1}^{\infty} p(x_i, y_j)$ as the *marginal probability distribution* of Y.

In the *continuous* case we proceed as follows: Let f be the joint pdf of the continuous two-dimensional random variable (X, Y). We define g and h, the *marginal probability density functions* of X and Y, respectively, as follows:

$$g(x) = \int_{-\infty}^{+\infty} f(x, y) \, dy; \qquad h(y) = \int_{-\infty}^{+\infty} f(x, y) \, dx.$$

These pdf's correspond to the basic pdf's of the one-dimensional random variables X and Y, respectively. For example

$$P(c \leq X \leq d) = P[c \leq X \leq d, -\infty < Y < \infty]$$

$$= \int_c^d \int_{-\infty}^{+\infty} f(x, y) \, dy \, dx$$

$$= \int_c^d g(x) \, dx.$$

EXAMPLE 6.5. Two characteristics of a rocket engine's performance are thrust X and mixture ratio Y. Suppose that (X, Y) is a two-dimensional continuous random variable with joint pdf:

$$f(x, y) = 2(x + y - 2xy), \qquad 0 \leq x \leq 1, \; 0 \leq y \leq 1,$$
$$= 0, \qquad \text{elsewhere.}$$

(The units have been adjusted in order to use values between 0 and 1.) The marginal pdf of X is given by

$$g(x) = \int_0^1 2(x + y - 2xy) \, dy = 2(xy + y^2/2 - xy^2)|_0^1$$
$$= 1, \qquad 0 \leq x \leq 1.$$

That is, X is uniformly distributed over $[0, 1]$.
    The marginal pdf of Y is given by

$$h(y) = \int_0^1 2(x + y - 2xy) \, dx = 2(x^2/2 + xy - x^2y)|_0^1$$
$$= 1, \qquad 0 \leq y \leq 1.$$

Hence Y is also uniformly distributed over $[0, 1]$.

Definition. We say that the two-dimensional continuous random variable is *uniformly distributed* over a region R in the Euclidean plane if

$$f(x, y) = \text{const} \qquad \text{for } (x, y) \in R,$$
$$= 0, \qquad \text{elsewhere.}$$

Because of the requirement $\int_{-\infty}^{+\infty} \int_{-\infty}^{+\infty} f(x, y) \, dx \, dy = 1$, the above implies that the constant equals $1/\text{area}(R)$. We are assuming that R is a region with finite, nonzero area.

Note: This definition represents the two-dimensional analog to the one-dimensional uniformly distributed random variable.

EXAMPLE 6.6. Suppose that the two-dimensional random variable (X, Y) is *uniformly distributed* over the shaded region R indicated in Fig. 6.5. Hence

$$f(x, y) = \frac{1}{\text{area}(R)}, \qquad (x, y) \in R.$$

We find that

$$\text{area}(R) = \int_0^1 (x - x^2)\, dx = \tfrac{1}{6}.$$

Therefore the pdf is given by

$$f(x, y) = 6, \qquad (x, y) \in R$$
$$= 0, \qquad (x, y) \notin R.$$

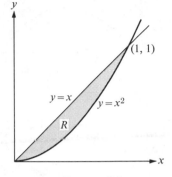

FIGURE 6.5

In the following equations we find the marginal pdf's of X and Y.

$$g(x) = \int_{-\infty}^{+\infty} f(x, y)\, dy = \int_{x^2}^{x} 6\, dy$$

$$= 6(x - x^2), \qquad 0 \le x \le 1;$$

$$h(y) = \int_{-\infty}^{+\infty} f(x, y)\, dx = \int_{y}^{\sqrt{y}} 6\, dx$$

$$= 6(\sqrt{y} - y), \qquad 0 \le y \le 1.$$

The graphs of these pdf's are sketched in Fig. 6.6.

The concept of conditional probability may be introduced in a very natural way.

EXAMPLE 6.7. Consider again Examples 6.1 and 6.4. Suppose that we want to evaluate the conditional probability $P(X = 2 \mid Y = 2)$. According to the definition of conditional probability we have

$$P(X = 2 \mid Y = 2) = \frac{P(X = 2, Y = 2)}{P(Y = 2)} = \frac{0.05}{0.25} = 0.20.$$

We can carry out such a computation quite generally for the discrete case. We have

$$p(x_i \mid y_j) = P(X = x_i \mid Y = y_j)$$
$$= \frac{p(x_i, y_j)}{q(y_j)} \qquad \text{if} \quad q(y_j) > 0, \qquad (6.5)$$
$$q(y_j \mid x_i) = P(Y = y_j \mid X = x_i)$$
$$= \frac{p(x_i, y_j)}{p(x_i)} \qquad \text{if} \quad p(x_i) > 0. \qquad (6.6)$$

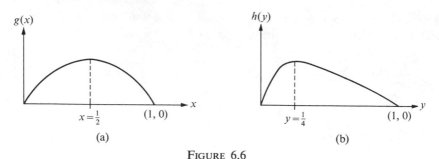

FIGURE 6.6

Note: For given j, $p(x_i \mid y_j)$ satisfies all the conditions for a probability distribution. We have $p(x_i \mid y_j) \geq 0$ and also

$$\sum_{i=1}^{\infty} p(x_i \mid y_j) = \sum_{i=1}^{\infty} \frac{p(x_i, y_j)}{q(y_j)} = \frac{q(y_j)}{q(y_j)} = 1.$$

In the *continuous case* the formulation of conditional probability presents some difficulty since for any given x_0, y_0, we have $P(X = x_0) = P(Y = y_0) = 0$. We make the following formal definitions.

Definition. Let (X, Y) be a continuous two-dimensional random variable with joint pdf f. Let g and h be the marginal pdf's of X and Y, respectively.

The *conditional* pdf of X for given $Y = y$ is defined by

$$g(x \mid y) = \frac{f(x, y)}{h(y)}, \qquad h(y) > 0. \tag{6.7}$$

The *conditional* pdf of Y for given $X = x$ is defined by

$$h(y \mid x) = \frac{f(x, y)}{g(x)}, \qquad g(x) > 0. \tag{6.8}$$

Notes: (a) The above conditional pdf's satisfy all the requirements for a one-dimensional pdf. Thus, for *fixed y*, we have $g(x \mid y) \geq 0$ and

$$\int_{-\infty}^{+\infty} g(x \mid y) \, dx = \int_{-\infty}^{+\infty} \frac{f(x, y)}{h(y)} \, dx = \frac{1}{h(y)} \int_{-\infty}^{+\infty} f(x, y) \, dx = \frac{h(y)}{h(y)} = 1.$$

An analogous computation may be carried out for $h(y \mid x)$. Hence Eqs. (6.7) and (6.8) *define* pdf's on R_X and R_Y, respectively.

(b) An intuitive interpretation of $g(x \mid y)$ is obtained if we consider slicing the surface represented by the joint pdf f with the plane $y = c$, say. The intersection of the plane with the surface $z = f(x, y)$ will result in a one-dimensional pdf, namely the pdf of X for $Y = c$. This will be precisely $g(x \mid c)$.

(c) Suppose that (X, Y) represents the height and weight of a person, respectively. Let f be the joint pdf of (X, Y) and let g be the marginal pdf of X (irrespective of Y).

Hence $\int_{5.8}^{6} g(x)\,dx$ would represent the probability of the event $\{5.8 \le X \le 6\}$ irrespective of the weight Y. And $\int_{5.8}^{6} g(x \mid 150)\,dx$ would be interpreted as $P(5.8 \le X \le 6 \mid Y = 150)$. Strictly speaking, this conditional probability is not defined in view of our previous convention with conditional probability, since $P(Y = 150) = 0$. However, we simply use the above integral to *define* this probability. Certainly on intuitive grounds this ought to be the meaning of this number.

EXAMPLE 6.8. Referring to Example 6.3, we have

$$g(x) = \int_0^2 \left(x^2 + \frac{xy}{3} \right) dy = 2x^2 + \frac{2}{3}x,$$

$$h(y) = \int_0^1 \left(x^2 + \frac{xy}{3} \right) dx = \frac{y}{6} + \frac{1}{3}.$$

Hence,

$$g(x \mid y) = \frac{x^2 + xy/3}{1/3 + y/6} = \frac{6x^2 + 2xy}{2 + y}, \qquad 0 \le x \le 1, \;\; 0 \le y \le 2;$$

$$h(y \mid x) = \frac{x^2 + xy/3}{2x^2 + 2/3(x)} = \frac{3x^2 + xy}{6x^2 + 2x} = \frac{3x + y}{6x + 2},$$

$$0 \le y \le 2, \;\; 0 \le x \le 1.$$

To check that $g(x \mid y)$ is a pdf, we have

$$\int_0^1 \frac{6x^2 + 2xy}{2 + y}\,dx = \frac{2 + y}{2 + y} = 1 \qquad \text{for all } y.$$

A similar computation can be carried out for $h(y \mid x)$.

6.3 Independent Random Variables

Just as we defined the concept of independence between two events A and B, we shall now define *independent random variables*. Intuitively, we intend to say that X and Y are independent random variables if the outcome of X, say, in no way influences the outcome of Y. This is an extremely important notion and there are many situations in which such an assumption is justified.

EXAMPLE 6.9. Consider two sources of radioactive material at some distance from each other which are emitting α-particles. Suppose that these two sources are observed for a period of two hours and the number of particles emitted is recorded. Assume that the following random variables are of interest: X_1 and X_2, the number of particles emitted from the first source during the first and second hour, respectively; and Y_1 and Y_2, the number of particles emitted from the second source during the first and second hour, respectively. It seems intuitively obvious that (X_1 and Y_1), or (X_1 and Y_2), or (X_2 and Y_1), or (X_2 and Y_2) are all pairs of independent random variables. For the X's depend only on the characteristics of source 1 while the Y's depend on the characteristics of source 2, and there is presumably no reason to assume that the two sources influence each other's behavior in any way. When we consider the possible independence of X_1 and X_2, however, the matter is not so clearcut. Is the number of particles emitted during the second hour influenced by the number that was emitted during the first hour? To answer this question we would have to obtain additional information about the mechanism of emission. We could certainly not assume, *a priori*, that X_1 and X_2 are independent.

Let us now make the above intuitive notion of independence more precise.

Definition. (a) Let (X, Y) be a two-dimensional discrete random variable. We say that X and Y are independent random variables if and only if $p(x_i, y_j) = p(x_i)q(y_j)$ for all i and j. That is, $P(X = x_i, Y = y_j) = P(X = x_i)P(Y = y_j)$, for all i and j.

(b) Let (X, Y) be a two-dimensional continuous random variable. We say that X and Y are independent random variables if and only if $f(x, y) = g(x)h(y)$ for all (x, y), where f is the joint pdf, and g and h are the marginal pdf's of X and Y, respectively.

Note: If we compare the above definition with that given for independent *events*, the similarity is apparent: we are essentially requiring that the joint probability (or joint pdf) can be factored. The following theorem indicates that the above definition is equivalent to another approach we might have taken.

Theorem 6.1. (a) Let (X, Y) be a two-dimensional discrete random variable. Then X and Y are independent if and only if $p(x_i \mid y_j) = p(x_i)$ for all i and j (or equivalently, if and only if $q(y_j \mid x_i) = q(y_j)$ for all i and j).

(b) Let (X, Y) be a two-dimensional continuous random variable. Then X and Y are independent if and only if $g(x \mid y) = g(x)$, or equivalently, if and only if $h(y \mid x) = h(y)$, for all (x, y).

Proof: See Problem 6.10.

EXAMPLE 6.10. Suppose that a machine is used for a particular task in the morning and for a different task in the afternoon. Let X and Y represent the number of times the machine breaks down in the morning and in the afternoon, respectively. Table 6.3 gives the joint probability distribution of (X, Y).

An easy computation reveals that for *all* the entries in Table 6.3 we have

$$p(x_i, y_j) = p(x_i)q(y_j).$$

Thus X and Y are independent random variables. (See also Example 3.7, for comparison.)

TABLE 6.3

X Y	0	1	2	$q(y_j)$
0	0.1	0.2	0.2	0.5
1	0.04	0.08	0.08	0.2
2	0.06	0.12	0.12	0.3
$p(x_i)$	0.2	0.4	0.4	1.0

EXAMPLE 6.11. Let X and Y be the life lengths of two electronic devices. Suppose that their joint pdf is given by

$$f(x, y) = e^{-(x+y)}, \qquad x \geq 0, \quad y \geq 0.$$

Since we can factor $f(x, y) = e^{-x}e^{-y}$, the independence of X and Y is established.

EXAMPLE 6.12. Suppose that $f(x, y) = 8xy$, $0 \leq x \leq y \leq 1$. (The domain is given by the shaded region in Fig. 6.7.) Although f is (already) written in factored form, X and Y are *not* independent, since the domain of definition

$$\{(x, y) \mid 0 \leq x \leq y \leq 1\}$$

is such that for given x, y may assume only values greater than that given x and less than 1. Hence X and Y are not independent.

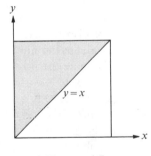

FIGURE 6.7

Note: From the definition of the marginal probability distribution (in either the discrete or the continuous case) it is clear that the joint probability distribution determines, uniquely, the marginal probability distribution. That is, from a knowledge of the joint pdf f, we can obtain the marginal pdf's g and h. However, the converse is not true! That is, in general, a knowledge of the marginal pdf's g and h do not determine the joint pdf f. Only when X and Y are independent is this true, for in this case we have $f(x, y) = g(x)h(y)$.

The following result indicates that our definition of independent random variables is consistent with our previous definition of independent events.

Theorem 6.2. Let (X, Y) be a two-dimensional random variable. Let A and B be events whose occurrence (or nonoccurrence) depends only on X and Y, respectively. (That is, A is a subset of R_X, the range space of X, while B is a subset of R_Y, the range space of Y.) Then, if X and Y are independent random variables, we have $P(A \cap B) = P(A)P(B)$.

Proof (continuous case only):

$$P(A \cap B) = \iint_{A \cap B} f(x, y) \, dx \, dy = \iint_{A \cap B} g(x)h(y) \, dx \, dy$$

$$= \int_A g(x) \, dx \int_B h(y) \, dy = P(A)P(B).$$

6.4 Functions of a Random Variable

In defining a random variable X we pointed out, quite emphatically, that X is a *function* defined from the sample space S to the real numbers. In defining a two-dimensional random variable (X, Y) we were concerned with a pair of functions, $X = X(s)$, $Y = Y(s)$, each of which is defined on the sample space of some experiment and each of which assigns a real number to every $s \in S$, thus yielding the two-dimensional vector $[X(s), Y(s)]$.

Let us now consider $Z = H_1(X, Y)$, a function of the two random variables X and Y. It should be clear that $Z = Z(s)$ is again a random variable. Consider the following sequence of steps:

(a) Perform the experiment ε and obtain the outcome s.
(b) Evaluate the numbers $X(s)$ and $Y(s)$.
(c) Evaluate the number $Z = H_1[X(s), Y(s)]$.

The value of Z clearly depends on s, the original outcome of the experiment. That is, $Z = Z(s)$ is a function assigning to every outcome $s \in S$ a real number, $Z(s)$. Hence Z is a random variable. Some of the important random variables we shall be interested in are $X + Y$, XY, X/Y, min (X, Y), max (X, Y), etc.

The problem we solved in the previous chapter for the one-dimensional random variable arises again: given the joint probability distribution of (X, Y), what is the probability distribution of $Z = H_1(X, Y)$? (It should be clear from the numerous previous discussions on this point, that a probability distribution is *induced* on R_Z, the sample space of Z.)

If (X, Y) is a discrete random variable, this problem is quite easily solved. Suppose that (X, Y) has the distribution given in Examples 6.1 and 6.4. The following (one-dimensional) random variables might be of interest:

$U = \min (X, Y) = $ least number of items produced by the two lines;

$V = \max (X, Y) = $ greatest number of items produced by the two lines;

$W = X + Y = $ total number of items produced by the two lines.

To obtain the probability distribution of U, say, we proceed as follows. The possible values of U are: 0, 1, 2, and 3. To evaluate $P(U = 0)$ we argue that $U = 0$ if and only if one of the following occurs: $X = 0$, $Y = 0$ or $X = 0$, $Y = 1$ or $X = 0$, $Y = 2$ or $X = 0$, $Y = 3$ or $X = 1$, $Y = 0$ or $X = 2$, $Y = 0$ or $X = 3$,

$Y = 0$ or $X = 4$, $Y = 0$ or $X = 5$, $Y = 0$. Hence $P(U = 0) = 0.28$. The rest of the probabilities associated with U may be obtained in a similar way. Hence the probability distribution of U may be summarized as follows: u: 0, 1, 2, 3; $P(U = u)$: 0.28, 0.30, 0.25, 0.17. The probability distribution of the random variables V and W as defined above may be obtained in a similar way. (See Problem 6.9.)

If (X, Y) is a continuous two-dimensional random variable and if $Z = H_1(X, Y)$ is a continuous function of (X, Y), then Z will be a continuous (one-dimensional) random variable and the problem of finding its pdf is somewhat more involved. In order to solve this problem we shall need a theorem which we state and discuss below. Before doing this, let us briefly outline the basic idea.

In finding the pdf of $Z = H_1(X, Y)$ it is often simplest to introduce a second random variable, say $W = H_2(X, Y)$, and first obtain the *joint* pdf of Z and W, say $k(z, w)$. From a knowledge of $k(z, w)$ we can then obtain the desired pdf of Z, say $g(z)$, by simply integrating $k(z, w)$ with respect to w. That is,

$$g(z) = \int_{-\infty}^{+\infty} k(z, w)\, dw.$$

The remaining problems are (1) how to find the joint pdf of Z and W, and (2) how to choose the appropriate random variable $W = H_2(X, Y)$. To resolve the latter problem, let us simply state that we usually make the simplest possible choice for W. In the present context, W plays only an intermediate role, and we are not really interested in it for its own sake. In order to find the joint pdf of Z and W we need Theorem 6.3.

Theorem 6.3. Suppose that (X, Y) is a two-dimensional continuous random variable with joint pdf f. Let $Z = H_1(X, Y)$ and $W = H_2(X, Y)$, and assume that the functions H_1 and H_2 satisfy the following conditions:

(a) The equations $z = H_1(x, y)$ and $w = H_2(x, y)$ may be uniquely solved for x and y in terms of z and w, say $x = G_1(z, w)$ and $y = G_2(z, w)$.

(b) The partial derivatives $\partial x/\partial z$, $\partial x/\partial w$, $\partial y/\partial z$, and $\partial y/\partial w$ exist and are continuous.

Then the joint pdf of (Z, W), say $k(z, w)$, is given by the following expression: $k(z, w) = f[G_1(z, w), G_2(z, w)]\,|J(z, w)|$, where $J(z, w)$ is the following 2×2 determinant:

$$J(z, w) = \begin{vmatrix} \dfrac{\partial x}{\partial z} & \dfrac{\partial x}{\partial w} \\[2ex] \dfrac{\partial y}{\partial z} & \dfrac{\partial y}{\partial w} \end{vmatrix}.$$

This determinant is called the *Jacobian* of the transformation $(x, y) \rightarrow (z, w)$ and is sometimes denoted by $\partial(x, y)/\partial(z, w)$. We note that $k(z, w)$ will be nonzero for those values of (z, w) corresponding to values of (x, y) for which $f(x, y)$ is nonzero.

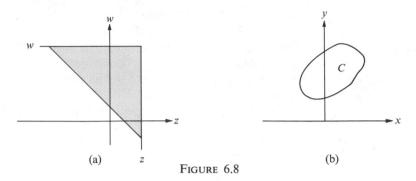

(a)   (b)

FIGURE 6.8

Notes: (a) Although we shall not prove this theorem, we will at least indicate what needs to be shown and where the difficulties lie. Consider the joint cdf of the two-dimensional random variable (Z, W), say

$$K(z, w) = P(Z \leq z, W \leq w) = \int_{-\infty}^{w} \int_{-\infty}^{z} k(s, t) \, ds \, dt,$$

where k is the sought pdf. Since the transformation $(x, y) \rightarrow (z, w)$ is assumed to be one to one [see assumption (a) above], we may find the event, equivalent to $\{Z \leq z, W \leq w\}$, in terms of X and Y. Suppose that this event is denoted by C. (See Fig. 6.8.) That is, $\{(X, Y) \in C\}$ if and only if $\{Z \leq z, W \leq w\}$. Hence

$$\int_{-\infty}^{w} \int_{-\infty}^{z} k(s, t) \, ds \, dt = \int_C \int f(x, y) \, dx \, dy.$$

Since f is assumed to be known, the integral on the right-hand side can be evaluated. Differentiating it with respect to z and w will yield the required pdf. In most texts on advanced calculus it is shown that these techniques lead to the result as stated in the above theorem.

(b) Note the striking similarity between the above result and the result obtained in the one-dimensional case treated in the previous chapter. (See Theorem 5.1.) The monotonicity requirement for the function $y = H(x)$ is replaced by the assumption that the correspondence between (x, y) and (z, w) is one to one. The differentiability condition is replaced by certain assumptions about the partial derivatives involved. The final solution obtained is also very similar to the one obtained in the one-dimensional case: the variables x and y are simply replaced by their equivalent expressions in terms of z and w, and the absolute value of dx/dy is replaced by the absolute value of the Jacobian.

EXAMPLE 6.13. Suppose that we are aiming at a circular target of radius one which has been placed so that its center is at the origin of a rectangular coordinate system (Fig. 6.9). Suppose that the coordinates (X, Y) of the point of impact are uniformly distributed over the circle. That is,

$f(x, y) = 1/\pi$ if (x, y) lies inside (or on) the circle,

    $= 0,$ elsewhere.

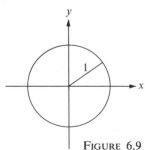

FIGURE 6.9

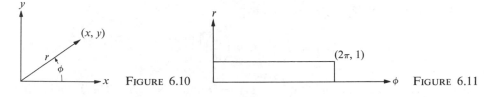

FIGURE 6.10   FIGURE 6.11

Suppose that we are interested in the random variable R representing the *distance* from the origin. (See Fig. 6.10.) That is, $R = \sqrt{X^2 + Y^2}$. We shall find the pdf of R, say g, as follows: Let $\Phi = \tan^{-1}(Y/X)$. Hence $X = H_1(R, \Phi)$ and $Y = H_2(R, \Phi)$ where $x = H_1(r, \phi) = r \cos \phi$ and $y = H_2(r, \phi) = r \sin \phi$. (We are simply introducing polar coordinates.)

The Jacobian is

$$J = \begin{vmatrix} \dfrac{\partial x}{\partial r} & \dfrac{\partial x}{\partial \phi} \\[2mm] \dfrac{\partial y}{\partial r} & \dfrac{\partial y}{\partial \phi} \end{vmatrix} = \begin{vmatrix} \cos \phi & -r \sin \phi \\[1mm] \sin \phi & r \cos \phi \end{vmatrix}$$

$$= r \cos^2 \phi + r \sin^2 \phi = r.$$

Under the above transformation the unit circle in the xy-plane is mapped into the rectangle in the ϕr-plane in Fig. 6.11. Hence the joint pdf of (Φ, R) is given by

$$g(\phi, r) = \frac{r}{\pi}, \qquad 0 \le r \le 1, \qquad 0 \le \phi < 2\pi.$$

Thus the required pdf of R, say h, is given by

$$h(r) = \int_0^{2\pi} g(\phi, r) \, d\phi = 2r, \qquad 0 \le r \le 1.$$

Note: This example points out the importance of obtaining a precise representation of the region of possible values for the new random variables introduced.

6.5 Distribution of Product and Quotient of Independent Random Variables

Among the most important functions of X and Y which we wish to consider are the sum $S = X + Y$, the product $W = XY$, and the quotient $Z = X/Y$. We can use the method in this section to obtain the pdf of each of these random variables under very general conditions.

We shall investigate the sum of random variables in much greater detail in Chapter 11. Hence we will defer discussion of the probability distribution of $X + Y$ until then. We shall, however, consider the product and quotient in the following two theorems.

Theorem 6.4. Let (X, Y) be a continuous two-dimensional random variable and assume that X and Y are *independent*. Hence the pdf f may be written as $f(x, y) = g(x)h(y)$. Let $W = XY$.

Then the pdf of W, say p, is given by

$$p(w) = \int_{-\infty}^{+\infty} g(u)h\left(\frac{w}{u}\right)\left|\frac{1}{u}\right| du. \tag{6.9}$$

Proof: Let $w = xy$ and $u = x$. Thus $x = u$ and $y = w/u$. The Jacobian is

$$J = \begin{vmatrix} 1 & 0 \\ -\dfrac{w}{u^2} & \dfrac{1}{u} \end{vmatrix} = \frac{1}{u}.$$

Hence the joint pdf of $W = XY$ and $U = X$ is

$$s(w, u) = g(u)h\left(\frac{w}{u}\right)\left|\frac{1}{u}\right|.$$

The marginal pdf of W is obtained by integrating $s(w, u)$ with respect to u, yielding the required result. The values of w for which $p(w) > 0$ would depend on the values of (x, y) for which $f(x, y) > 0$.

Note: In evaluating the above integral we may use the fact that

$$\int_{-\infty}^{+\infty} g(u)h\left(\frac{w}{u}\right)\left|\frac{1}{u}\right| du = \int_{0}^{\infty} g(u)h\left(\frac{w}{u}\right)\frac{1}{u} du - \int_{-\infty}^{0} g(u)h\left(\frac{w}{u}\right)\frac{1}{u} du.$$

EXAMPLE 6.14. Suppose that we have a circuit in which both the current I and the resistance R vary in some random way. Specifically, assume that I and R are independent continuous random variables with the following pdf's.

$$I: \quad g(i) = 2i, \qquad 0 \le i \le 1 \qquad \text{and 0 elsewhere;}$$
$$R: \quad h(r) = r^2/9, \qquad 0 \le r \le 3 \qquad \text{and 0 elsewhere.}$$

Of interest is the random variable $E = IR$ (the voltage in the circuit). Let p be the pdf of E.

By Theorem 6.4 we have

$$p(e) = \int_{-\infty}^{+\infty} g(i)h\left(\frac{e}{i}\right)\left|\frac{1}{i}\right| di.$$

Some care must be taken in evaluating this integral. First, we note that the variable of integration cannot assume negative values. Second, we note that in order for the integrand to be positive, *both* the pdf's appearing in the integrand must be positive. Noting the values for which g and h are not equal to zero we find that the following conditions must be satisfied:

$$0 \le i \le 1 \qquad \text{and} \qquad 0 \le e/i \le 3.$$

These *two* inequalities are, in turn, equivalent to
$e/3 \leq i \leq 1$. Hence the above integral becomes $p(e)$

$$p(e) = \int_{e/3}^{1} 2i \frac{e^2}{9i^2} \frac{1}{i} \, di$$

$$= -\tfrac{2}{9}e^2 \frac{1}{i} \Big|_{e/3}^{1}$$

$$= \tfrac{2}{9}e(3 - e), \qquad 0 \leq e \leq 3.$$

An easy computation shows that $\int_0^3 p(e)\, de = 1$.
(See Fig. 6.12.)

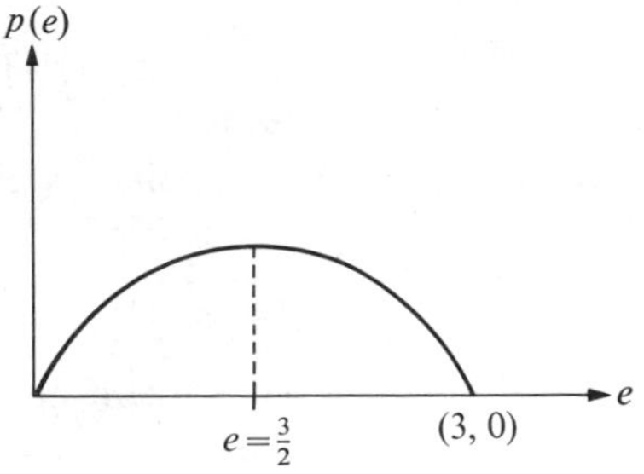

$e = \tfrac{3}{2}$ (3, 0)

FIGURE 6.12

Theorem 6.5. Let (X, Y) be a continuous two-dimensional random variable
and assume that X and Y are independent. [Hence the pdf of (X, Y) may be
written as $f(x, y) = g(x)h(y)$.] Let $Z = X/Y$. Then the pdf of Z, say q,
is given by

$$q(z) = \int_{-\infty}^{+\infty} g(vz)h(v)|v| \, dv. \tag{6.10}$$

Proof: Let $z = x/y$ and let $v = y$. Hence $x = vz$ and $y = v$. The Jacobian is

$$J = \begin{vmatrix} v & z \\ 0 & 1 \end{vmatrix} = v.$$

Hence the joint pdf of $Z = X/Y$ and $V = Y$ equals

$$t(z, v) = g(vz)h(v)|v|.$$

Integrating this joint pdf with respect to v yields the required marginal pdf of Z.

EXAMPLE 6.15. Let X and Y represent the life lengths of two light bulbs manufactured by different processes. Assume that X and Y are independent random
variables with the pdf's f and g, respectively, where

$$f(x) = e^{-x}, \qquad x \geq 0, \qquad \text{and 0 elsewhere;}$$
$$g(y) = 2e^{-2y}, \qquad y \geq 0, \qquad \text{and 0 elsewhere.}$$

Of interest might be the random variable X/Y, representing the ratio of the two
life lengths. Let q be the pdf of Z.

By Theorem 6.5 we have $q(z) = \int_{-\infty}^{+\infty} g(vz)h(v)|v| \, dv$. Since X and Y can assume
only nonnegative quantities, the above integration need only be carried out over
the positive values of the variable of integration. In addition, the integrand will
be positive only when *both* the pdf's appearing are positive. This implies that we
must have $v \geq 0$ and $vz \geq 0$. Since $z > 0$, these inequalities imply that $v \geq 0$.

Thus the above becomes

$$q(z) = \int_0^\infty e^{-vz} 2e^{-2v} v \, dv = 2 \int_0^\infty v e^{-v(2+z)} \, dv.$$

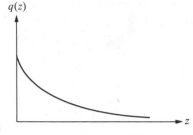

$q(z)$

An easy integration by parts yields

$$q(z) = \frac{2}{(z+2)^2}, \quad z \geq 0.$$

(See Fig. 6.13.) It is again an easy exercise to check that $\int_0^\infty q(z) \, dz = 1$.

FIGURE 6.13

6.6 n-Dimensional Random Variables

Our discussion so far has been entirely restricted to two-dimensional random variables. As we indicated at the beginning of this chapter, however, we may have to be concerned with three or more simultaneous numerical characteristics.

We shall give only the briefest discussion of n-dimensional random variables. Most of the concepts introduced above for the two-dimensional case can be extended to the n-dimensional one. We shall restrict ourselves to the continuous case. (See the note at the end of this chapter.)

Suppose then that $(X_1, \ldots, X_n)$ may assume all values in some region of n-dimensional space. That is, the value is an n-dimensional vector

$$(X_1(s), \ldots, X_n(s)).$$

We characterize the probability distribution of $(X_1, \ldots, X_n)$ as follows.

There exists a joint probability density function f satisfying the following conditions:

(a) $f(x_1, \ldots, x_n) \geq 0$ for all $(x_1, \ldots, x_n)$.
(b) $\int_{-\infty}^{+\infty} \cdots \int_{-\infty}^{+\infty} f(x_1, \ldots, x_n) \, dx_1 \cdots dx_n = 1$.

With the aid of this pdf we *define*

$$P[(X_1, \ldots, X_n) \in C] = \int \cdots \int_C f(x_1, \ldots, x_n) \, dx_1 \cdots dx_n,$$

where C is a subset of the range space of $(X_1, \ldots, X_n)$.

With each n-dimensional random variable we can associate a number of lower-dimensional random variables. For example, if $n = 3$, then

$$\int_{-\infty}^{+\infty} \int_{-\infty}^{+\infty} f(x_1, x_2, x_3) \, dx_1 \, dx_2 = g(x_3),$$

where g is the marginal pdf of the one-dimensional random variable X_3, while

$$\int_{-\infty}^{+\infty} f(x_1, x_2, x_3) \, dx_3 = h(x_1, x_2),$$

where h represents the joint pdf of the two-dimensional random variable (X_1, X_2), etc. The concept of independent random variables is also extended in a natural way. We say that $(X_1, \ldots, X_n)$ are independent random variables if and only if their joint pdf $f(x_1, \ldots, x_n)$ can be factored into

$$g_1(x_1) \cdots g_n(x_n).$$

There are many situations in which we wish to consider n-dimensional random variables. We shall give a few examples.

(a) Suppose that we study the pattern of precipitation due to a particular storm system. If we have a network of, say, 5 observing stations and if we let X_i be the rainfall at station i due to a particular frontal system, we might wish to consider the five-dimensional random variable $(X_1, X_2, X_3, X_4, X_5)$.

(b) One of the most important applications of n-dimensional random variables occurs when we deal with repeated measurements on some random variable X. Suppose that information about the life length, say X, of an electron tube is required. A large number of these tubes are produced by a certain manufacturer, and we test n of these. Let X_i be the life length of ith tube, $i = 1, \ldots, n$. Hence $(X_1, \ldots, X_n)$ is an n-dimensional random variable. If we assume that each X_i has the same probability distribution (since all tubes are produced in the same manner), and if we assume that the X_i's are all independent random variables (since, presumably, the production of one tube does not affect the production of other tubes), we may suppose that the n-dimensional random variable $(X_1, \ldots, X_n)$ is composed of the independent identically distributed components $X_1, \ldots, X_n$. (It should be obvious that although X_1 and X_2 have the same distribution, they need not assume the same value.)

(c) Another way in which n-dimensional random variables arise is the following. Let $X(t)$ represent the power required by a certain industrial concern at time t. For fixed t, $X(t)$ is a one-dimensional random variable. However, we may be interested in describing the power required at certain n specified times, say $t_1 < t_2 < \cdots < t_n$. Thus we wish to study the n-dimensional random variable

$$[X(t_1), X(t_2), \ldots, X(t_n)].$$

Problems of this type are studied at a more advanced level. (An excellent reference for this subject area is "Stochastic Processes" by Emanuel Parzen, Holden-Day, San Francisco, 1962.)

Note: In some of our discussion we have referred to the concept of "n-space." Let us summarize a few of the basic ideas required.

With each real number x we may associate a point on the real number line, and conversely. Similarly, with each pair of real numbers (x_1, x_2) we may associate a point in the rectangular coordinate plane, and conversely. Finally, with each set of three real numbers (x_1, x_2, x_3), we may associate a point in the three-dimensional, rectangular coordinate space, and conversely.

In many of the problems with which we are concerned we deal with a set of n real numbers, $(x_1, x_2, \ldots, x_n)$, also called an n-tuple. Although we cannot draw any sketches if $n > 3$, we can continue to adopt geometric terminology as suggested by the lower-dimensional cases referred to above. Thus, we shall speak of a "point" in n-dimensional space determined by the n-tuple $(x_1, \ldots, x_n)$. We shall define as n-space (sometimes called Euclidean n-space) the set of all $(x_1, \ldots, x_n)$, where x_i may be any real number.

Although we will not actually need to evaluate n-dimensional integrals, we shall find the concept a very useful one and will occasionally need to express a quantity as a multiple integral. If we recall the definition of

$$\iint_A f(x, y)\, dx\, dy,$$

where A is a region in the (x, y)-plane, then the extension of this concept to

$$\int \cdots \int_R f(x_1, \ldots, x_n) dx_1 \cdots dx_n,$$

where R is a region in n-space should be clear. If f represents the joint pdf of the two-dimensional random variable (X, Y) then

$$\iint_A f(x, y)\, dx\, dy$$

represents $P[(X, Y) \in A]$. Similarly, if f represents the joint pdf of $(X_1, \ldots, X_n)$ then

$$\int \cdots \int_R f(x_1, \ldots, x_n)\, dx_1 \cdots dx_n$$

represents

$$P[(X_1, \ldots, X_n) \in R].$$

PROBLEMS

6.1. Suppose that the following table represents the joint probability distribution of the discrete random variable (X, Y). Evaluate all the marginal and conditional distributions.

Y \ X	1	2	3
1	$\frac{1}{12}$	$\frac{1}{6}$	0
2	0	$\frac{1}{9}$	$\frac{1}{5}$
3	$\frac{1}{18}$	$\frac{1}{4}$	$\frac{2}{15}$

6.2. Suppose that the two-dimensional random variable (X, Y) has joint pdf

$$f(x, y) = kx(x - y), \quad 0 < x < 2, \quad -x < y < x,$$
$$= 0, \quad \text{elsewhere.}$$

(a) Evaluate the constant k.
(b) Find the marginal pdf of X.
(c) Find the marginal pdf of Y.

6.3. Suppose that the joint pdf of the two-dimensional random variable (X, Y) is given by

$$f(x, y) = x^2 + \frac{xy}{3}, \qquad 0 < x < 1, \qquad 0 < y < 2,$$

$$= 0, \qquad \text{elsewhere.}$$

Compute the following.

(a) $P(X > \frac{1}{2})$;   (b) $P(Y < X)$;   (c) $P(Y < \frac{1}{2} \mid X < \frac{1}{2})$.

6.4. Suppose that two cards are drawn at random from a deck of cards. Let X be the number of aces obtained and let Y be the number of queens obtained.

(a) Obtain the joint probability distribution of (X, Y).
(b) Obtain the marginal distribution of X and of Y.
(c) Obtain the conditional distribution of X (given Y) and of Y (given X).

6.5. For what value of k is $f(x, y) = ke^{-(x+y)}$ a joint pdf of (X, Y) over the region $0 < x < 1, 0 < y < 1$?

6.6. Suppose that the continuous two-dimensional random variable (X, Y) is uniformly distributed over the square whose vertices are $(1, 0)$, $(0, 1)$, $(-1, 0)$, and $(0, -1)$. Find the marginal pdf's of X and of Y.

6.7. Suppose that the dimensions, X and Y, of a rectangular metal plate may be considered to be independent continuous random variables with the following pdf's.

$$X: \quad g(x) = x - 1, \qquad 1 < x \le 2,$$
$$= -x + 3, \qquad 2 < x < 3,$$
$$= 0, \qquad \text{elsewhere.}$$

$$Y: \quad h(y) = \frac{1}{2}, \qquad 2 < y < 4,$$
$$= 0, \qquad \text{elsewhere.}$$

Find the pdf of the area of the plate, $A = XY$.

6.8. Let X represent the life length of an electronic device and suppose that X is a continuous random variable with pdf

$$f(x) = \frac{1000}{x^2}, \qquad x > 1000,$$

$$= 0, \qquad \text{elsewhere.}$$

Let X_1 and X_2 be two independent determinations of the above random variable X. (That is, suppose that we are testing the life length of two such devices.) Find the pdf of the random variable $Z = X_1/X_2$.

6.9. Obtain the probability distribution of the random variables V and W introduced on p. 95.

6.10. Prove Theorem 6.1.

6.11.   The magnetizing force H at a point P, X units from a wire carrying a current I, is given by $H = 2I/X$. (See Fig. 6.14.) Suppose that P is a variable point. That is, X is a continuous random variable uniformly distributed over $(3, 5)$. Assume that the current I is also a continuous random variable, uniformly distributed over $(10, 20)$. Suppose, in addition, that the random variables X and I are independent. Find the pdf of the random variable H.

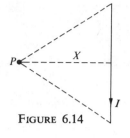

FIGURE 6.14

6.12.   The intensity of light at a given point is given by the relationship $I = C/D^2$, where C is the candlepower of the source and D is the distance that the source is from the given point. Suppose that C is uniformly distributed over $(1, 2)$, while D is a continuous random variable with pdf $f(d) = e^{-d}$, $d > 0$. Find the pdf of I, if C and D are independent. [*Hint:* First find the pdf of D^2 and then apply the results of this chapter.]

6.13.   When a current I (amperes) flows through a resistance R (ohms), the power generated is given by $W = I^2R$ (watts). Suppose that I and R are independent random variables with the following pdf's.

$$I: \quad f(i) = 6i(1 - i), \quad 0 \le i \le 1,$$
$$= 0, \quad \text{elsewhere.}$$

$$R: \quad g(r) = 2r, \quad 0 < r < 1,$$
$$= 0, \quad \text{elsewhere.}$$

Determine the pdf of the random variable W and sketch its graph.

6.14.   Suppose that the joint pdf of (X, Y) is given by

$$f(x, y) = e^{-y}, \quad \text{for } x > 0, \quad y > x,$$
$$= 0, \quad \text{elsewhere.}$$

(a) Find the marginal pdf of X.
(b) Find the marginal pdf of Y.
(c) Evaluate $P(X > 2 \mid Y < 4)$.

7

Further Characteristics of Random Variables

7.1 The Expected Value of a Random Variable

Consider the deterministic relationship $ax + by = 0$. We recognize this as a linear relationship between x and y. The constants a and b are the *parameters* of this relationship in the sense that for any particular choice of a and b we obtain a specific linear function. In other cases one or more parameters may characterize the relationship under consideration. For example, if $y = ax^2 + bx + c$, three parameters are needed. If $y = e^{-kx}$, one parameter is sufficient. Not only is a particular relationship characterized by parameters but, conversely, from a certain relationship we may define various pertinent parameters. For example, if $ay + bx = 0$, then $m = -b/a$ represents the slope of the line. Again, if $y = ax^2 + bx + c$, then $-b/2a$ represents the value at which a relative maximum or relative minimum occurs.

In the nondeterministic or random mathematical models which we have been considering, parameters may also be used to characterize the probability distribution. With each probability distribution we may associate certain parameters which yield valuable information about the distribution (just as the slope of a line yields valuable information about the linear relationship it represents).

EXAMPLE 7.1. Suppose that X is a continuous random variable with pdf $f(x) = ke^{-kx}$, $x \geq 0$. To check that this is a pdf note that $\int_0^\infty ke^{-kx}\,dx = 1$ for all $k > 0$, and that $ke^{-kx} > 0$ for $k > 0$. This distribution is called an exponential distribution, which we shall study in greater detail later. It is a particularly useful distribution for representing the life length, say X, of certain types of equipment or components. The interpretation of k, in this context, will also be discussed subsequently.

EXAMPLE 7.2. Assume that items are produced indefinitely on an assembly line. The probability of an item being defective is p, and this value is the same for all items. Suppose also that the successive items are defective (D) or nondefective (N) independently of each other. Let the random variable X be the number of items inspected until the first defective item is found. Thus a typical outcome of the experiment would be of the form $NNNND$. Here $X(NNNND) = 5$. The possible values of X are: $1, 2, \ldots, n, \ldots$ Since $X = k$ if and only if the first $(k - 1)$ items are nondefective and the kth item is defective, we assign the following probability to the event $\{X = k\}$: $P(X = k) = p(1 - p)^{k-1}, k = 1, 2, \ldots, n, \ldots$ To check that this is a legitimate probability distribution we note that

$$\sum_{k=1}^{\infty} p(1 - p)^{k-1} = p[1 + (1 - p) + (1 - p)^2 + \cdots]$$

$$= p \frac{1}{1 - (1 - p)} = 1 \quad \text{if} \quad 0 < |p| < 1.$$

Thus the parameter p may be any number satisfying $0 < p < 1$.

Suppose that a random variable and its probability distribution is specified. Is there some way of characterizing this distribution in terms of a few pertinent numerical parameters?

Before pursuing the above question, let us motivate our discussion by considering the following example.

EXAMPLE 7.3. A wire cutting machine cuts wire to a specified length. Due to certain inaccuracies of the cutting mechanism, the length of the cut wire (in inches), say X, may be considered as a uniformly distributed random variable over [11.5, 12.5]. The specified length is 12 inches. If $11.7 \leq X < 12.2$, the wire can be sold for a profit of $0.25. If $X \geq 12.2$, the wire can be recut, and an eventual profit of $0.10 is realized. And if $X < 11.7$, the wire is discarded with a loss of $0.02. An easy computation shows that $P(X \geq 12.2) = 0.3$, $P(11.7 \leq X < 12.2) = 0.5$, and $P(X < 11.7) = 0.2$.

Suppose that a large number of wire specimens are cut, say N. Let N_S be the number of specimens for which $X < 11.7$, N_R the number of specimens for which $11.7 \leq X < 12.2$, and N_L the number of specimens for which $X \geq 12.2$. Hence the total profit realized from the production of the N specimens equals $T = N_S(-0.02) + N_R(0.25) + N_L(0.10)$. The *total profit per wire cut*, say W, equals $W = (N_S/N)(-0.02) + (N_R/N)(0.25) + (N_L/N)(0.1)$. (Note that W is a random variable, since N_S, N_R, and N_L are random variables.)

We have already mentioned that the relative frequency of an event is close to the probability of that event if the number of repetitions on which the relative frequency is based is large. (We shall discuss this more precisely in Chapter 12.) Hence, if N is large, we would expect N_S/N to be close to 0.2, N_R/N to be close to 0.5, and N_L/N to be close to 0.3. Therefore, for large N, W could be approximated as follows:

$$W \simeq (0.2)(-0.02) + 0.5(0.25) + (0.3)(0.1) = \$0.151.$$

Thus, if a large number of wires were produced, we would expect to make a profit of $0.151 per wire. The number 0.151 is called the *expected value* of the random variable W.

Definition. Let X be a discrete random variable with possible values $x_1, \ldots, x_n, \ldots$ Let $p(x_i) = P(X = x_i)$, $i = 1, 2, \ldots, n, \ldots$ Then the *expected value* of X (or mathematical expectation of X), denoted by $E(X)$, is defined as

$$E(X) = \sum_{i=1}^{\infty} x_i p(x_i) \tag{7.1}$$

if the series $\sum_{i=1}^{\infty} x_i p(x_i)$ converges absolutely, i.e., if $\sum_{i=1}^{\infty} |x_i| p(x_i) < \infty$. This number is also referred to as the *mean value* of X.

Notes: (a) If X assumes only a finite number of values, the above expression becomes $E(X) = \sum_{i=1}^{n} p(x_i) x_i$. This may be considered as a "weighted average" of the possible values $x_1, \ldots, x_n$. If all these possible values are equally probable, $E(X) = (1/n)\sum_{i=1}^{n} x_i$, which represents the ordinary arithmetic average of the n possible values.

(b) If a fair die is tossed and the random variable X designates the number of points showing, then $E(X) = \frac{1}{6}(1 + 2 + 3 + 4 + 5 + 6) = \frac{7}{2}$. This simple example illustrates, strikingly, that $E(X)$ is not the outcome we would expect when X is observed a single time. In fact, in the above situation, $E(X) = \frac{7}{2}$ is not even a *possible* value for X! Rather, it turns out that if we obtain a large number of independent observations of X, say $x_1, \ldots, x_n$, and compute the arithmetic mean of these outcomes, then, under fairly general conditions, the arithmetic mean will be close to $E(X)$ in a probabilistic sense. For example, in the above situation, if we were to throw the die a large number of times and then compute the arithmetic mean of the various outcomes, we would expect this average to become closer to $\frac{7}{2}$ the more often the die were tossed.

(c) We should note the similarity between the notion of expected value as defined above (particularly if X may assume only a finite number of values) and the notion of an average of a set of numbers say, $z_1, \ldots, z_n$. We usually define $\bar{z} = (1/n)\sum_{i=1}^{n} z_i$ as the arithmetic mean of the numbers $z_1, \ldots, z_n$. Suppose, furthermore, that we have numbers $z_1', \ldots, z_k'$, where z_i' occurs n_i times, $\sum_{i=1}^{k} n_i = n$. Letting $f_i = n_i/n$, $\sum_{i=1}^{k} f_i = 1$, we define the weighted mean of the numbers $z_1', \ldots, z_k'$ as

$$\frac{1}{n} \sum_{i=1}^{k} n_i z_i' = \sum_{i=1}^{k} f_i z_i'.$$

Although there is a strong resemblance between the above weighted average and the definition of $E(X)$, it is important to realize that the latter is a number (parameter) associated with a theoretical probability distribution, while the former is simply the result of combining a set of numbers in a particular way. However, there is more than just a superficial resemblance. Consider a random variable X and let $x_1, \ldots, x_n$ be the values obtained when the experiment giving rise to X was performed n times independently. (That is, $x_1, \ldots, x_n$ simply represent the outcomes of n repeated measurements of the numerical characteristic X.) Let $\bar{x}$ be the arithmetic mean of these n numbers. Then, as we shall discuss much more precisely in Chapter 12, if n is sufficiently large, $\bar{x}$ will be "close" to $E(X)$ in a certain sense. This result is very much related to the idea (also to be discussed in Chapter 12) that the relative frequency f_A associated with n repetitions of an experiment will be close to the probability $P(A)$ if f_A is based on a large number of repetitions of $\mathcal{E}$.

EXAMPLE 7.4. A manufacturer produces items such that 10 percent are defective and 90 percent are nondefective. If a defective item is produced, the manufacturer loses $1 while a nondefective item brings a profit of $5. If X is the net profit per item, then X is a random variable whose expected value is computed as $E(X) = -1(0.1) + 5(0.9) = \$4.40$. Suppose that a large number of such items are produced. Then, since the manufacturer will lose $1 about 10 percent of the time and earn $5 about 90 percent of the time, he will expect to gain about $4.40 per item in the long run.

Theorem 7.1. Let X be a binomially distributed random variable with parameter p, based on n repetitions of an experiment. Then

$$E(X) = np.$$

Proof: Since $P(X = k) = \binom{n}{k}p^k(1 - p)^{n-k}$, we have

$$E(X) = \sum_{k=0}^{n} k \frac{n!}{k!(n - k)!} p^k(1 - p)^{n-k}$$

$$= \sum_{k=1}^{n} \frac{n!}{(k - 1)!(n - k)!} p^k(1 - p)^{n-k}$$

(since the term with $k = 0$ equals zero). Let $s = k - 1$ in the above sum. As k assumes values from one through n, s assumes values from zero through $(n - 1)$. Replacing k everywhere by $(s + 1)$ we obtain

$$E(X) = \sum_{s=0}^{n-1} n \binom{n - 1}{s} p^{s+1}(1 - p)^{n-s-1}$$

$$= np \sum_{s=0}^{n-1} \binom{n - 1}{s} p^s(1 - p)^{n-1-s}.$$

The sum in the last expression is simply the sum of the binomial probabilities with n replaced by $(n - 1)$ [that is, $(p + (1 - p))^{n-1}$] and hence equals one. This establishes the result.

Note: The above result certainly corresponds to our intuitive notion. For suppose that the probability of some event A is, say 0.3, when an experiment is performed. If we repeat this experiment, say 100 times, we would expect A to occur about $100(0.3) = 30$ times. The concept of expected value, introduced above for the discrete random variable, will shortly be extended to the continuous case.

EXAMPLE 7.5. A printing machine has a constant probability of 0.05 of breaking down on any given day. If the machine has no breakdowns during the week, a profit of $\$S$ is realized. If 1 or 2 breakdowns occur, a profit of $\$R$ is realized

$(R < S)$. If 3 or more breakdowns occur, a profit of $\$(-L)$ is realized. (We assume that R, S, and L are greater than zero; we also suppose that if the machine breaks down on any given day, it stays shut down for the remainder of that day.) Let X be the profit realized per five-day week. The possible values of X are R, S, and $(-L)$. Let B be the number of breakdowns per week. We have

$$P(B = k) = \binom{5}{k}(0.05)^k(0.95)^{5-k}, \qquad k = 0, 1, \ldots, 5.$$

Since $X = S$ if and only if $B = 0$, $X = R$ if and only if $B = 1$ or 2, and $X = (-L)$ if and only if $B = 3, 4,$ or 5, we find that,

$$E(X) = SP(B = 0) + RP(B = 1 \text{ or } 2) + (-L)P(B = 3, 4, \text{ or } 5)$$
$$= S(0.95)^5 + R[5(0.05)(0.95)^4 + 10(0.05)^2(0.95)^3]$$
$$+ (-L)[10(0.05)^3(0.95)^2 + 5(0.05)^4(0.95) + (0.05)^5] \text{ dollars.}$$

Definition. Let X be a continuous random variable with pdf f. The *expected value* of X is defined as
$$E(X) = \int_{-\infty}^{+\infty} xf(x)\,dx. \tag{7.2}$$

Again it may happen that this (improper) integral does not converge. Hence we say that $E(X)$ exists if and only if
$$\int_{-\infty}^{\infty} |x|\,f(x)\,dx$$
is finite.

Note: We should observe the analogy between the expected value of a random variable and the concept of "center of mass" in mechanics. If a unit mass is distributed along the line at the discrete points $x_1, \ldots, x_n, \ldots$ and if $p(x_i)$ is the mass at x_i, then we see that $\sum_{i=1}^{\infty} x_i p(x_i)$ represents the center of mass (about the origin). Similarly, if a unit mass is distributed continuously over a line, and if $f(x)$ represents the mass density at x, then $\int_{-\infty}^{+\infty} xf(x)\,dx$ may again be interpreted as the center of mass. In the above sense, $E(X)$ can represent "a center" of the probability distribution. Also, $E(X)$ is sometimes called a measure of central tendency and is in the *same units* as X.

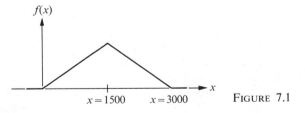

FIGURE 7.1

EXAMPLE 7.6. Let the random variable X be defined as follows. Suppose that X is the time (in minutes) during which electrical equipment is used at maximum load in a certain specified time period. Suppose that X is a continuous random variable with the following pdf:

$$f(x) = \frac{1}{(1500)^2} x, \qquad 0 \le x \le 1500,$$

$$= \frac{-1}{(1500)^2} (x - 3000), \qquad 1500 \le x \le 3000,$$

$$= 0, \qquad \text{elsewhere.}$$

Thus

$$E(X) = \int_{-\infty}^{+\infty} x f(x) \, dx$$

$$= \frac{1}{(1500)(1500)} \left[\int_{0}^{1500} x^2 \, dx - \int_{1500}^{3000} x(x - 3000) \, dx \right]$$

$$= 1500 \text{ minutes.}$$

(See Fig. 7.1.)

EXAMPLE 7.7. The ash content in coal (percentage), say X, may be considered as a continuous random variable with the following pdf: $f(x) = \frac{1}{4875} x^2$, $10 \le x \le 25$. Hence $E(X) = \frac{1}{4875} \int_{10}^{25} x^3 \, dx = 19.5$ percent. Thus the expected ash content in the particular coal specimen being considered is 19.5 percent.

Theorem 7.2. Let X be uniformly distributed over the interval $[a, b]$. Then

$$E(X) = \frac{a + b}{2}.$$

Proof: The pdf of X is given by $f(x) = 1/(b - a)$, $a \le x \le b$. Hence

$$E(X) = \int_{a}^{b} \frac{x}{b - a} \, dx = \frac{1}{b - a} \frac{x^2}{2} \Big|_{a}^{b} = \frac{a + b}{2}.$$

(Observe that this represents the *midpoint* of the interval $[a, b]$, as we would expect intuitively.)

Note: It might be worthwhile to recall at this juncture that a random variable X is a function from a sample space S to the range space R_X. As we have pointed out repeatedly, for most purposes we are only concerned with the range space and with probabilities defined on it. This notion of expected value was defined entirely in terms of the range space [see Eqs. (7.1) and (7.2).] However, occasionally we should observe the functional nature of X. For example, how do we express Eq. (7.1) in terms of the outcome $s \in S$, assuming S to be finite? Since $x_i = X(s)$ for some $s \in S$ and since

$$p(x_i) = P[s: X(s) = x_i],$$

we may write

$$E(X) = \sum_{i=1}^{n} x_i p(x_i) = \sum_{s \in S} X(s)P(s), \qquad (7.3)$$

where $P(s)$ is the probability of the event $\{s\} \subset S$. For instance, if the experiment consists of classifying three items as defective (D) or nondefective (N), a sample space for this experiment would be

$$S = \{NNN, NND, NDN, DNN, NDD, DND, DDN, DDD\}.$$

If X is defined as the number of defectives, and if all the above outcomes are assumed to be equally likely, we have according to Eq. (7.3),

$$E(X) = \sum_{s \in S} X(s)P(s)$$
$$= 0 \cdot (\tfrac{1}{8}) + 1(\tfrac{1}{8}) + 1(\tfrac{1}{8}) + 1(\tfrac{1}{8}) + 2(\tfrac{1}{8}) + 2(\tfrac{1}{8}) + 2(\tfrac{1}{8}) + 3(\tfrac{1}{8})$$
$$= \tfrac{3}{2}.$$

Of course, this result could have been obtained more easily by applying Eq. (7.1) directly. However, it is well to remember that in order to use Eq. (7.1) we needed to know the numbers $p(x_i)$, which in turn meant that a computation such as the one used above had to be carried out. The point is that once the probability distribution over R_X is known [in this case the values of the numbers $p(x_i)$], we can suppress the functional relationship between R_X and S.

7.2 Expectation of a Function of a Random Variable

As we have discussed previously, if X is a random variable and if $Y = H(X)$ is a function of X, then Y is also a random variable with some probability distribution. Hence it will be of interest and meaningful to evaluate $E(Y)$. There are two ways of evaluating $E(Y)$ which turn out to be equivalent. To show that they are in general equivalent is not trivial, and we shall only prove a special case. However, it is important that the reader understands the two approaches discussed below.

Definition. Let X be a random variable and let $Y = H(X)$.

(a) If Y is a discrete random variable with possible values $y_1, y_2, \ldots$ and if $q(y_i) = P(Y = y_i)$, we define

$$E(Y) = \sum_{i=1}^{\infty} y_i q(y_i). \qquad (7.4)$$

(b) If Y is a continuous random variable with pdf g, we define

$$E(Y) = \int_{-\infty}^{+\infty} y g(y)\, dy. \qquad (7.5)$$

Note: Of course, these definitions are completely consistent with the previous definition given for the expected value of a random variable. In fact, the above simply represents a restatement in terms of Y. One "disadvantage" of applying the above definition in order to obtain $E(Y)$ is that the probability distribution of Y (that is, the probability distribution over the range space R_Y) is required. We discussed, in the previous chapter, methods by which we may obtain either the point probabilities $q(y_i)$ or g, the pdf of Y. However, the question arises as to whether we can obtain $E(Y)$ without first finding the probability distribution of Y, simply from the knowledge of the probability distribution of X. The answer is in the affirmative as the following theorem indicates.

Theorem 7.3. Let X be a random variable and let $Y = H(X)$.

(a) If X is a discrete random variable and $p(x_i) = P(X = x_i)$, we have

$$E(Y) = E(H(X)) = \sum_{j=1}^{\infty} H(x_j)p(x_j). \tag{7.6}$$

(b) If X is a continuous random variable with pdf f, we have

$$E(Y) = E(H(X)) = \int_{-\infty}^{+\infty} H(x)f(x)\,dx. \tag{7.7}$$

Note: This theorem makes the evaluation of $E(Y)$ much simpler, for it means that we need not find the probability distribution of Y in order to evaluate $E(Y)$. The knowledge of the probability distribution of X suffices.

Proof: [We shall only prove Eq. (7.6). The proof of Eq. (7.7) is somewhat more intricate.]   Consider the sum $\sum_{j=1}^{\infty} H(x_j)p(x_j) = \sum_{j=1}^{\infty} (\sum_i H(x_i)p(x_i))$, where the inner sum is taken over all indices i for which $H(x_i) = y_j$, for some fixed y_j. Hence all the terms $H(x_i)$ are constant in the inner sum.   Hence

$$\sum_{j=1}^{\infty} H(x_j)p(x_j) = \sum_{j=1}^{\infty} y_j \sum_i p(x_i).$$

However,

$$\sum_i p(x_i) = \sum_i P[x \mid H(x_i) = y_j] = q(y_j).$$

Therefore, $\sum_{j=1}^{\infty} H(x_j)p(x_j) = \sum_{j=1}^{\infty} y_j q(y_j)$, which establishes Eq. (7.6).

Note: The method of proof is essentially equivalent to the method of counting in which we put together all items having the same value. Thus, if we want to find the total sum of the values 1, 1, 2, 3, 5, 3, 2, 1, 2, 2, 3, we could either add directly or point out that since there are 3 one's, 4 two's, 3 three's, and 1 five, the total sum equals

$$3(1) + 4(2) + 3(3) + 1(5) = 25.$$

EXAMPLE 7.8. Let V be the wind velocity (mph) and suppose that V is uniformly distributed over the interval [0, 10]. The pressure, say W (in lb/ft^2), on the surface of an airplane wing is given by the relationship: $W = 0.003V^2$. To find the expected value of W, $E(W)$, we can proceed in two ways:

(a) Using Theorem 7.3, we have

$$E(W) = \int_0^{10} 0.003v^2 f(v)\, dv$$

$$= \int_0^{10} 0.003v^2\, \frac{1}{10}\, dv$$

$$= 0.1\ \text{lb/ft}^2.$$

(b) Using the definition of $E(W)$, we first need to find the pdf of W, say g, and then evaluate $\int_{-\infty}^{+\infty} wg(w)\, dw$. To find $g(w)$, we note that $w = 0.003v^2$ is a monotone function of v, for $v \geq 0$. We may apply Theorem 5.1 and obtain

$$g(w) = \frac{1}{10}\left|\frac{dv}{dw}\right|$$

$$= \tfrac{1}{2}\sqrt{\tfrac{10}{3}}\, w^{-1/2}, \qquad 0 \leq w \leq 0.3,$$

$$= 0, \qquad \text{elsewhere.}$$

Hence

$$E(W) = \int_0^{0.3} wg(w)\, dw = 0.1$$

after a simple computation. Thus, as the theorem stated, the two evaluations of $E(W)$ yield the same result.

EXAMPLE 7.9. In many problems we are interested only in the *magnitude* of a random variable without regard to its algebraic sign. That is, we are concerned with $|X|$. Suppose that X is a continuous random variable with the following pdf:

$$f(x) = \frac{e^x}{2} \qquad \text{if}\ \ x \leq 0,$$

$$= \frac{e^{-x}}{2} \qquad \text{if}\ \ x > 0.$$

Let $Y = |X|$. To obtain $E(Y)$ we may proceed in one of two ways.
(a) Using Theorem 7.3, we have

$$E(Y) = \int_{-\infty}^{+\infty} |x| f(x)\, dx$$

$$= \tfrac{1}{2}\left[\int_{-\infty}^0 (-x)e^x\, dx + \int_0^\infty (x)e^{-x}\, dx\right]$$

$$= \tfrac{1}{2}[1 + 1] = 1.$$

(b) To evaluate $E(Y)$ using the definition, we need to obtain the pdf of $Y = |X|$, say g. Let G be the cdf of Y. Hence

$$G(y) = P(Y \leq y) = P[|X| \leq y] = P[-y \leq X \leq y] = 2P(0 \leq X \leq y),$$

since the pdf of X is symmetric about zero. Therefore

$$G(y) = 2\int_0^y f(x)\,dx = 2\int_0^y \frac{e^{-x}}{2}\,dx = -e^{-y} + 1.$$

Thus we have for g, the pdf of Y, $g(y) = G'(y) = e^{-y}$, $y \geq 0$. Hence $E(Y) = \int_0^\infty yg(y)\,dy = \int_0^\infty ye^{-y}\,dy = 1$, as above.

EXAMPLE 7.10. In many problems we can use the expected value of a random variable in order to make a certain decision in an optimum way.

Suppose that a manufacturer produces a certain type of lubricating oil which loses some of its special attributes if it is not used within a certain period of time. Let X be the number of units of oil ordered from the manufacturer during each year. (One unit equals 1000 gallons.) Suppose that X is a continuous random variable, uniformly distributed over [2, 4]. Hence the pdf f has the form,

$$f(x) = \tfrac{1}{2}, \qquad 2 \leq x \leq 4,$$
$$= 0, \qquad \text{elsewhere.}$$

Suppose that for each unit sold a profit of $300 is earned, while for each unit not sold (during any specified year) a loss of $100 is taken, since a unit not used will have to be discarded. Assume that the manufacturer must decide a few months prior to the beginning of each year how much he will produce, and that he decides to manufacture Y units. (Y is *not* a random variable; it is specified by the manufacturer.) Let Z be the profit per year (in dollars). Here Z is clearly a random variable since it is a function of the random variable X. Specifically, $Z = H(X)$ where

$$H(X) = 300\,Y \qquad \text{if} \quad X \geq Y,$$
$$= 300\,X + (-100)(Y - X) \qquad \text{if} \quad X < Y.$$

(The last expression may be written as $400X - 100\,Y$.)

In order for us to obtain $E(Z)$ we apply Theorem 7.3 and write

$$E(Z) = \int_{-\infty}^{+\infty} H(x)f(x)\,dx$$

$$= \frac{1}{2}\int_2^4 H(x)\,dx.$$

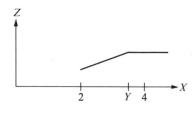

FIGURE 7.2

To evaluate this integral we must consider three cases: $Y < 2$, $2 \leq Y \leq 4$, and $Y > 4$. With the aid of Fig. 7.2 and after some simplification we obtain

$$E(Z) = 300\,Y \qquad \text{if} \quad Y \leq 2$$
$$= -100\,Y^2 + 700\,Y - 400$$
$$\text{if} \quad 2 < Y < 4$$
$$= 1200 - 100\,Y \qquad \text{if} \quad Y \geq 4.$$

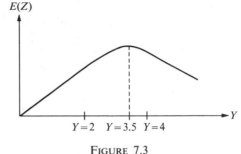

FIGURE 7.3

The following question is of interest. How should the manufacturer choose the value of Y in order to maximize his expected profit? We can answer this question easily by simply setting $dE(Z)/dY = 0$. This yields $Y = 3.5$. (See Fig. 7.3.)

7.3 Two-Dimensional Random Variables

The concepts discussed above for the one-dimensional case also hold for higher-dimensional random variables. In particular, for the two-dimensional case, we make the following definition.

Definition. Let (X, Y) be a two-dimensional random variable and let $Z = H(X, Y)$ be a real-valued function of (X, Y). Hence Z is a (one-dimensional) random variable and we define $E(Z)$ as follows:

(a) If Z is a discrete random variable with possible values $z_1, z_2, \ldots$ and with

$$p(z_i) = P(Z = z_i),$$

then

$$E(Z) = \sum_{i=1}^{\infty} z_i p(z_i). \tag{7.8}$$

(b) If Z is a continuous random variable with pdf f, we have

$$E(Z) = \int_{-\infty}^{+\infty} z f(z)\, dz. \tag{7.9}$$

As in the one-dimensional case, the following theorem (analogous to Theorem 7.3) may be proved.

Theorem 7.4. Let (X, Y) be a two-dimensional random variable and let $Z = H(X, Y)$.

(a) If (X, Y) is a discrete random variable and if

$$p(x_i, y_j) = P(X = x_i, Y = y_j), \qquad i, j = 1, 2, \ldots,$$

we have

$$E(Z) = \sum_{j=1}^{\infty} \sum_{i=1}^{\infty} H(x_i, y_j)p(x_i, y_j). \tag{7.10}$$

(b) If (X, Y) is a continuous random variable with joint pdf f, we have

$$E(Z) = \int_{-\infty}^{+\infty} \int_{-\infty}^{+\infty} H(x, y)f(x, y) \, dx \, dy. \tag{7.11}$$

Note: We shall not prove Theorem 7.4. Again, as in the one-dimensional case, this is an extremely useful result since it states that we need *not* find the probability distribution of the random variable Z in order to evaluate its expectation. We can find $E(Z)$ directly from the knowledge of the joint distribution of (X, Y).

EXAMPLE 7.11. Let us reconsider Example 6.14 and find $E(E)$ where $E = IR$. We found that I and R were independent random variables with the following pdf's g and h, respectively:

$$g(i) = 2i, \quad 0 \le i \le 1; \qquad h(r) = r^2/9, \quad 0 \le r \le 3.$$

We also found that the pdf of E is $p(e) = \frac{2}{9}e(3 - e)$, $0 \le e \le 3$. Since I and R are independent random variables, the joint pdf of (I, R) is simply the product of the pdf of I and R: $f(i, r) = \frac{2}{9}ir^2$, $0 \le i \le 1, 0 \le r \le 3$. To evaluate $E(E)$ using Theorem 7.4 we have

$$E(E) = \int_0^3 \int_0^1 ir f(i, r) \, di \, dr = \int_0^3 \int_0^1 ir\frac{2}{9}ir^2 \, di \, dr$$

$$= \frac{2}{9} \int_0^1 i^2 \, di \int_0^3 r^3 \, dr = \frac{3}{2}.$$

Using the definition (7.9) directly, we have

$$E(E) = \int_0^3 ep(e) \, de = \int_0^3 e\frac{2}{9}e(3 - e) \, de$$

$$= \frac{2}{9} \int_0^3 (3e^2 - e^3) \, de = \frac{3}{2}.$$

7.4 Properties of Expected Value

We shall list a number of important properties of the expected value of a random variable which will be very useful for subsequent work. In each case we shall assume that all the expected values to which we refer exist. The proofs will be given only for the continuous case. The reader should be able to supply the argument for the discrete case by simply replacing integrals by summations.

Property 7.1. If $X = C$ where C is a constant, then $E(X) = C$.

Proof

$$E(X) = \int_{-\infty}^{+\infty} Cf(x)\, dx$$

$$= C \int_{-\infty}^{+\infty} f(x)\, dx = C.$$

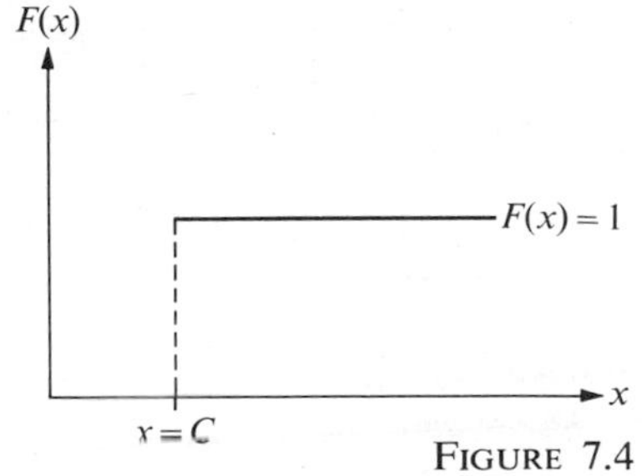

FIGURE 7.4

Note: The meaning of X equals C is the following. Since X is a function from the sample space to R_X, the above means that R_X consists of the single value C. Hence X equals C if and only if $P[X(s) = C] = 1$. This notion is best explained in terms of the cdf of X. Namely, $F(x) = 0$, if $x < C$; $F(x)$ equals 1, if $x \geq C$ (Fig. 7.4). Such a random variable is sometimes called *degenerate*.

Property 7.2. Suppose that C is a constant and X is a random variable. Then $E(CX) = CE(X)$.

Proof: $E(CX) = \int_{-\infty}^{+\infty} Cxf(x)\, dx = C \int_{-\infty}^{+\infty} xf(x)\, dx = CE(X).$

Property 7.3. Let (X, Y) be a two-dimensional random variable with a joint probability distribution. Let $Z = H_1(X, Y)$ and $W = H_2(X, Y)$. Then $E(Z + W) = E(Z) + E(W)$.

Proof

$$E(Z + W) = \int_{-\infty}^{+\infty} \int_{-\infty}^{+\infty} [H_1(x, y) + H_2(x, y)] f(x, y)\, dx\, dy$$

[where f is the joint pdf of (X, Y)]

$$= \int_{-\infty}^{+\infty} \int_{-\infty}^{+\infty} H_1(x, y) f(x, y)\, dx\, dy + \int_{-\infty}^{+\infty} \int_{-\infty}^{+\infty} H_2(x, y) f(x, y)\, dx\, dy$$

$$= E(Z) + E(W).$$

Property 7.4. Let X and Y be any two random variables. Then $E(X + Y) = E(X) + E(Y)$.

Proof: This follows immediately from Property 7.3 by letting $H_1(X, Y) = X$, and $H_2(X, Y) = Y$.

Notes: (a) Combining Properties 7.1, 7.2, and 7.4 we observe the following important fact: If $Y = aX + b$, where a and b are constants, then $E(Y) = aE(X) + b$. In words:

The expectation of a linear function is that same linear function of the expectation. This is *not* true unless a linear function is involved, and it is a common error to believe otherwise. For instance, $E(X^2) \neq (E(X))^2$, $E(\ln X) \neq \ln E(X)$, etc. Thus if X assumes the values -1 and $+1$, each with probability $\frac{1}{2}$, then $E(X) = 0$. However,

$$E(X^2) = (-1)^2(\tfrac{1}{2}) + (1)^2(\tfrac{1}{2}) = 1 \neq 0^2.$$

(b) In general, it is difficult to obtain expressions for $E(1/X)$ or $E(X^{1/2})$, say, in terms of $1/E(X)$ or $(E(X))^{1/2}$. However, some inequalities are available, which are very easy to derive. (See articles by Fleiss, Murthy and Pillai, and Gurland in the February 1966, December 1966, and April 1967 issues, respectively, of *The American Statistician.*)
For instance, we have:
(1) If X assumes only positive values and has finite expectation, then $E(1/X) \geq 1/E(X)$.
(2) Under the same hypotheses as in (1), $E(X^{1/2}) \leq (E(X))^{1/2}$.

Property 7.5. Let $X_1, \ldots, X_n$ be n random variables. Then

$$E(X_1 + \cdots + X_n) = E(X_1) + \cdots + E(X_n).$$

Proof: This follows immediately from Property 7.4 by applying mathematical induction.

Note: Combining this property with the above, we obtain

$$E\left(\sum_{i=1}^{n} a_i X_i\right) = \sum_{i=1}^{n} a_i E(X_i),$$

where the a_i's are constants.

Property 7.6. Let (X, Y) be a two-dimensional random variable and suppose that X and Y are *independent*. Then $E(XY) = E(X)E(Y)$.

Proof

$$E(XY) = \int_{-\infty}^{+\infty} \int_{-\infty}^{+\infty} xy f(x, y) \, dx \, dy$$

$$= \int_{-\infty}^{+\infty} \int_{-\infty}^{+\infty} xy g(x) h(y) \, dx \, dy$$

$$= \int_{-\infty}^{+\infty} x g(x) \, dx \int_{-\infty}^{+\infty} y h(y) \, dy = E(X)E(Y).$$

Note: The additional hypothesis of independence is required to establish Property 7.6, whereas no such assumption was needed to obtain Property 7.4.

EXAMPLE 7.12. (This example is based on a problem in *An Introduction to Probability Theory and Its Applications* by W. Feller, p. 225.)

Suppose that we need to test a large number of persons for some characteristic, with either positive or negative results. Furthermore, suppose that one can take specimens from several persons and test the combined specimen as a unit, such as may be the case in certain types of blood tests.

Assume: The combined specimen will give a negative result if and only if all contributing specimens are negative.

Thus, in the case of a positive result (of the combined specimen), *all* specimens must be retested individually to determine which are positive. If the N persons are divided into n groups of k persons (assume $N = kn$) then the following choices arise:

(a) Test all N persons individually, requiring N tests.

(b) Test groups of k specimens which may require as few as $n = N/k$ or as many as $(k + 1)n = N + n$ tests.

It shall be our purpose to study the expected number of tests required under (b) and then to compare this with N.

Assume: The probability that the results of the test are positive equals p and is the same for all persons. Furthermore, test outcomes for persons within the same group being tested are independent. Let $X = $ number of tests required to determine the characteristic being studied for all N persons, and let $X_i = $ number of tests required for testing persons in the ith group, $i = 1, \ldots, n$.

Hence $X = X_1 + \cdots + X_n$, and therefore $E(X) = E(X_1) + \cdots + E(X_n)$, which equals $nE(X_1)$, say, since all of the X_i's have the same expectation. Now X_1 assumes just two values: 1 and $k + 1$. Furthermore,

$$P(X_1 = 1) = P(\text{all } k \text{ persons in group 1 are negative})$$
$$= (1 - p)^k.$$

Therefore

$$P(X_1 = k + 1) = 1 - (1 - p)^k$$

and hence

$$E(X_1) = 1 \cdot (1 - p)^k + (k + 1)[1 - (1 - p)^k]$$
$$= k[1 - (1 - p)^k + k^{-1}].$$

Thus

$$E(X) = nE(X_1) = N[1 - (1 - p)^k + k^{-1}].$$

(The above formula is valid only for $k > 1$, since for $k = 1$ it yields $E(X) = N + pn$, which is obviously false!)

One question of interest is the choice of k for which the above $E(X)$ is smallest. This could easily be handled by some numerical procedure. (See Problem 7.11a.)

Finally, note that in order for "group testing" to be preferable to individual testing, we should have $E(X) < N$, that is, $1 - (1 - p)^k + k^{-1} < 1$, which is equivalent to $k^{-1} < (1 - p)^k$. This *cannot* occur if $(1 - p) < \frac{1}{2}$. For, in that case, $(1 - p)^k < \frac{1}{2}^k < 1/k$, the last inequality following from the fact that $2^k > k$. Thus we obtain the following interesting conclusion: If p, the probability of a positive test on any given individual, is greater than $\frac{1}{2}$, then it is *never preferable* to group specimens before testing. (See Problem 7.11b.)

EXAMPLE 7.13. Let us apply some of the above properties to derive (again) the expectation of a binomially distributed random variable. The method used may be applied to advantage in many similar situations.

Consider n independent repetitions of an experiment and let X be the number of times some event, say A, occurs. Let p equal $P(A)$ and assume that this number is constant for all repetitions considered.

Define the auxiliary random variables $Y_1, \ldots, Y_n$ as follows:

$$Y_i = 1 \qquad \text{if the event } A \text{ occurs on the } i\text{th repetition,}$$
$$= 0, \qquad \text{elsewhere.}$$

Hence

$$X = Y_1 + Y_2 + \cdots + Y_n,$$

and applying Property 7.5, we obtain

$$E(X) = E(Y_1) + \cdots + E(Y_n).$$

However,

$$E(Y_i) = 1(p) + 0(1 - p) = p, \qquad \text{for all } i.$$

Thus $E(X) = np$, which checks with the previous result.

Note: Let us reinterpret this important result. Consider the random variable X/n. This represents the relative frequency of the event A among the n repetitions of $\mathcal{E}$. Using Property 7.2, we have $E(X/n) = (np)/n = p$. This is, intuitively, as it should be, for it says that the expected relative frequency of the event A is p, where $p = P(A)$. It represents the first theoretical verification of the fact that there is a connection between the relative frequency of an event and the probability of that event. In a later chapter we shall obtain further results yielding a much more precise relationship between relative frequency and probability.

EXAMPLE 7.14. Suppose that the demand D, per week, of a certain product is a random variable with a certain probability distribution, say $P(D = n) = p(n)$, $n = 0, 1, 2, \ldots$ Suppose that the cost to the supplier is C_1 dollars per item, while he sells the item for C_2 dollars. Any item which is not sold at the end of the week must be stored at a cost of C_3 dollars per item. If the supplier decides to produce

N items at the beginning of the week, what is his expected profit per week? For what value of N is the expected profit maximized? If T is the profit per week, we have

$$T = NC_2 - NC_1 \quad \text{if} \quad D > N,$$
$$= DC_2 - C_1N - C_3(N - D) \quad \text{if} \quad D \leq N.$$

Rewriting the above, we obtain

$$T = N(C_2 - C_1) \quad \text{if} \quad D > N,$$
$$= (C_2 + C_3)D - N(C_1 + C_3) \quad \text{if} \quad D \leq N.$$

Hence the expected profit is obtained as follows:

$$E(T) = N(C_2 - C_1)P(D > N) + (C_2 + C_3)\sum_{n=0}^{N} np(n)$$

$$- N(C_1 + C_3)P(D \leq N)$$

$$= N(C_2 - C_1)\sum_{n=N+1}^{\infty} p(n) + (C_2 + C_3)\sum_{n=0}^{N} np(n)$$

$$- N(C_1 + C_3)\sum_{n=0}^{N} p(n)$$

$$= N(C_2 - C_1) + (C_2 + C_3)\left[\sum_{n=0}^{N} np(n) - N\sum_{n=0}^{N} p(n)\right]$$

$$= N(C_2 - C_1) + (C_2 + C_3)\sum_{n=0}^{N} p(n)(n - N).$$

Suppose that the following probability distribution is known to be appropriate for D: $P(D = n) = \frac{1}{5}, n = 1, 2, 3, 4, 5$. Hence

$$E(T) = N(C_2 - C_1) + \frac{(C_2 + C_3)}{5}[N(N + 1)/2 - N^2] \quad \text{if} \quad N \leq 5,$$

$$= N(C_2 - C_1) + (C_2 + C_3)\tfrac{1}{5}(15 - 5N) \quad \text{if} \quad N > 5.$$

Suppose that $C_2 = \$9$, $C_1 = \$3$, and $C_3 = \$1$. Therefore

$$E(T) = 6N + 2\left[\frac{N(N + 1)}{2} - N^2\right]$$

$$\text{if} \quad N \leq 5,$$
$$= 6N + 2(15 - 5N) \quad \text{if} \quad N > 5,$$
$$= 7N - N^2 \quad \text{if} \quad N \leq 5,$$
$$= 30 - 4N \quad \text{if} \quad N > 5.$$

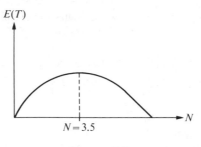

FIGURE 7.5

Hence the maximum occurs for $N = 3.5$. (See Fig. 7.5.) For $N = 3$ or 4, we have $E(T) = 12$, which is the maximum attainable since N is an integer.

7.5 The Variance of a Random Variable

Suppose that for a random variable X we find that $E(X)$ equals 2. What is the significance of this? It is important that we do not attribute more meaning to this information than is warranted. It simply means that if we consider a large number of determinations of X, say $x_1, \ldots, x_n$, and average these values of X, this average would be close to 2 if n is large. However, it is very crucial that we should not put too much meaning into an expected value. For example, suppose that X represents the life length of light bulbs being received from a manufacturer, and that $E(X) = 1000$ hours. This could mean one of several things. It could mean that most of the bulbs would be expected to last somewhere between 900 hours and 1100 hours. It could also mean that the bulbs being supplied are made up of two entirely different types of bulbs: about half are of very high quality and will last about 1300 hours, while the other half are of very poor quality and will last about 700 hours.

There is an obvious need to introduce a quantitative measure which will distinguish between such situations. Various measures suggest themselves, but the following is the most commonly used quantity.

Definition. Let X be a random variable. We define the *variance* of X, denoted by $V(X)$ or σ_X^2, as follows:

$$V(X) = E[X - E(X)]^2. \tag{7.12}$$

The positive square root of $V(X)$ is called the *standard deviation* of X and is denoted by σ_X.

Notes: (a) The *number* $V(X)$ is expressed in *square units* of X. That is, if X is measured in hours, then $V(X)$ is expressed in (hours)². This is one reason for considering the standard deviation. It is expressed in the *same* units as X.

(b) Another possible measure might have been $E|X - E(X)|$. For a number of reasons, one of which is that X^2 is a "better-behaved" function than $|X|$, the variance is preferred.

(c) If we interpret $E(X)$ as the center of a unit mass distributed over a line, we may interpret $V(X)$ as the moment of inertia of this mass about a perpendicular axis through the center of mass.

(d) $V(X)$ as defined in Eq. (7.12) is a special case of the following more general notion. The *kth moment* of the random variable X about its expectation is defined as $\mu_k = E[X - E(X)]^k$. Clearly for $k = 2$, we obtain the variance.

The evaluation of $V(X)$ may be simplified with the aid of the following result.

Theorem 7.5
$$V(X) = E(X^2) - [E(X)]^2.$$

Proof: Expanding $E[X - E(X)]^2$ and using the previously established properties for expectation, we obtain

$$
\begin{aligned}
V(X) &= E[X - E(X)]^2 \\
&= E\{X^2 - 2XE(X) + [E(X)]^2\} \\
&= E(X^2) - 2E(X)E(X) + [E(X)]^2 \qquad \text{[Recall that } E(X) \text{ is a constant.]} \\
&= E(X^2) - [E(X)]^2.
\end{aligned}
$$

EXAMPLE 7.15. The weather bureau classifies the type of sky that is visible in terms of "degrees of cloudiness." A scale of 11 categories is used: 0, 1, 2, ..., 10, where 0 represents a perfectly clear sky, 10 represents a completely overcast sky, while the other values represent various intermediate conditions. Suppose that such a classification is made at a particular weather station on a particular day and time. Let X be the random variable assuming one of the above 11 values. Suppose that the probability distribution of X is

$$p_0 = p_{10} = 0.05;$$
$$p_1 = p_2 = p_8 = p_9 = 0.15;$$
$$p_3 = p_4 = p_5 = p_6 = p_7 = 0.06.$$

Hence
$$
\begin{aligned}
E(X) &= 1(0.15) + 2(0.15) + 3(0.06) + 4(0.06) + 5(0.06) \\
&\quad + 6(0.06) + 7(0.06) + 8(0.15) + 9(0.15) \\
&\quad + 10(0.05) = 5.0.
\end{aligned}
$$

In order to compute $V(X)$ we need to evaluate $E(X^2)$.

$$
\begin{aligned}
E(X^2) &= 1(0.15) + 4(0.15) + 9(0.06) + 16(0.06) + 25(0.06) \\
&\quad + 36(0.06) + 49(0.06) + 64(0.15) + 81(0.15) \\
&\quad + 100(0.05) = 35.6.
\end{aligned}
$$

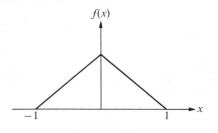

FIGURE 7.6

Hence

$$V(X) = E(X^2) - (E(X))^2 = 35.6 - 25 = 10.6,$$

and the standard deviation $\sigma = 3.25$.

EXAMPLE 7.16. Suppose that X is a continuous random variable with pdf

$$f(x) = 1 + x, \qquad -1 \le x \le 0,$$
$$= 1 - x, \qquad 0 \le x \le 1.$$

(See Fig. 7.6.) Because of the symmetry of the pdf, $E(X) = 0$. (See Note below.)
Furthermore,

$$E(X^2) = \int_{-1}^{0} x^2(1 + x)\, dx$$

$$+ \int_{0}^{1} x^2(1 - x)\, dx = \tfrac{1}{6}.$$

Hence $V(X) = \tfrac{1}{6}$.

Note: Suppose that a continuous random variable has a pdf which is symmetric about
$x = 0$. That is, $f(-x) = f(x)$ for all x. Then, provided $E(X)$ exists, $E(X) = 0$, which
is an immediate consequence of the definition of $E(X)$. This may be extended to an
arbitrary point of symmetry $x = a$, in which case $E(X) = a$. (See Problem 7.33.)

7.6 Properties of the Variance of a Random Variable

There are various important properties, in part analogous to those discussed
for the expectation of a random variable, which hold for the variance.

Property 7.7. If C is a constant,

$$V(X + C) = V(X). \tag{7.13}$$

Proof

$$V(X + C) = E[(X + C) - E(X + C)]^2 = E[(X + C) - E(X) - C]^2$$
$$= E[X - E(X)]^2 = V(X).$$

Note: This property is intuitively clear, for adding a constant to an outcome X does not change its variability, which is what the variance measures. It simply "shifts" the values of X to the right or to the left, depending on the sign of C.

Property 7.8. If C is a constant,

$$V(CX) = C^2 V(X). \tag{7.14}$$

Proof

$$V(CX) = E(CX)^2 - (E(CX))^2 = C^2 E(X^2) - C^2 (E(X))^2$$
$$= C^2 [E(X^2) - (E(X))^2] = C^2 V(X).$$

Property 7.9. If (X, Y) is a two-dimensional random variable, and if X and Y are *independent* then

$$V(X + Y) = V(X) + V(Y). \tag{7.15}$$

Proof

$$V(X + Y) = E(X + Y)^2 - (E(X + Y))^2$$
$$= E(X^2 + 2XY + Y^2) - (E(X))^2 - 2E(X)E(Y) - (E(Y))^2$$
$$= E(X^2) - (E(X))^2 + E(Y^2) - (E(Y))^2 = V(X) + V(Y).$$

Note: It is important to realize that the variance is *not additive*, in general, as is the expected value. With the additional assumption of independence, Property 7.9 is valid. Nor does the variance possess the linearity property which we discussed for the expectation, that is, $V(aX + b) \neq aV(X) + b$. Instead we have $V(aX + b) = a^2 V(X)$.

Property 7.10. Let $X_1, \ldots, X_n$ be n independent random variables. Then

$$V(X_1 + \cdots + X_n) = V(X_1) + \cdots + V(X_n). \tag{7.16}$$

Proof: This follows from Property 7.9 by mathematical induction.

Property 7.11. Let X be a random variable with finite variance. Then for any real number α,

$$V(X) = E[(X - \alpha)^2] - [E(X) - \alpha]^2. \tag{7.17}$$

Proof: See Problem 7.36.

Notes: (a) This is an obvious extension of Theorem 7.5, for by letting $\alpha = 0$ we obtain Theorem 7.5.

(b) If we interpret $V(X)$ as the moment of inertia and $E(X)$ as the center of a unit mass, then the above property is a statement of the well-known *parallel-axis theorem* in mechanics: The moment of inertia about an arbitrary point equals the moment of inertia about the center of mass plus the square of the distance of this arbitrary point from the center of mass.

(c) $E[X - \alpha]^2$ is minimized if $\alpha = E(X)$. This follows immediately from the above property. Thus the moment of inertia (of a unit mass distributed over a line) about an axis through an arbitrary point is minimized if this point is chosen as the center of mass.

EXAMPLE 7.17. Let us compute the variance of a binomially distributed random variable with parameter p.

To compute $V(X)$ we can proceed in two ways. Since we already know that $E(X) = np$, we must simply compute $E(X^2)$ and then evaluate $V(X)$ as $E(X^2) - (E(X))^2$. To compute $E(X^2)$ we use the fact that $P(X = k) = \binom{n}{k} p^k (1 - p)^{n-k}$, $k = 0, 1, \ldots, n$. Hence $E(X^2) = \sum_{k=0}^{n} k^2 \binom{n}{k} p^k (1 - p)^{n-k}$. This sum may be evaluated fairly easily, but rather than do this, we shall employ a simpler method.

We shall again use the representation of X introduced in Example 7.13, namely $X = Y_1 + Y_2 + \cdots + Y_n$. We now note that the Y_i's are *independent* random variables since the value of Y_i depends only on the outcome of the ith repetition, and the successive repetitions are assumed to be independent. Hence we may apply Property 7.10 and obtain

$$V(X) = V(Y_1 + \cdots + Y_n) = V(Y_1) + \cdots + V(Y_n).$$

But $V(Y_i) = E(Y_i)^2 - [E(Y_i)]^2$. Now

$$E(Y_i) = 1(p) + 0(1 - p) = p, \qquad E(Y_i)^2 = 1^2(p) + 0^2(1 - p) = p.$$

Therefore $V(Y_i) = p - p^2 = p(1 - p)$ for all i. Thus $V(X) = np(1 - p)$.

Note: Let us consider $V(X) = np(1 - p)$ as a function of p, for given n. We sketch a graph as shown in Fig. 7.7.

Solving $(d/dp)np(1 - p) = 0$ we find that the maximum value for $V(X)$ occurs for $p = \frac{1}{2}$. The minimum value of $V(X)$ obviously occurs at the endpoints of the interval at $p = 0$ and $p = 1$. This is intuitively as it should be. Recalling that the variance is a measure of the variation of the random variable X defined as the number of times the event A occurs in n repetitions, we find that this variation is zero if $p = 0$ or 1 (that is, if A occurs with probability 0 or 1) and is maximum when we are as "uncertain as we can be" about the occurrence or nonoccurrence of A, that is, when $P(A) = \frac{1}{2}$.

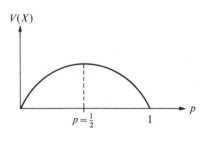

FIGURE 7.7

EXAMPLE 7.18. Suppose that the random variable X is uniformly distributed over $[a, b]$. As we have computed previously, $E(X) = (a + b)/2$.

To compute $V(X)$ we evaluate $E(X^2)$:

$$E(X^2) = \int_a^b x^2 \frac{1}{b - a} dx = \frac{b^3 - a^3}{3(b - a)}.$$

Hence

$$V(X) = E(X^2) - [E(X)]^2 = \frac{(b-a)^2}{12}$$

after a simple computation.

Notes: (a) This result is intuitively meaningful. It states that the variance of X does not depend on a and b individually but only on $(b-a)^2$, that is, on the square of their *difference*. Hence two random variables each of which is uniformly distributed over some interval (not necessarily the same) will have equal variances so long as the *lengths* of the intervals are the same.

(b) It is a well-known fact that the moment of inertia of a slim rod of mass M and length L about a transverse axis through the center is given by $ML^2/12$.

7.7 Approximate Expressions for Expectation and Variance

We have already noted that in order to evaluate $E(Y)$ or $V(Y)$, where $Y = H(X)$, we need not find the probability distribution of Y, but may work directly with the probability distribution of X. Similarly, if $Z = H(X, Y)$, we can evaluate $E(Z)$ and $V(Z)$ without first obtaining the distribution of Z.

If the function H is quite involved, the evaluation of the above expectations and variances may lead to integrations (or summations) which are quite difficult. Hence the following approximations are very useful.

Theorem 7.6. Let X be a random variable with $E(X) = \mu$, and $V(X) = \sigma^2$. Suppose that $Y = H(X)$. Then

$$E(Y) \simeq H(\mu) + \frac{H''(\mu)}{2}\sigma^2, \tag{7.18}$$

$$V(Y) \simeq [H'(\mu)]^2\sigma^2. \tag{7.19}$$

(In order to make the above approximations meaningful, we obviously require that H be at least twice differentiable at $x = \mu$.)

Proof (outline only): In order to establish Eq. (7.18), we expand the function H in a Taylor series about $x = \mu$ to two terms. Thus

$$Y = H(\mu) + (X - \mu)H'(\mu) + \frac{(X-\mu)^2 H''(\mu)}{2} + R_1,$$

where R_1 is a remainder. If we discard the remainder term R_1, then, taking the expected value of both sides, we have

$$E(Y) \simeq H(\mu) + \frac{H''(\mu)}{2}\sigma^2,$$

since $E(X - \mu) = 0$. In order to establish Eq. (7.19), we expand H in a Taylor series about $x = \mu$ to *one* term. Then $Y = H(\mu) + (X - \mu)H'(\mu) + R_2$. If we discard the remainder R_2 and take the variance of both sides, we have

$$V(Y) \simeq [H'(\mu)]^2\sigma^2.$$

EXAMPLE 7.19. Under certain conditions, the surface tension of a liquid (dyn/cm) is given by the formula $S = 2(1 - 0.005T)^{1.2}$, where T is the temperature of the liquid (degrees centigrade).

Suppose that T is a continuous random variable with the following pdf,

$$f(t) = 3000t^{-4}, \quad t \geq 10,$$
$$= 0, \quad \text{elsewhere.}$$

Hence

$$E(T) = \int_{10}^{\infty} 3000t^{-3} \, dt = 15 \text{ (degrees centigrade).}$$

And

$$V(T) = E(T^2) - (15)^2$$
$$= \int_{10}^{\infty} 3000t^{-2} \, dt - 225 = 75 \text{ (degrees centigrade)}^2.$$

To compute $E(S)$ and $V(S)$ we have to evaluate the following integrals:

$$\int_{10}^{\infty} (1 - 0.005t)^{1.2} t^{-4} \, dt$$

and

$$\int_{10}^{\infty} (1 - 0.005t)^{2.4} t^{-4} \, dt.$$

Rather than evaluate these expressions, we shall obtain approximations for $E(S)$ and $V(S)$ by using Eqs. (7.18) and (7.19). In order to use these formulas we have to compute $H'(15)$ and $H''(15)$, where $H(t) = 2(1 - 0.005t)^{1.2}$. We have

$$H'(t) = 2.4(1 - 0.005t)^{0.2}(-0.005) = -0.012(1 - 0.005t)^{0.2}.$$

Hence

$$H(15) = 1.82, \, H'(15) = 0.01.$$

Similarly,

$$H''(t) = -0.0024(1 - 0.005t)^{-0.8}(-0.005) = 0.000012(1 - 0.005t)^{-0.8}.$$

Therefore

$$H''(15) = \frac{0.000012}{(0.925)^{0.8}} = 0^+.$$

Thus we have

$$E(S) \simeq H(15) + 75H''(15) = 1.82 \text{ (dyne/cm),}$$
$$V(S) \simeq 75[H''(15)]^2 = 0.87 \text{ (dyne/cm)}^2.$$

If Z is a function of two random variables, say $Z = H(X, Y)$, an analogous result is available.

Theorem 7.7. Let (X, Y) be a two-dimensional random variable. Suppose that $E(X) = \mu_x$, $E(Y) = \mu_y$; $V(X) = \sigma_x^2$ and $V(Y) = \sigma_y^2$. Let $Z = H(X, Y)$. [We shall assume that the various derivatives of H exist at (μ_x, μ_y).] Then if X and Y are independent, we have

$$E(Z) \simeq H(\mu_x, \mu_y) + \frac{1}{2}\left[\frac{\partial^2 H}{\partial x^2}\sigma_x^2 + \frac{\partial^2 H}{\partial y^2}\sigma_y^2\right],$$

$$V(Z) \simeq \left[\frac{\partial H}{\partial x}\right]^2 \sigma_x^2 + \left[\frac{\partial H}{\partial y}\right]^2 \sigma_y^2,$$

where all the partial derivatives are evaluated at (μ_x, μ_y).

Proof: The proof involves the expansion of H in a Taylor series about the point (μ_x, μ_y) to one and two terms, discarding the remainder, and then taking the expectation and variance of both sides as was done in the proof of Theorem 7.6. We shall leave the details to the reader. (If X and Y are not independent, a slightly more complicated formula may be derived.)

Note: The above result may be extended to a function of n independent random variables, say $Z = H(X_1, \ldots, X_n)$. If $E(X_i) = \mu_i$, $V(X_i) = \sigma_i^2$, we have the following approximations, assuming that all the derivatives exist:

$$E(Z) \simeq H(\mu_1, \ldots, \mu_n) + \frac{1}{2}\sum_{i=1}^{n}\frac{\partial^2 H}{\partial x_i^2}\sigma_i^2,$$

$$V(Z) \simeq \sum_{i=1}^{n}\left(\frac{\partial H}{\partial x_i}\right)^2 \sigma_i^2,$$

where all the partial derivatives are evaluated at the point $(\mu_1, \ldots, \mu_n)$.

EXAMPLE 7.20. Suppose that we have a simple circuit for which the voltage, say M, is expressed by Ohm's Law as $M = IR$, where I and R are the current and resistance of the circuit, respectively. If I and R are independent random variables, then M is a random variable, and using Theorem 7.7, we may write

$$E[M] \simeq E(I)E(R), \qquad V[M] \simeq [E(R)]^2 V(I) + [E(I)]^2 V(R).$$

7.8 Chebyshev's Inequality

There is a well-known inequality due to the Russian mathematician Chebyshev which will play an important role in our subsequent work. In addition, it will give us a means of understanding precisely how the variance measures variability about the expected value of a random variable.

If we know the probability distribution of a random variable X (either the pdf in the continuous case or the point probabilities in the discrete case), we may then compute $E(X)$ and $V(X)$, if these exist. However, the converse is *not* true. That is, from a knowledge of $E(X)$ and $V(X)$ we cannot reconstruct the probability distribution of X and hence cannot compute quantities such as $P[|X - E(X)| \leq C]$.

Nonetheless, it turns out that although we cannot evaluate such probabilities [from a knowledge of $E(X)$ and $V(X)$], we can give a very useful upper (or lower) bound to such probabilities. This result is contained in what is known as **Chebyshev's inequality**.

Chebyshev's inequality. Let X be a random variable with $E(X) = \mu$ and let c be any real number. Then, if $E(X - c)^2$ is finite and ϵ is any positive number, we have

$$P[|X - c| \geq \epsilon] \leq \frac{1}{\epsilon^2} E(X - c)^2. \qquad (7.20)$$

The following forms, equivalent to (7.20), are immediate:
(a) By considering the complementary event we obtain

$$P[|X - c| < \epsilon] \geq 1 - \frac{1}{\epsilon^2} E(X - c)^2. \qquad (7.20a)$$

(b) Choosing $c = \mu$ we obtain

$$P[|X - \mu| \geq \epsilon] \leq \frac{\text{Var } X}{\epsilon^2}. \qquad (7.20b)$$

(c) Choosing $c = \mu$ and $\epsilon = k\sigma$, where $\sigma^2 = \text{Var } X > 0$, we obtain

$$P[|X - \mu| \geq k\sigma] \leq k^{-2}. \qquad (7.21)$$

This last form (7.21) is particularly indicative of how the variance measures the "degree of concentration" of probability near $E(X) = \mu$.

Proof (We shall prove only 7.20 since the others follow as indicated. We shall deal only with the continuous case. In the discrete case the argument is very similar with integrals replaced by sums. However, some care must be taken with endpoints of intervals.):

Consider

$$P([|X - c| \geq \epsilon] = \int_{x:|x-c|\geq\epsilon} f(x)\, dx.$$

(The limit on the integral says that we are integrating between $-\infty$ and $c - \epsilon$ and between $c + \epsilon$ and $+\infty$.)

Now $|x - c| \geq \epsilon$ is equivalent to $(x - c)^2/\epsilon^2 \geq 1$. Hence the above integral is

$$\leq \int_R \frac{(x - c)^2}{\epsilon^2} f(x)\, dx,$$

where

$$R = \{x: |x - c| \geq \epsilon\}.$$

This integral is, in turn,

$$\leq \int_{-\infty}^{+\infty} \frac{(x - c)^2}{\epsilon^2} f(x)\, dx$$

which equals

$$\frac{1}{\epsilon^2} E[X - c]^2,$$

as was to be shown.

Notes: (a) It is important to realize that the above result is remarkable precisely because so little is assumed about the probabilistic behavior of the random variable X.

(b) As we might suspect, additional information about the distribution of the random variable X will enable us to improve on the inequality derived. For example, if $C = \frac{3}{2}$ we have, from Chebyshev's inequality,

$$P[|X - \mu| \geq \tfrac{3}{2}\sigma] \leq \tfrac{4}{9} = 0.44.$$

Suppose that we also *know* that X is uniformly distributed over $(1 - 1/\sqrt{3}, 1 + 1/\sqrt{3})$. Hence $E(X) = 1$, $V(X) = \frac{1}{9}$ and thus

$$P[|X - \mu| \geq \tfrac{3}{2}\sigma] = P[|X - 1| \geq \tfrac{1}{2}] = 1 - P[|X - 1| < \tfrac{1}{2}]$$

$$= 1 - P[\tfrac{1}{2} < X < \tfrac{3}{2}] = 1 - \frac{\sqrt{3}}{2} = 0.134.$$

Observe that although the statement obtained from Chebyshev's inequality is consistent with this result, the latter is a more precise statement. However, in many problems no assumption concerning the specific distribution of the random variable is justified, and in such cases Chebyshev's inequality can give us important information about the behavior of the random variable.

As we note from Eq. (7.21), if $V(X)$ is small, most of the probability distribution of X is "concentrated" near $E(X)$. This may be expressed more precisely in the following theorem.

Theorem 7.8. Suppose that $V(X) = 0$. Then $P[X = \mu] = 1$, where $\mu = E(X)$. (Informally, $X = \mu$, with "probability 1.")

Proof: From Eq. (7.20b) we find that

$$P[|X - \mu| \geq \epsilon] = 0 \qquad \text{for any } \epsilon > 0.$$

Hence

$$P[|X - \mu| < \epsilon] = 1 \qquad \text{for any } \epsilon > 0.$$

Since ϵ may be chosen arbitrarily small, the theorem is established.

Notes: (a) This theorem shows that zero variance does imply that all the probability is concentrated at a single point, namely at $E(X)$.

(b) If $E(X) = 0$, then $V(X) = E(X^2)$, and hence in this case, $E(X^2) = 0$ implies the same conclusion.

(c) It is in the above sense that we say that a random variable X is *degenerate*: It assumes only one value with probability 1.

7.9 The Correlation Coefficient

So far we have been concerned with associating parameters such as $E(X)$ and $V(X)$ with the distribution of one-dimensional random variables. These parameters measure, in a sense described previously, certain characteristics of the distribution. If we have a two-dimensional random variable (X, Y), an analogous problem is encountered. Of course, we may again discuss the one-dimensional random variables X and Y associated with (X, Y). However, the question arises whether there is a meaningful parameter which measures in some sense the "degree of association" between X and Y. This rather vague notion will be made precise shortly. We state the following formal definition.

Definition. Let (X, Y) be a two-dimensional random variable. We define ρ_{xy}, the *correlation coefficient*, between X and Y, as follows:

$$\rho_{xy} = \frac{E\{[X - E(X)][Y - E(Y)]\}}{\sqrt{V(X)V(Y)}}. \qquad (7.22)$$

Notes: (a) We assume that all the expectations exist and that both $V(X)$ and $V(Y)$ are nonzero. When there is no question as to which random variables are involved we shall simply write ρ instead of ρ_{xy}.

(b) The numerator of ρ, $E\{[X - E(X)][Y - E(Y)]\}$, is called the *covariance* of X and Y, and is sometimes denoted by σ_{xy}.

(c) The correlation coefficient is a dimensionless quantity.

(d) Before the above definition can be very meaningful we must discover exactly what ρ measures. This we shall do by considering a number of properties of ρ.

Theorem 7.9

$$\rho = \frac{E(XY) - E(X)E(Y)}{\sqrt{V(X)V(Y)}}.$$

Proof: Consider

$$
\begin{aligned}
E\{[X - E(X)][Y - E(Y)]\} &= E[XY - XE(Y) - YE(X) + E(X)E(Y)] \\
&= E(XY) - E(X)E(Y) - E(Y)E(X) + E(X)E(Y) \\
&= E(XY) - E(X)E(Y).
\end{aligned}
$$

Theorem 7.10. If X and Y are independent, then $\rho = 0$.

Proof: This follows immediately from Theorem 7.9, since

$$E(XY) = E(X)E(Y)$$

if X and Y are independent.

Note: The converse of Theorem 7.10 is in general *not* true. (See Problem 7.39.) That is, we may have $\rho = 0$, and yet X and Y need not be independent. If $\rho = 0$, we say that X and Y are *uncorrelated.* Thus, being uncorrelated and being independent are, in general, not equivalent. The following example illustrates this point.*

Let X and Y be any random variables having the *same* distribution. Let $U = X - Y$ and $V = X + Y$. Hence $E(U) = 0$ and $\text{cov}(U, V) = E[(X - Y)(X + Y)] = E(X^2 - Y^2) = 0$. Thus U and V are uncorrelated. Even if X and Y are independent, U and V may be dependent, as the following choice of X and Y indicates. Let X and Y be the numbers appearing on the first and second fair dice, respectively, which have been tossed. We now find, for example, that $P[V = 4 \mid U = 3] = 0$ (since if $X - Y = 3$, $X + Y$ cannot equal 4), while $P(V = 4) = 3/36$. Thus U and V are dependent.

Theorem 7.11. $-1 \le \rho \le 1$.   (That is, ρ assumes the values between -1 and $+1$ inclusive.)

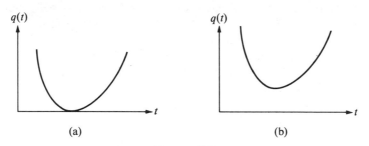

$$\text{(a)} \qquad\qquad\qquad\qquad\qquad\qquad \text{(b)}$$

FIGURE 7.8

Proof: Consider the following function of the real variable t:

$$q(t) = E[V + tW]^2,$$

where $V = X - E(X)$ and $W = Y - E(Y)$. Since $[V + tW]^2 \ge 0$, we have that $q(t) \ge 0$ for all t. Expanding, we obtain

$$q(t) = E[V^2 + 2tVW + t^2W^2] = E(V^2) + 2tE(VW) + t^2E(W^2).$$

Thus $q(t)$ is a quadratic expression in t. In general, if a quadratic expression $q(t) = at^2 + bt + c$ has the property that $q(t) \ge 0$ for all t, it means that its graph touches the t-axis at just one place or not at all, as indicated in Fig. 7.8. This, in turn, means that its discriminant $b^2 - 4ac$ must be ≤ 0, since $b^2 - 4ac > 0$ would mean that $q(t)$ has *two distinct* real roots. Applying this conclusion to the function $q(t)$ under consideration above, we obtain

$$4[E(VW)]^2 - 4E(V^2)E(W^2) \le 0.$$

---

* The example in this note is taken from a discussion appearing in an article entitled "Mutually Exclusive Events, Independence and Zero Correlation," by J. D. Gibbons, appearing in *The American Statistician,* **22**, No. 5, December 1968, pp. 31–32.

This implies

$$\frac{[E(VW)]^2}{E(V^2)E(W^2)} \le 1, \quad \text{and hence} \quad \frac{\{E[X - E(X)][Y - E(Y)]\}^2}{V(X)V(Y)} = \rho^2 \le 1.$$

Thus $-1 \le \rho \le 1$.

Theorem 7.12. Suppose that $\rho^2 = 1$. Then (with probability 1 in the sense of Theorem 7.8), $Y = AX + B$, where A and B are constants. In words: If the correlation coefficient ρ is ± 1, then Y is a linear function of X (with probability 1).

Proof: Consider again the function $q(t)$ described in the proof of Theorem 7.11. It is a simple matter to observe in the proof of that theorem that if $q(t) > 0$ for all t, then $\rho^2 < 1$. Hence the hypothesis of the present theorem, namely $\rho^2 = 1$, implies that there must exist at least one value of t, say t_0, such that $q(t_0) = E(V + t_0 W)^2 = 0$. Since $V + t_0 W = [X - E(X)] + t_0[Y - E(Y)]$, we have that $E(V + t_0 W) = 0$ and hence variance $(V + t_0 W) = E(V + t_0 W)^2$. Thus we find that the hypothesis of Theorem 7.12 leads to the conclusion that the variance of $(V + t_0 W) = 0$. Hence, from Theorem 7.8 we may conclude that the random variable $(V + t_0 W) = 0$ (with probability 1). Therefore $[X - E(X)] + t_0[Y - E(Y)] = 0$. Rewriting this, we find that $Y = AX + B$ (with probability 1), as was to be proved.

Note: The converse of Theorem 7.12 also holds as is shown in Theorem 7.13.

Theorem 7.13. Suppose that X and Y are two random variables for which $Y = AX + B$, where A and B are constants. Then $\rho^2 = 1$. If $A > 0$, $\rho = +1$; if $A < 0, \rho = -1$.

Proof: Since $Y = AX + B$, we have $E(Y) = AE(X) + B$ and $V(Y) = A^2 V(X)$. Also,

$$E(XY) = E[X(AX + B)] = AE(X^2) + BE(X).$$

Hence

$$\begin{aligned}
\rho^2 &= \frac{[E(XY) - E(X)E(Y)]^2}{V(X)V(Y)} \\
&= \frac{\{AE(X^2) + BE(X) - E(X)[AE(X) + B]\}^2}{V(X)A^2 V(X)} \\
&= \frac{[AE(X^2) + BE(X) - A(E(X))^2 - BE(X)]^2}{A^2 (V(X))^2} \\
&= \frac{A^2 \{E(X^2) - [E(X)]^2\}^2}{A^2 (V(X))^2} = 1.
\end{aligned}$$

(The second statement of the theorem follows by noting that $\sqrt{A^2} = |A|$.)

Note: Theorems 7.12 and 7.13 establish the following important characteristic of the correlation coefficient: The correlation coefficient is a measure of the *degree of linearity* between X and Y. Values of ρ near $+1$ or -1 indicate a high degree of linearity while values of ρ near 0 indicate a lack of such linearity. Positive values of ρ show that Y tends to increase with increasing X, while negative values of ρ show that Y tends to decrease with increasing values of X. There is considerable misunderstanding about the interpretation of the correlation coefficient. A value of ρ close to zero only indicates the absence of a *linear* relationship between X and Y. It does not preclude the possibility of some *nonlinear* relationship.

EXAMPLE 7.21.   Suppose that the two-dimensional random variable (X, Y) is uniformly distributed over the triangular region

$$R = \{(x, y) \mid 0 < x < y < 1\}.$$

(See Fig. 7.9.)   Hence the pdf is given as

$$f(x, y) = 2, \qquad (x, y) \in R,$$
$$ = 0, \qquad \text{elsewhere.}$$

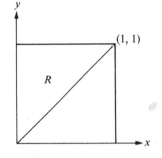

FIGURE 7.9

Thus the marginal pdf's of X and of Y are

$$g(x) = \int_x^1 (2)\, dy = 2(1 - x), \qquad 0 \le x \le 1;$$

$$h(y) = \int_0^y (2)\, dx = 2y, \qquad 0 \le y \le 1.$$

Therefore

$$E(X) = \int_0^1 x2(1 - x)\, dx = \tfrac{1}{3}, \qquad E(Y) = \int_0^1 y2y\, dy = \tfrac{2}{3};$$

$$E(X^2) = \int_0^1 x^2 2(1 - x)\, dx = \tfrac{1}{6}, \qquad E(Y^2) = \int_0^1 y^2 2y\, dy = \tfrac{1}{2};$$

$$V(X) = E(X^2) - [E(X)]^2 = \tfrac{1}{18}, \qquad V(Y) = E(Y^2) - [E(Y)]^2 = \tfrac{1}{18};$$

$$E(XY) = \int_0^1 \int_0^y xy2\, dx\, dy = \tfrac{1}{4}.$$

Hence

$$\rho = \frac{E(XY) - E(X)E(Y)}{\sqrt{V(X)V(Y)}} = \frac{1}{2}.$$

As we have noted, the correlation coefficient is a dimensionless quantity. Its value is not affected by a change of scale. The following theorem may easily be proved. (See Problem 7.41.)

Theorem 7.14.   If ρ_{XY} is the correlation coefficient between X and Y, and if $V = AX + B$ and $W = CY + D$, where A, B, C, and D are constants, then $\rho_{vw} = (AC/|AC|)\rho_{xy}$. (We suppose that $A \neq 0$, $C \neq 0$.)

7.10 Conditional Expectation

Just as we defined the expected value of a random variable X (in terms of its probability distribution) as $\int_{-\infty}^{+\infty} xf(x)\,dx$ or $\sum_{i=1}^{\infty} x_i p(x_i)$, so we can define the conditional expectation of a random variable (in terms of its conditional probability distribution) as follows.

> **Definition.** (a) If (X, Y) is a two-dimensional continuous random variable we define the *conditional expectation* of X for given $Y = y$ as
>
> $$E(X \mid y) = \int_{-\infty}^{+\infty} xg(x \mid y)\,dx. \tag{7.23}$$
>
> (b) If (X, Y) is a two-dimensional discrete random variable we define the conditional expectation of X for given $Y = y_j$ as
>
> $$E(X \mid y_j) = \sum_{i=1}^{\infty} x_i p(x_i \mid y_j). \tag{7.24}$$

The conditional expectation of Y for given X is defined analogously.

Notes: (a) The interpretation of conditional expectation is as follows. Since $g(x \mid y)$ represents the conditional pdf of X for given $Y = y$, $E(X \mid y)$ is the expectation of X conditioned on the event $\{Y = y\}$. For example, if (X, Y) represents the tensile strength and hardness of a specimen of steel, then $E(X \mid y = 52.7)$ is the expected tensile strength of a specimen of steel chosen at random from the population of specimens whose hardness (measured on the Rockwell scale) is 52.7.

(b) It is important to realize that in general $E(X \mid y)$ is a function of y and hence is a *random variable*. Similarly $E(Y \mid x)$ is a function of x and is also a random variable. [Strictly speaking, $E(X \mid y)$ is the *value* of the random variable $E(X \mid Y)$.]

(c) Since $E(Y \mid X)$ and $E(X \mid Y)$ are random variables, it will be meaningful to speak of *their* expectations. Thus we may consider $E[E(X \mid Y)]$, for instance. It is important to realize that the inner expectation is taken with respect to the conditional distribution of X given Y equals y, while the outer expectation is taken with respect to the probability distribution of Y.

Theorem 7.15

$$E[E(X \mid Y)] = E(X), \tag{7.25}$$
$$E[E(Y \mid X)] = E(Y). \tag{7.26}$$

Proof (continuous case only): By definition,

$$E(X \mid y) = \int_{-\infty}^{+\infty} xg(x \mid y)\,dx = \int_{-\infty}^{+\infty} x\,\frac{f(x, y)}{h(y)}\,dx,$$

where f is the joint pdf of (X, Y) and h is the marginal pdf of Y.

Hence

$$E[E(X \mid Y)] = \int_{-\infty}^{+\infty} E(X \mid y)h(y)\, dy = \int_{-\infty}^{+\infty} \left[\int_{-\infty}^{+\infty} x\, \frac{f(x, y)}{h(y)}\, dx \right] h(y)\, dy.$$

If all the expectations exist, it is permissible to write the above iterated integral with the order of integration reversed. Thus

$$E[E(X \mid Y)] = \int_{-\infty}^{+\infty} x \left[\int_{-\infty}^{+\infty} f(x, y)\, dy \right] dx = \int_{-\infty}^{+\infty} xg(x)\, dx = E(X).$$

[A similar argument may be used to establish Eq. (7.26).] This theorem is very useful as the following example illustrates.

EXAMPLE 7.22. Suppose that shipments involving a varying number of parts arrive each day. If N is the number of items in the shipment, the probability distribution of the random variable N is given as follows:

n:	10	11	12	13	14	15
$P(N=n)$:	0.05	0.10	0.10	0.20	0.35	0.20

The probability that any particular part is defective is the same for all parts and equals 0.10. If X is the number of defective parts arriving each day, what is the expected value of X? For *given* N equals n, X has a binomial distribution. Since N is itself a random variable, we proceed as follows.

We have $E(X) = E[E(X \mid N)]$. However, $E(X \mid N) = 0.10N$, since for given N, X has a binomial distribution. Hence

$$E(X) = E(0.10N) = 0.10E(N)$$
$$= 0.10[10(0.05) + 11(0.10) + 12(0.10) + 13(0.20) + 14(0.35) + 15(0.20)]$$
$$= 1.33.$$

Theorem 7.16. Suppose that X and Y are independent random variables. Then

$$E(X \mid Y) = E(X) \qquad \text{and} \quad E(Y \mid X) = E(Y).$$

Proof: See Problem 7.43.

EXAMPLE 7.23. Suppose that the power supply (kilowatts) to a hydroelectric company during a specified time period is a random variable X, which we shall assume to have a uniform distribution over $[10, 30]$. The demand for power (kilowatts), say Y, also is a random variable which we shall assume to be uniformly distributed over $[10, 20]$. (Thus, on the average, more power is supplied than is demanded since $E(X) = 20$, while $E(Y) = 15$.) For every kilowatt supplied, the company makes a profit of $0.03. If the demand exceeds the supply, the company gets additional power from another source making a profit on this power of $0.01 per kilowatt supplied. What is the expected profit during the specified time considered?

Let T be this profit. We have

$$T = 0.03\,Y \quad \text{if} \quad Y < X,$$
$$= 0.03X + 0.01(Y - X) \quad \text{if} \quad Y > X.$$

To evaluate $E(T)$ write it as $E[E(T \mid X)]$. We have

$$E(T \mid x) = \begin{cases} \int_{10}^{x} 0.03y\tfrac{1}{10}\,dy + \int_{x}^{20} (0.01y + 0.02x)\tfrac{1}{10}\,dy & \text{if} \quad 10 < x < 20, \\ \int_{10}^{20} 0.03y\tfrac{1}{10}\,dy & \text{if} \quad 20 < x < 30, \end{cases}$$

$$= \begin{cases} \tfrac{1}{10}[0.015x^2 - 1.5 + 2 + 0.4x - 0.005x^2 - 0.02x^2] \\ \qquad\qquad\qquad\qquad\qquad\qquad\qquad \text{if} \quad 10 < x < 20, \\ \tfrac{9}{20} \qquad \text{if} \quad 20 < x < 30, \end{cases}$$

$$= \begin{cases} 0.05 + 0.04x - 0.001x^2 & \text{if} \quad 10 < x < 20, \\ 0.45 & \text{if} \quad 20 < x < 30. \end{cases}$$

Therefore

$$E[E(T \mid X)] = \tfrac{1}{20}\int_{10}^{20} (0.05 + 0.04x - 0.001x^2)\,dx + \tfrac{1}{20}\int_{20}^{30} 0.45\,dx = \$0.43.$$

7.11 Regression of the Mean

As we have pointed out in the previous section, $E(X \mid y)$ is the value of the random variable $E(X \mid Y)$ and is a *function of y*. The graph of this function of y is known as the *regression curve* (of the mean) of X on Y. Analogously, the graph of the function of x, $E(Y \mid x)$ is called the regression curve (of the mean) of Y on X. For each fixed y, $E(X \mid y)$ is the expected value of the (one-dimensional) random variable whose probability distribution is defined by Eq. (6.5) or (6.7). (See Fig. 7.10.) In general, this expected value will depend on y. [Analogous interpretations may be made for $E(Y \mid x)$.]

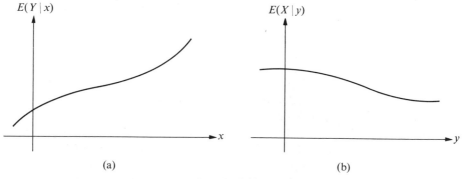

$E(Y \mid x)$

$E(X \mid y)$

(a)   (b)

FIGURE 7.10

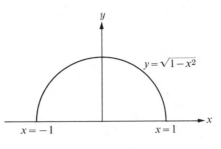

$$\text{FIGURE 7.11}$$

EXAMPLE 7.24. Suppose that (X, Y) is uniformly distributed over the semicircle indicated in Fig. 7.11. Then $f(x, y) = 2/\pi$, $(x, y) \in$ semicircle. Thus

$$g(x) = \int_0^{\sqrt{1-x^2}} \frac{2}{\pi} \, dy = \frac{2}{\pi} \sqrt{1 - x^2}, \qquad -1 \le x \le 1;$$

$$h(y) = \int_{-\sqrt{1-y^2}}^{\sqrt{1-y^2}} \frac{2}{\pi} \, dx = \frac{4}{\pi} \sqrt{1 - y^2}, \qquad 0 \le y \le 1.$$

Therefore

$$g(x \mid y) = \frac{1}{2\sqrt{1 - y^2}}, \qquad -\sqrt{1 - y^2} \le x \le \sqrt{1 - y^2};$$

$$h(y \mid x) = \frac{1}{\sqrt{1 - x^2}}, \qquad 0 \le y \le \sqrt{1 - x^2}.$$

Hence

$$E(Y \mid x) = \int_0^{\sqrt{1-x^2}} y h(y \mid x) \, dy$$

$$= \int_0^{\sqrt{1-x^2}} y \frac{1}{\sqrt{1 - x^2}} \, dy = \frac{1}{\sqrt{1 - x^2}} \frac{y^2}{2} \Big|_0^{\sqrt{1-x^2}} = \tfrac{1}{2}\sqrt{1 - x^2}.$$

Similarly

$$E(X \mid y) = \int_{-\sqrt{1-y^2}}^{+\sqrt{1-y^2}} x g(x \mid y) \, dx$$

$$= \int_{-\sqrt{1-y^2}}^{+\sqrt{1-y^2}} x \frac{1}{2\sqrt{1 - y^2}} \, dx = \frac{1}{2\sqrt{1 - y^2}} \frac{x^2}{2} \Big|_{-\sqrt{1-y^2}}^{+\sqrt{1-y^2}}$$

$$= 0.$$

It may happen that either or both of the regression curves are in fact straight lines (Fig. 7.12). That is, $E(Y \mid x)$ may be a *linear* function of x and/or $E(X \mid y)$ may be a linear function of y. In this case we say that the regression of the mean of Y on X (say) is linear.

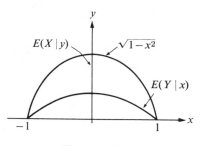

FIGURE 7.12

EXAMPLE 7.25. Suppose that (X, Y) is uniformly distributed over the triangle indicated in Fig. 7.13. Then $f(x, y) = 1$, $(x, y) \in T$. The following expressions for the marginal and conditional pdf's are easily verified:

$$g(x) = 2x, \quad 0 \le x \le 1; \qquad h(y) = \frac{2 - y}{2}, \quad 0 \le y \le 2.$$

$$g(x \mid y) = \frac{2}{2 - y}, \quad y/2 \le x \le 1; \qquad h(y \mid x) = \frac{1}{2x}, \quad 0 \le y \le 2x.$$

Thus $E(Y \mid x) = \int_0^{2x} yh(y \mid x)\, dy = \int_0^{2x} y(1/2x)\, dy = x$. Similarly,

$$E(X \mid y) = \int_{y/2}^{1} xg(x \mid y)\, dx = \int_{y/2}^{1} x\frac{2}{2 - y}\, dx = \frac{y}{4} + \frac{1}{2}.$$

Thus *both* the regression of Y on X and of X on Y are linear (Fig. 7.14).

It turns out that *if* the regression of the mean of Y on X is linear, say $E(Y \mid x) = \alpha x + \beta$, then we can easily express the coefficients α and β in terms of certain parameters of the joint distribution of (X, Y). We have the following theorem.

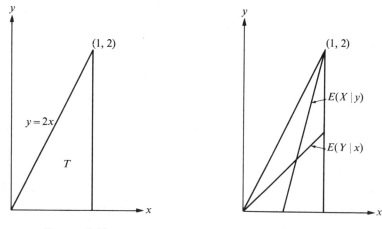

FIGURE 7.13   FIGURE 7.14

Theorem 7.17. Let (X, Y) be a two-dimensional random variable and suppose that

$$E(X) = \mu_x, \quad E(Y) = \mu_y, \quad V(X) = \sigma_x^2, \quad \text{and} \quad V(Y) = \sigma_y^2.$$

Let ρ be the correlation coefficient between X and Y. *If* the regression of Y on X is linear, we have

$$E(Y \mid x) = \mu_y + \rho \frac{\sigma_y}{\sigma_x} (x - \mu_x). \tag{7.27}$$

If the regression of X on Y is linear, we have

$$E(X \mid y) = \mu_x + \rho \frac{\sigma_x}{\sigma_y} (y - \mu_y). \tag{7.28}$$

Proof: The proof of this theorem is outlined in Problem 7.44.

Notes: (a) As is suggested by the above wording, it is possible that *one* of the regressions of the mean is linear while the other one is not.

(b) Note the crucial role played by the correlation coefficient in the above expressions. *If* the regression of X on Y, say, is linear, and if $\rho = 0$, then we find (again) that $E(X \mid y)$ does not depend on y. Also observe that the algebraic sign of ρ determines the sign of the slope of the regression line.

(c) If both regression functions are linear, we find, upon solving Eqs. (7.27) and (7.28) simultaneously, that the regression lines intersect at the "center" of the distribution, (μ_x, μ_y).

As we have noted (Example 7.23, for instance), the regression functions need not be linear. However we might still be interested in trying to *approximate* the regression curve with a linear function. This is usually done by appealing to the *principle of least squares*, which in the present context is as follows: Choose the constants a and b so that $E[E(Y \mid X) - (aX + b)]^2$ is minimized. Similarly, choose the constants c and d so that $E[E(X \mid Y) - (cY + d)]^2$ is minimized.

The lines $y = ax + b$ and $x = cy + d$ are called the *least-squares approximations* to the corresponding regression curves $E(Y \mid x)$ and $E(X \mid y)$, respectively. The following theorem relates these regression lines to those discussed earlier.

Theorem 7.18. If $y = ax + b$ is the least-squares approximation to $E(Y \mid x)$ and if $E(Y \mid x)$ *is* in fact a linear function of x, that is

$$E(Y \mid x) = a'x + b',$$

then $a = a'$ and $b = b'$. An analogous statement holds for the regression of X on Y.

Proof: See Problem 7.45.

PROBLEMS

7.1.   Find the expected value of the following random variables.
(a) The random variable X defined in Problem 4.1.
(b) The random variable X defined in Problem 4.2.
(c) The random variable T defined in Problem 4.6.
(d) The random variable X defined in Problem 4.18.

7.2.   Show that $E(X)$ does not exist for the random variable X defined in Problem 4.25.

7.3.   The following represents the probability distribution of D, the daily demand of a certain product. Evaluate $E(D)$.

$$d: \ 1, \ 2, \ 3, \ 4, \ 5,$$
$$P(D=d): \ 0.1, \ 0.1, \ 0.3, \ 0.3, \ 0.2.$$

7.4.   In the manufacture of petroleum, the distilling temperature, say T (degrees centigrade), is crucial in determining the quality of the final product. Suppose that T is considered as a random variable uniformly distributed over (150,300).
    Suppose that it costs C_1 dollars to produce one gallon of petroleum. If the oil distills at a temperature less than 200°C, the product is known as naphtha and sells for C_2 dollars per gallon. If it is distilled at a temperature greater than 200°C, it is known as refined oil distillate and sells for C_3 dollars per gallon. Find the expected net profit (per gallon).

7.5.   A certain alloy is formed by combining the melted mixture of two metals. The resulting alloy contains a certain percent of lead, say X, which may be considered as a random variable. Suppose that X has the following pdf:

$$f(x) \ = \ \tfrac{3}{5}10^{-5}x(100 \ - \ x), \qquad 0 \le x \le 100.$$

Suppose that P, the net profit realized in selling this alloy (per pound), is the following function of the percent content of lead: $P \ = \ C_1 + C_2X$. Compute the expected profit (per pound).

7.6.   Suppose that an electronic device has a life length X (in units of 1000 hours) which is considered as a continuous random variable with the following pdf:

$$f(x) \ = \ e^{-x}, \qquad x > 0.$$

Suppose that the cost of manufacturing one such item is $2.00. The manufacturer sells the item for $5.00, but guarantees a total refund if $X \le 0.9$. What is the manufacturer's expected profit per item?

7.7.   The first 5 repetitions of an experiment cost $10 each. All subsequent repetitions cost $5 each. Suppose that the experiment is repeated until the first successful outcome occurs. If the probability of a successful outcome always equals 0.9, and if the repetitions are independent, what is the expected cost of the entire operation?

7.8.   A lot is known to contain 2 defective and 8 nondefective items. If these items are inspected at random, one after another, what is the expected number of items that must be chosen *for inspection* in order to remove all the defective ones?

7.9. A lot of 10 electric motors must either be totally rejected or is sold, depending on the outcome of the following procedure: Two motors are chosen at random and inspected. If one or more are defective, the lot is rejected. Otherwise it is accepted. Suppose that each motor costs $75 and is sold for $100. If the lot contains 1 defective motor, what is the manufacturer's expected profit?

7.10. Suppose that D, the daily demand for an item, is a random variable with the following probability distribution:

$$P(D = d) = C2^d/d!, \qquad d = 1, 2, 3, 4.$$

(a) Evaluate the constant C.
(b) Compute the expected demand.
(c) Suppose that an item is sold for $5.00. A manufacturer produces K items daily. Any item which is not sold at the end of the day must be discarded at a loss of $3.00. (i) Find the probability distribution of the daily profit, as a function of K. (ii) How many items should be manufactured to maximize the expected daily profit?

7.11. (a) With $N = 50$, $p = 0.3$, perform some computations to find that value of k which minimizes $E(X)$ in Example 7.12.
(b) Using the above values of N and p and using $k = 5, 10, 25$, determine for each of these values of k whether "group testing" is preferable.

7.12. Suppose that X and Y are independent random variables with the following pdf's:

$$f(x) = 8/x^3, \quad x > 2; \qquad g(y) = 2y, \quad 0 < y < 1.$$

(a) Find the pdf of $Z = XY$.
(b) Obtain $E(Z)$ in two ways: (i) using the pdf of Z as obtained in (a). (ii) Directly, without using the pdf of Z.

7.13. Suppose that X has pdf

$$f(x) = 8/x^3, \qquad x > 2.$$

Let $W = \frac{1}{3}X$.
(a) Evaluate $E(W)$ using the pdf of W.
(b) Evaluate $E(W)$ without using the pdf of W.

7.14. A fair die is tossed 72 times. Given that X is the number of times six appears, evaluate $E(X^2)$.

7.15. Find the expected value and variance of the random variables Y and Z of Problem 5.2.

7.16. Find the expected value and variance of the random variable Y of Problem 5.3.

7.17. Find the expected value and variance of the random variables Y and Z of Problem 5.5.

7.18. Find the expected value and variance of the random variables Y, Z, and W of Problem 5.6.

7.19. Find the expected value and variance of the random variables V and S of Problem 5.7.

7.20.   Find the expected value and variance of the random variable Y of Problem 5.10 for each of the three cases.

7.21.   Find the expected value and variance of the random variable A of Problem 6.7.

7.22.   Find the expected value and variance of the random variable H of Problem 6.11.

7.23.   Find the expected value and variance of the random variable W of Problem 6.13.

7.24.   Suppose that X is a random variable for which $E(X) = 10$ and $V(X) = 25$. For what positive values of a and b does $Y = aX - b$ have expectation 0 and variance 1?

7.25.   Suppose that S, a random voltage, varies between 0 and 1 volt and is uniformly distributed over that interval. Suppose that the signal S is perturbed by an additive, independent random noise N which is uniformly distributed between 0 and 2 volts.

(a) Find the expected voltage of the signal, taking noise into account.

(b) Find the expected power when the perturbed signal is applied to a resistor of 2 ohms.

7.26.   Suppose that X is uniformly distributed over $[-a, 3a]$. Find the variance of X.

7.27.   A target is made of three concentric circles of radii $1/\sqrt{3}$, 1, and $\sqrt{3}$ feet. Shots within the inner circle count 4 points, within the next ring 3 points, and within the third ring 2 points. Shots outside the target count zero. Let R be the random variable representing the distance of the hit from the center. Suppose that the pdf of R is $f(r) = 2/\pi(1 + r^2)$, $r > 0$. Compute the expected value of the score after 5 shots.

7.28.   Suppose that the continuous random variable X has pdf

$$f(x) = 2xe^{-x^2}, \qquad x \geq 0.$$

Let $Y = X^2$. Evaluate $E(Y)$:

(a) directly without first obtaining the pdf of Y,

(b) by first obtaining the pdf of Y.

7.29.   Suppose that the two-dimensional random variable (X, Y) is uniformly distributed over the triangle in Fig. 7.15. Evaluate $V(X)$ and $V(Y)$.

7.30.   Suppose that (X, Y) is uniformly distributed over the triangle in Fig. 7.16.

(a) Obtain the marginal pdf of X and of Y.

(b) Evaluate $V(X)$ and $V(Y)$.

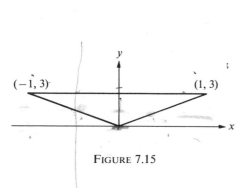

FIGURE 7.15

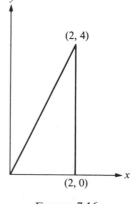

FIGURE 7.16

7.31.   Suppose that X and Y are random variables for which $E(X) = \mu_x$, $E(Y) = \mu_y$, $V(X) = \sigma_x^2$, and $V(Y) = \sigma_y^2$. Using Theorem 7.7, obtain an approximation for $E(Z)$ and $V(Z)$, where $Z = X/Y$.

7.32.   Suppose that X and Y are independent random variables, each uniformly distributed over $(1, 2)$. Let $Z = X/Y$.

(a) Using Theorem 7.7, obtain approximate expressions for $E(Z)$ and $V(Z)$.

(b) Using Theorem 6.5, obtain the pdf of Z and then find the exact value of $E(Z)$ and $V(Z)$. Compare with (a).

7.33.   Show that if X is a continuous random variable with pdf f having the property that the graph of f is symmetric about $x = a$, then $E(X) = a$, provided that $E(X)$ exists. (See Example 7.16.)

7.34.   (a) Suppose that the random variable X assumes the values -1 and 1 each with probability $\frac{1}{2}$. Consider $P[|X - E(X)| \geq k\sqrt{V(X)}]$ as a function of k, $k > 0$. Plot this function of k and, on the same coordinate system, plot the upper bound of the above probability as given by Chebyshev's inequality.

(b) Same as (a) except that $P(X = -1) = \frac{1}{3}$, $P(X = 1) = \frac{2}{3}$.

7.35.   Compare the upper bound on the probability $P[|X - E(X)| \geq 2\sqrt{V(X)}]$ obtained from Chebyshev's inequality with the exact probability if X is uniformly distributed over $(-1, 3)$.

7.36.   Verify Eq. (7.17).

7.37.   Suppose that the two-dimensional random variable (X, Y) is uniformly distributed over R, where R is defined by $\{(x, y) \mid x^2 + y^2 \leq 1, y \geq 0\}$. (See Fig. 7.17.) Evaluate ρ_{xy}, the correlation coefficient.

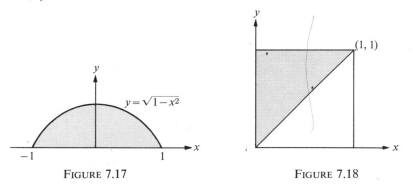

FIGURE 7.17   FIGURE 7.18

7.38.   Suppose that the two-dimensional random variable (X, Y) has pdf given by

$$f(x, y) = ke^{-y}, \qquad 0 < x < y < 1$$
$$= 0, \qquad \text{elsewhere.}$$

(See Fig. 7.18.) Find the correlation coefficient ρ_{xy}.

7.39.   The following example illustrates that $\rho = 0$ does not imply independence. Suppose that (X, Y) has a joint probability distribution given by Table 7.1.

(a) Show that $E(XY) = E(X)E(Y)$ and hence $\rho = 0$.

(b) Indicate why X and Y are not independent.

(c) Show that this example may be generalized as follows. The choice of the number $\frac{1}{8}$ is not crucial. What is important is that all the circled values are the same, all the boxed values are the same, and the center value equals zero.

TABLE 7.1

Y \ X	−1	0	1
−1	$\frac{1}{8}$	$\frac{1}{8}$	$\frac{1}{8}$
0	$\frac{1}{8}$	0	$\frac{1}{8}$
1	$\frac{1}{8}$	$\frac{1}{8}$	$\frac{1}{8}$

7.40. Suppose that A and B are two events associated with an experiment $\mathcal{E}$. Suppose that $P(A) > 0$ and $P(B) > 0$. Let the random variables X and Y be defined as follows.

$$X = 1 \text{ if } A \text{ occurs and } 0 \text{ otherwise,}$$
$$Y = 1 \text{ if } B \text{ occurs and } 0 \text{ otherwise.}$$

Show that $\rho_{xy} = 0$ implies that X and Y are independent.

7.41. Prove Theorem 7.14.

7.42. For the random variable (X, Y) defined in Problem 6.15, evaluate $E(X \mid y)$, $E(Y \mid x)$, and check that $E(X) = E[E(X \mid Y)]$ and $E(Y) = E[E(Y \mid X)]$.

7.43. Prove Theorem 7.16.

7.44. Prove Theorem 7.17. [*Hint:* For the continuous case, multiply the equation $E(Y \mid x) = Ax + B$ by $g(x)$, the pdf of X, and integrate from $-\infty$ to ∞. Do the same thing, using $xg(x)$ and then solve the resulting two equations for A and for B.]

7.45. Prove Theorem 7.18.

7.46. If X, Y, and Z are uncorrelated random variables with standard deviations 5, 12, and 9, respectively and if $U = X + Y$ and $V = Y + Z$, evaluate the correlation coefficient between U and V.

7.47. Suppose that both of the regression curves of the mean are in fact linear. Specifically, assume that $E(Y \mid x) = -\frac{3}{2}x - 2$ and $E(X \mid y) = -\frac{3}{5}y - 3$.

(a) Determine the correlation coefficient ρ.
(b) Determine $E(X)$ and $E(Y)$.

7.48. Consider weather forecasting with two alternatives: "rain" or "no rain" in the next 24 hours. Suppose that $p = \text{Prob(rain in next 24 hours)} > 1/2$. The forecaster scores 1 point if he is correct and 0 points if not. In making n forecasts, a forecaster with no ability whatsoever chooses at random r days $(0 \le r \le n)$ to say "rain" and the remaining $n - r$ days to say "no rain." His total point score is S_n. Compute $E(S_n)$ and $\text{Var}(S_n)$ and find that value of r for which $E(S_n)$ is largest. [*Hint:* Let $X_i = 1$ or 0 depending on whether the ith forecast is correct or not. Then $S_n = \sum_{i=1}^{n} X_i$. Note that the X_i's are *not* independent.]

8

The Poisson and Other Discrete Random Variables

8.1 The Poisson Distribution

As for deterministic models, in which certain functional relationships play an important role (such as linear, quadratic, exponential, trigonometric, etc.) so we find that, in constructing nondeterministic models for observable phenomena, certain probability distributions arise more often than others. One reason for this is that as in the deterministic case, certain relatively simple mathematical models seem to be capable of describing a very large class of phenomena.

In this chapter we shall discuss a number of discrete random variables in considerable detail. In the next chapter we shall do the same for continuous random variables.

Let us formally introduce the following random variable. Subsequently we shall indicate under what circumstances this random variable might represent the outcome of a random experiment.

Definition. Let X be a discrete random variable assuming the possible values: $0, 1, \ldots, n, \ldots$ If

$$P(X = k) = \frac{e^{-\alpha}\alpha^k}{k!}, \qquad k = 0, 1, \ldots, n, \ldots, \tag{8.1}$$

we say that X has a *Poisson distribution* with parameter $\alpha > 0$.

To check that the above represents a legitimate probability distribution, simply observe that $\sum_{k=0}^{\infty} P(X = k) = \sum_{k=0}^{\infty} (e^{-\alpha}\alpha^k/k!) = e^{-\alpha}e^{\alpha} = 1$.

Note: Since we are defining the random variable directly in terms of its range space and probability distribution, without reference to any underlying sample space S, we may suppose that the sample space S has been identified with R_X and that $X(s) = s$. That is, the outcomes of the experiment are simply the numbers $0, 1, 2, \ldots$ and the probabilities associated with each of these outcomes are given by Eq. (8.1).

Theorem 8.1. If X has a Poisson distribution with parameter α, then $E(X) = \alpha$ and $V(x) = \alpha$.

Proof
$$E(X) = \sum_{k=0}^{\infty} \frac{ke^{-\alpha}\alpha^k}{k!} = \sum_{k=1}^{\infty} \frac{e^{-\alpha}\alpha^k}{(k-1)!}.$$

Letting $s = k - 1$, we find that this becomes

$$E(X) = \sum_{s=0}^{\infty} \frac{e^{-\alpha}\alpha^{s+1}}{s!} = \alpha \sum_{s=0}^{\infty} \frac{e^{-\alpha}\alpha^s}{s!} = \alpha.$$

Similarly,

$$E(X^2) = \sum_{k=0}^{\infty} \frac{k^2 e^{-\alpha}\alpha^k}{k!} = \sum_{k=1}^{\infty} k \frac{e^{-\alpha}\alpha^k}{(k-1)!}.$$

Again letting $s = k - 1$, we obtain

$$E(X^2) = \sum_{s=0}^{\infty} (s+1) \frac{e^{-\alpha}\alpha^{s+1}}{s!} = \alpha \sum_{s=0}^{\infty} s \frac{e^{-\alpha}\alpha^s}{s!} + \alpha \sum_{s=0}^{\infty} \frac{e^{-\alpha}\alpha^s}{s!} = \alpha^2 + \alpha$$

[since the first sum represents $E(X)$ while the second sum equals one]. Hence

$$V(X) = E(X^2) - (E(X))^2 = \alpha^2 + \alpha - \alpha^2 = \alpha.$$

Note: Observe the interesting property which a Poisson random variable possesses: its expectation equals its variance.

8.2 The Poisson Distribution as an Approximation to the Binomial Distribution

The Poisson distribution plays a very important role in its own right as an appropriate probabilistic model for a large number of random phenomena. We shall discuss this in the next section. Here we shall be concerned with the importance of this distribution for approximating to the binomial probabilities.

EXAMPLE 8.1.   Suppose that telephone calls come into a large exchange and that in a particular three-hour period (180 minutes) a total of 270 calls have been received, or 1.5 calls per minute. Suppose we want, based on the above evidence, to compute the probability of receiving 0, 1, 2, etc. calls during the next three minutes.

In considering the phenomena of incoming calls we might reach the conclusion that at *any* instant a phone call is as likely to occur as at any other instant. That is, the probability remains constant from "time point" to "time point." The difficulty is that even in a very short time interval the number of points is not only infinite but cannot even be enumerated.   Thus we are led to a series of approximations which we shall now describe.

To begin with, we might consider subdividing the three-minute interval into nine subintervals of 20 seconds each. We might then treat each of these nine intervals as a Bernoulli trial during which we observed a call (success) or no call (failure) with $P(\text{success}) = (1.5) \; 20/60 = 0.5$. Thus we might be tempted to say that the probability of two calls during the three-minute interval (i.e., 2 successes in 9 trials with $P(\text{success}) = 0.5$) is equal to $\binom{9}{2}(1/2)^9 = 9/128$.

The trouble with this approximation is that we are ignoring the possibility of, say, two or three, etc., calls during one of our 20-second trial periods. If this possibility were taken into account, the above use of the binomial distribution would not be legitimate, for that distribution is applicable only when a dichotomy exists—a call or no call.

It is to avoid this difficulty that we turn to the next approximation and, in fact, are led to an entire sequence of approximations. One way of being fairly certain that at most one call is received at the exchange during a small time interval is to make that interval very short. Thus, instead of considering nine intervals of 20-second duration, let us consider next 18 intervals, each 10 seconds long. Now we may represent our experiment as 18 Bernoulli trials with $P(\text{success}) = P(\text{incoming call during subinterval}) = (1.5)10/60 = 0.25$. Hence $P(\text{two calls during the three-minute interval}) = \binom{18}{2}(0.25)^2(0.75)^{16}$. Note that although now we are dealing with a different binomial distribution than before (i.e., having parameters $n = 18$, $p = 0.25$ instead of $n = 9$, $p = 0.5$), the expected value np is the same, namely, $np = 18(0.25) = 9(0.5) = 4.5$.

If we continue in this manner, increasing the number of subintervals (i.e., n), we shall at the same time decrease the probability of an incoming call (i.e., p) in such a way that np remains constant.

Thus the preceding example leads us to ask the following question: What happens to the binomial probabilities $\binom{n}{k}p^k(1 - p)^{n-k}$ if $n \to \infty$ and $p \to 0$ in such a manner that np remains constant, say $np = \alpha$?

The following calculation yields the answer to this very important question. Consider the general expression for the binomial probability,

$$P(X = k) = \binom{n}{k}p^k(1 - p)^{n-k} = \frac{n!}{k!(n - k)!}p^k(1 - p)^{n-k}$$

$$= \frac{n(n - 1)(n - 2) \cdots (n - k + 1)}{k!}p^k(1 - p)^{n-k}.$$

Let $np = \alpha$. Hence $p = \alpha/n$, and $1 - p = 1 - \alpha/n = (n - \alpha)/n$. Replacing all terms involving p by their equivalent expression in terms of α, we obtain

$$P(X = k) = \frac{n(n - 1) \cdots (n - k + 1)}{k!}\left(\frac{\alpha}{n}\right)^k \left(\frac{n - \alpha}{n}\right)^{n-k}$$

$$= \frac{\alpha^k}{k!}\left[(1)\left(1 - \frac{1}{n}\right)\left(1 - \frac{2}{n}\right) \cdots \left(1 - \frac{k - 1}{n}\right)\right]\left[1 - \frac{\alpha}{n}\right]^{n-k}$$

$$= \frac{\alpha^k}{k!}\left[(1)\left(1 - \frac{1}{n}\right)\left(1 - \frac{2}{n}\right) \cdots \left(1 - \frac{k - 1}{n}\right)\right]$$

$$\times \left(1 - \frac{\alpha}{n}\right)^n \left(1 - \frac{\alpha}{n}\right)^{-k}.$$

Now let $n \to \infty$ in such a way that $np = \alpha$ remains fixed. This obviously means that $p \to 0$ as $n \to \infty$, for otherwise np could not remain constant. (Equivalently we could require that $n \to \infty$ and $p \to 0$ in such a way that $np \to \alpha$.)

In the above expression the terms of the form $(1 - 1/n)$, $(1 - 2/n)$, ... approach one as n approaches infinity, as does $(1 - \alpha/n)^{-k}$. It is well known (from the definition of the number e) that $(1 - \alpha/n)^n \to e^{-\alpha}$ as $n \to \infty$.

Thus, $\lim_{n \to \infty} P(X = k) = e^{-\alpha} \alpha^k / k!$. That is, in the limit we obtain the Poisson distribution with parameter α. Let us summarize this important result in the following theorem.

Theorem 8.2. Let X be a binomially distributed random variable with parameter p (based on n repetitions of an experiment). That is,

$$P(X = k) = \binom{n}{k} p^k (1 - p)^{n-k}.$$

Suppose that as $n \to \infty$, $np = \alpha$ (const), or equivalently, as $n \to \infty$, $p \to 0$ such that $np \to \alpha$. Under these conditions we have

$$\lim_{n \to \infty} P(X = k) = \frac{e^{-\alpha} \alpha^k}{k!},$$

the Poisson distribution with parameter α.

Notes: (a) The above theorem says, essentially, that we may approximate the binomial probabilities with the probabilities of the Poisson distribution whenever n is large and p is small.

(b) We have already verified that if X has a binomial distribution, $E(X) = np$, while if X has a Poisson distribution (with parameter α), $E(X) = \alpha$.

(c) The binomial distribution is characterized by two parameters, n and p, while the Poisson distribution is characterized by a single parameter, $\alpha = np$, which represents the expected number of successes per unit time (or per unit space in some other context). This parameter is also referred to as the *intensity* of the distribution. It is important to distinguish between the expected number of occurrences per *unit* time and the expected number of occurrences in the specified time. For instance, in Example 8.1 the intensity is 1.5 calls per minute and hence the expected number of calls in, say, a 10-minute period would be 15.

(d) We may also consider the following argument for evaluating the variance of a Poisson random variable X, with parameter α: X may be considered as a limiting case of a binomially distributed random variable Y with parameters n and p, where $n \to \infty$ and $p \to 0$ such that $np \to \alpha$. Since $E(Y) = np$ and $\text{Var}(Y) = np(1 - p)$, we observe that in the limit $\text{Var}(Y) \to \alpha$.

Extensive tables for the Poisson distribution are available. (E.C. Molina, *Poisson's Exponential Binomial Limit*, D. Van Nostrand Company, Inc., New York, 1942.) A brief tabulation of this distribution is given in the appendix.

Let us consider three additional examples, illustrating the previously mentioned application of the Poisson distribution.

EXAMPLE 8.2.   At a busy traffic intersection the probability p of an individual car having an accident is very small, say $p = 0.0001$. However, during a certain part of the day, say between 4 p.m. and 6 p.m., a large number of cars pass through the intersection, say 1000. Under these conditions, what is the probability of two or more accidents occurring during that period?

Let us make a few assumptions. Suppose that the above value of p is the same for each car. Secondly, suppose that whether a car does or does not have an accident does not depend on what happens to any other car. (This assumption is obviously not realistic but we shall make it nevertheless.) Thus we may assume that if X is the number of accidents among the 1000 cars which arrive, then X has a binomial distribution with $p = 0.0001$. (Another assumption, not explicitly stated, is that n, the number of cars passing through the intersection between 4 p.m. and 6 p.m., is predetermined at 1000. Obviously, a more realistic approach would be to consider n itself as a random variable whose value depends on a random mechanism. However, we shall not do so here, but shall consider n as fixed.) Hence we can obtain the exact value of the sought-after probability:

$$P(X \geq 2) = 1 - P(X = 0) - P(X = 1)$$
$$= 1 - (0.9999)^{1000} - 1000(0.0001)(0.9999)^{999}.$$

The evaluation of the above numbers gives rise to considerable difficulty. Since n is large and p small, we shall apply Theorem 8.2 and obtain the following approximation:

$$P(X = k) \cong \frac{e^{-0.1}(0.1)^k}{k!}.$$

Hence,

$$P(X \geq 2) \cong 1 - e^{-0.1}(1 + 0.1) = 0.0045.$$

EXAMPLE 8.3.   Suppose that a manufacturing process turns out items in such a way that a certain (constant) proportion of items, say p, are defective. If a lot of n such items is obtained, the probability of obtaining exactly k defectives may be computed from the binomial distribution as $P(X = k) = \binom{n}{k}p^k(1 - p)^{n-k}$, where X is the number of defectives in the lot. If n is large and p is small (as is often the case), we may approximate the above probability by

$$P(X = k) \simeq \frac{e^{-np}(np)^k}{k!}.$$

Suppose, for instance, that a manufacturer produces items of which about 1 in 1000 are defective. That is, $p = 0.001$. Hence using the binomial distribution, we find that in a lot of 500 items the probability that none of the items are defective is $(0.999)^{500} = 0.609$. If we apply the Poisson approximation, this probability may be written as $e^{-0.5} = 0.61$. The probability of finding 2 or more defective items is, according to the Poisson approximation, $1 - e^{-0.5}(1 + 0.5) = 0.085$.

EXAMPLE 8.4. [Suggested by a discussion in A. Renyi's *Calculus of Probability* (in German), VEB Deutscher Verlag der Wissenschaft, Berlin, 1962.]

In the manufacture of glass bottles, small, hard particles are found in the molten glass from which the bottles are produced. If such a single particle appears in a bottle, the bottle cannot be used and must be discarded. The particles may be assumed to be randomly scattered in the molten glass. We shall assume that the molten glass is produced in such a way that the number of particles is (on the average) the same for a constant quantity of molten glass. Assume in particular that in 100 kg of molten glass, x such particles are found and that 1 kg of molten glass is required to make one such bottle.

Question: What percentage of the bottles will have to be discarded because they are defective? At first glance the "solution" to this problem might be as follows. Since the material for 100 bottles contains x particles, there will be approximately x percent of the bottles that must be discarded. A bit of reflection will indicate, however, that the above solution is not correct. For, a defective bottle may have more than 1 particle, thus lowering the percentage of defective bottles obtained from the remaining material.

In order to obtain a "correct" solution, let us make the following simplifying assumptions: (a) Every particle may appear in the material of every bottle with equal probability, and (b) the distribution of any particular particle is independent of any other particular particle. With these assumptions we may reduce our problem to the following "urn" model. Among N urns, n balls are distributed at random. What is the probability that in a randomly chosen urn, exactly k balls are found? (The urns obviously correspond to the bottles, while the balls correspond to the particles.)

Letting Z be the number of balls found in a randomly chosen urn, it follows from the above assumptions, that Z is binomially distributed with parameter $1/N$. Hence

$$P(Z = k) = \binom{n}{k}\left(\frac{1}{N}\right)^k\left(1 - \frac{1}{N}\right)^{n-k}.$$

Suppose now that the molten glass is prepared in very large quantities. In fact, suppose that it is prepared in units of 100 kg and that M such units have been supplied. Hence $N = 100M$ and $n = xM$. Let $\alpha = x/100$, which equals the proportion of particles per bottle. Thus $N = n/\alpha$ and the above probability may be written as

$$P(Z = k) = \binom{n}{k}\left(\frac{\alpha}{n}\right)^k\left(1 - \frac{\alpha}{n}\right)^{n-k}.$$

Thus, as the production process continues (that is, $M \to \infty$ and hence $n \to \infty$), we obtain

$$P(Z = k) \simeq \frac{e^{-\alpha}\alpha^k}{k!} \qquad \text{where} \quad \alpha = \frac{x}{100} \, .$$

Let us compute the probability that a bottle must be discarded. This is equal to $1 - P(Z = 0)$. Hence P (defective bottle) $\simeq 1 - e^{-x/100}$. If the number of bottles produced is quite large, we may essentially identify the probability of a defective bottle with the relative frequency of defective bottles. Therefore the percentage of defective bottles is approximately $100(1 - e^{-x/100})$. If we expand $100(1 - e^{-x/100})$ in a Maclaurin series, we obtain

$$100\left[1 - \left(1 - \frac{x}{100} + \frac{x^2}{2(100)^2} - \frac{x^3}{3!(100)^3} + \cdots\right)\right]$$

$$= x - \frac{x^2}{2(100)} + \frac{x^3}{6(100)^2} - \cdots$$

Thus, if x is small, the proportion of discarded bottles *is* approximately x, as we first suggested. However, for large x this does not hold anymore. In the case in which $x = 100$, the percentage of discarded bottles is *not* 100, but rather $100(1 - e^{-1}) = 63.21$ percent. This is, of course, an extreme case and would not be encountered in a reasonably controlled process. Suppose that $x = 30$ (a more realistic number). Then, instead of discarding 30 percent (our initial solution again), we would discard only $100(1 - e^{-0.3}) = 25.92$ percent. We might note that if x is reasonably large, it is more economical to produce smaller bottles. For example, if we require only 0.25 kg of molten glass per bottle instead of 1 kg, and if $x = 30$, then the percent discarded is reduced from 25.92 percent to 7.22 percent.

8.3 The Poisson Process

In the previous section the Poisson distribution was used as a means for approximating a known distribution, namely the binomial. However, the Poisson distribution plays an extremely important role in its own right, since it represents an appropriate probabilistic model for a large number of observational phenomena.

Although we shall not give a completely rigorous derivation of some of the results we are about to discuss, the general approach is of such importance that it will be worthwhile to understand it, even though we may not justify every step.

In order to refer to a specific example as we go through the mathematical details, let us consider a source of radioactive material which is emitting α-particles. Let X_t be defined as the number of particles emitted during a specified time period $[0, t)$. We shall make some assumptions about the (discrete) random variable X_t which will enable us to determine the probability distribution of X_t. The plausibility of these assumptions (recalling what X_t represents) is substantiated by the fact that empirical evidence supports to a very considerable degree the theoretical results we shall derive.

It might be worthwhile to point out that in deducing any mathematical result, we must accept some underlying postulates or axioms. In seeking axioms to describe observational phenomena, some axioms may be far more plausible (and less arbitrary) than others. For example, in describing the motion of an object propelled upward with some initial velocity, we might assume that the distance above the ground, say s, is a quadratic function of time t; that is, $s = at^2 + bt + c$. This would hardly be a very intuitive assumption, in terms of our experience. Instead, we could assume that the acceleration is a constant and then *deduce* from this that s *must* be a quadratic function of t. The point is, of course, that if we must assume something in order to construct our mathematical model, we would prefer to assume that which is plausible, rather than that which is less so.

The same point of view guides us here in constructing a probabilistic model for the emission of α-particles from a radioactive source. The random variable X_t defined above may assume the values $0, 1, 2, \ldots$ Let $p_n(t) = P[X_t = n], n = 0, 1, 2, \ldots$

The following five *assumptions* will now be made.

$\mathbf{A_1}$: The number of particles emitted during *nonoverlapping* time intervals are *independent* random variables.

$\mathbf{A_2}$: If X_t is defined as above and if Y_t equals the number of particles emitted during $[t_1, t_1 + t)$, then for any $t_1 > 0$, the random variables X_t and Y_t have the same probability distribution. (In other words, the *distribution* of the number of particles emitted during any interval depends only on the *length* of that interval and not on the endpoints.)

$\mathbf{A_3}$: $p_1(\Delta t)$ is approximately equal to $\lambda \Delta t$, if Δt is sufficiently small, where λ is a positive constant. We shall write this as $p_1(\Delta t) \sim \lambda \Delta t$. Throughout *this* section $a(\Delta t) \sim b(\Delta t)$ means that $a(\Delta t)/b(\Delta t) \to 1$ as $\Delta t \to 0$. We shall also suppose that $\Delta t > 0$. (This assumption says that if the interval is sufficiently small, the probability of obtaining exactly one emission during that interval is directly proportional to the length of that interval.)

$\mathbf{A_4}$: $\sum_{k=2}^{\infty} p_k(\Delta t) \to 0$. (This implies that $p_k(\Delta t) \sim 0, k \geq 2$.) This says that the probability of obtaining two or more emissions during a sufficiently small interval is negligible.

$\mathbf{A_5}$: $X_0 = 0$, or equivalently, $p_0(0) = 1$. This amounts to an initial condition for the model we are describing.

As we shall shortly demonstrate, the five assumptions listed above will make it possible for us to derive an expression for $p_n(t) = P[X_t = n]$. Let us now draw a number of conclusions from the above assumptions.

(a) Assumptions $\mathbf{A_1}$ and $\mathbf{A_2}$ together imply that the random variables X_t and $[X_{t+\Delta t} - X_t]$ are independent random variables with the *same* probability distribution. (See Fig. 8.1.)

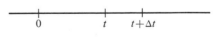

FIGURE 8.1

(b) From Assumptions A_3 and A_4 we may conclude that

$$p_0(\Delta t) = 1 - p_1(\Delta t) - \sum_{k=2}^{\infty} p_k(\Delta t) \sim 1 - \lambda t + p(\Delta t) \quad \text{as } t \to 0. \qquad (8.2)$$

(c) We may write

$$
\begin{aligned}
p_0(t + \Delta t) &= P[X_{t+\Delta t} = 0] \\
&= P[X_t = 0 \quad \text{and} \quad (X_{t+\Delta t} - X_t) = 0] \\
&= p_0(t)p_0(\Delta t). \qquad \text{[See conclusion (a).]} \\
&\sim p_0(t)[1 - \lambda \Delta t]. \qquad \text{[See Eq. (8.2).]}
\end{aligned}
$$

(d) Hence we have

$$\frac{p_0(t + \Delta t) - p_0(t)}{\Delta t} \sim -\lambda p_0(t).$$

Letting $\Delta t \to 0$, and observing that the left-hand side represents the difference quotient of the function p_0 and hence approaches $p_0'(t)$ (more properly, the right-hand derivative, since $\Delta t > 0$), we have

$$p_0'(t) = -\lambda p_0(t) \quad \text{or, equivalently,} \quad \frac{p_0'(t)}{p_0(t)} = -\lambda.$$

Integrating both sides with respect to t yields $\ln p_0(t) = -\lambda t + C$, where C is a constant of integration. From Assumption A_5 we find, by letting $t = 0$, that $C = 0$. Hence

$$p_0(t) = e^{-\lambda t}. \qquad (8.3)$$

Thus our assumptions have led us to an expression for $P[X_t = 0]$. Using essentially the same approach, we shall now obtain $p_n(t)$ for $n \geq 1$.

(e) Consider $p_n(t + \Delta t) = P[X_{t+\Delta t} = n]$.

Now $X_{t+\Delta t} = n$ if and only if $X_t = x$ and $[X_{t+\Delta t} - X_t] = n - x$, $x = 0, 1, 2, \ldots, n$. Using Assumptions A_1 and A_2, we have

$$p_n(t + \Delta t) = \sum_{x=0}^{n} p_x(t)p_{n-x}(\Delta t)$$

$$= \sum_{x=0}^{n-2} p_x(t)p_{n-x}(\Delta t) + p_{n-1}(t)p_1(\Delta t) + p_n(t)p_0(\Delta t).$$

Using Assumptions A_3 and A_4 and also Eq. (8.2), we obtain

$$p_n(t + \Delta t) \sim p_{n-1}(t)\lambda \Delta t + p_n(t)[1 - \lambda \Delta t].$$

Hence

$$\frac{p_n(t + \Delta t) - p_n(t)}{\Delta t} \sim \lambda p_{n-1}(t) - \lambda p_n(t).$$

Again letting $\Delta t \to 0$, and again observing that the left-hand side represents the difference quotient of the function p_n we obtain

$$p_n'(t) = -\lambda p_n(t) + \lambda p_{n-1}(t), \qquad n = 1, 2, \ldots$$

This represents an infinite system of (linear) difference differential equations. The interested reader may verify that if we define the function q_n by the relation $q_n(t) = e^{\lambda t} p_n(t)$, the above system becomes $q_n'(t) = \lambda q_{n-1}(t), n = 1, 2, \ldots$ Since $p_0(t) = e^{-\lambda t}$, we find that $q_0(t) = 1$. [Also note that $q_n(0) = 0$ for $n > 0$.] Thus we obtain, recursively,

$$q_1'(t) = \lambda, \qquad\qquad \text{and hence} \quad q_1(t) = \lambda t;$$

$$q_2'(t) = \lambda q_1(t) = \lambda^2 t, \qquad \text{and hence} \quad q_2(t) = \frac{(\lambda t)^2}{2}.$$

In general, $q_n'(t) = \lambda q_{n-1}(t)$ and hence $q_n(t) = (\lambda t)^n/n!$ Recalling the definition of q_n, we finally obtain

$$p_n(t) = e^{-\lambda t}(\lambda t)^n/n!, \qquad n = 0, 1, 2, \ldots \tag{8.4}$$

We have thus shown that the number of particles emitted during the time interval $[0, t)$ from a radioactive source, subject to the assumptions made above, is a random variable, with Poisson distribution with parameter (λt).

Notes: (a) It is important to realize that the Poisson distribution appeared as a *consequence* of certain assumptions we made. This means that whenever these assumptions are valid (or at least approximately so), the Poisson distribution should be used as an appropriate model. It turns out that there is a large class of phenomena for which the Poisson model is appropriate.

(i) Let X_t represent the number of telephone calls arriving at a telephone switchboard during a time period of length t. The above assumptions are approximately satisfied, particularly during the "busy period" of the day. Hence X_t has a Poisson distribution.

(ii) Let X_t represent the number of electrons released from the cathode of a vacuum tube. Again the above assumptions are appropriate, and hence X_t has a Poisson distribution.

(iii) The following example (from astronomy) indicates that the above reasoning may be applied not only to the number of occurrences of some event during a fixed *time* period, but also to the number of occurrences of an event in the confines of a fixed *area* or *volume*. Suppose that an astronomer investigates a portion of the Milky Way and assume that in the portion considered, the density of stars, say λ, is constant. (This means that in a volume of, say V (cubic units), one would find, on the average, λV stars.) Let X_V be equal to the number of stars found in a portion of the Milky Way, having volume V. If the above assumptions are fulfilled (with "volume" replacing "time"), then $P[X_V = n] = (\lambda V)^n e^{-\lambda V}/n!$. (The assumptions, interpreted in the present context, would essentially state that the number of stars appearing in non-overlapping portions of the sky represent independent random variables and that the probability of more than one star appearing in a very small portion of the sky is zero.)

(iv) Another application, from the field of biology, comes to mind if we let X_A be the number of blood cells visible under the microscope, where the visible surface area under the microscope is given by A square units.

(b) The constant λ originally appeared as a constant of proportionality in Assumption A_3. The following interpretations of λ are worthwhile to note: If X_t represents the number of occurrences of some event during a time interval of length t, then $E(X_t) = \lambda t$, and hence $\lambda = [E(X_t)]/t$ represents the expected *rate* at which particles are emitted. If X_V represents the number of occurrences of some event within a specified volume V, then $E(X_V) = \lambda V$, and hence $\lambda = [E(X_V)]/V$ represents the expected *density* at which stars appear.

(c) It is important to realize that our discussion in Section 8.3 did not deal with just *a* random variable X possessing a Poisson distribution. Rather, for every $t > 0$, say, we found that X_t had a Poisson distribution with a parameter depending on t. Such an (infinite) collection of random variables is also known as a *Poisson process*. (Equivalently, a Poisson process is generated whenever an event occurs in some time interval such that assumptions A_1 through A_5 are satisfied.) In an analogous way we can define a *Bernoulli process*: If $X_1, X_2, \ldots, X_n, \ldots$ are the number of occurrences of success in $1, 2, \ldots, n, \ldots$ Bernoulli trials, then the collection of random variables $X_1, \ldots X_n, \ldots$ is called a Bernoulli process.

EXAMPLE 8.5. A complicated piece of machinery, when running properly, can bring a profit of C dollars per hour ($C > 2$) to a firm. However, this machine has a tendency to break down at unexpected and unpredictable times. Suppose that the number of breakdowns during any period of length t hours is a random variable with a Poisson distribution with parameter t. If the machine breaks down x times during the t hours, the loss incurred (shutdown of machine plus repair) equals $(x^2 + x)$ dollars. Thus the total profit P during any period of t hours equals $P = Ct - (X^2 + X)$, where X is the random variable representing the number of breakdowns of the machine. Therefore P is a random variable, and it might be of interest to choose t (which is at our disposal) in such a manner that the *expected profit* is maximized. We have

$$E(P) = Ct - E(X^2 + X).$$

From Theorem 8.1 we find that $E(X) = t$ and $E(X^2) = t + (t)^2$. It then follows that $E(P) = Ct - 2t - t^2$. To find the value of t for which $E(P)$ is maximized, we differentiate $E(P)$ and set the resulting expression equal to zero. We obtain $C - 2 - 2t = 0$, yielding $t = \frac{1}{2}(C - 2)$ hours.

EXAMPLE 8.6. Let X_t be equal to the number of particles emitted from a radioactive source during a time interval of length t. Assume that X_t has a Poisson distribution with parameter αt. A counting device is set up to keep count of the number of particles emitted. Suppose that there is a constant probability p that any emitted particle is not counted. If R_t equals the number of particles counted during the specified interval, what is the probability distribution of R_t?

For *given* $X_t = x$, the random variable R_t has a binomial distribution based on x repetitions with parameter $(1 - p)$. That is,

$$P(R_t = k \mid X_t = x) = \binom{x}{k}(1 - p)^k p^{x-k}.$$

Using the formula for total probability (Eq. 3.4), we have

$$P(R_t = k) = \sum_{x=k}^{\infty} P(R_t = k \mid X_t = x)P(X_t = x)$$

$$= \sum_{x=k}^{\infty} \binom{x}{k}(1 - p)^k p^{x-k} e^{-\alpha t}(\alpha t)^x/x!$$

$$= \left(\frac{1 - p}{p}\right)^k \frac{e^{-\alpha t}}{k!} \sum_{x=k}^{\infty} \frac{1}{(x - k)!}(p\alpha t)^x.$$

Let $i = x - k$. Then

$$P(R_t = k) = \left(\frac{1 - p}{p}\right)^k \frac{e^{-\alpha t}}{k!} \sum_{i=0}^{\infty} \frac{(p\alpha t)^{i+k}}{i!}$$

$$= \left(\frac{1 - p}{p}\right)^k \frac{e^{-\alpha t}}{k!} (p\alpha t)^k e^{p\alpha t}$$

$$= \frac{e^{-\alpha(1-p)t}[(1 - p)\alpha t]^k}{k!}.$$

Thus we find that R_t also has a Poisson distribution with parameter $(1 - p)\alpha t$.

8.4 The Geometric Distribution

Suppose that we perform an experiment $\mathcal{E}$ and are concerned only about the occurrence or nonoccurrence of some event A. Assume, as in the discussion of the binomial distribution, that we perform $\mathcal{E}$ repeatedly, that the repetitions are independent, and that on each repetition $P(A) = p$ and $P(\bar{A}) = 1 - p = q$ remain the same. Suppose that we repeat the experiment until A occurs for the first time. (Here we depart from the assumptions leading to the binomial distribution. There the number of repetitions was predetermined, whereas here it is a random variable.)

Define the random variable X as the number of repetitions required up to and including the first occurrence of A. Thus X assumes the possible values 1, 2, . . . Since $X = k$ if and only if the first $(k - 1)$ repetitions of $\mathcal{E}$ result in $\bar{A}$ while the kth repetition results in A, we have

$$P(X = k) = q^{k-1}p, \qquad k = 1, 2, \ldots \tag{8.5}$$

A random variable with probability distribution Eq. (8.5) is said to have a *geometric* distribution.

An easy calculation shows that Eq. (8.5) defines a legitimate probability distribution. We obviously have $P(X = k) \geq 0$. And

$$\sum_{k=1}^{\infty} P(X = k) = p(1 + q + q^2 + \cdots) = p\left[\frac{1}{1 - q}\right] = 1.$$

We can obtain the expected value of X as follows.

$$E(X) = \sum_{k=1}^{\infty} kpq^{k-1} = p \sum_{k=1}^{\infty} \frac{d}{dq} q^k$$

$$= p \frac{d}{dq} \sum_{k=1}^{\infty} q^k = p \frac{d}{dq}\left[\frac{q}{1 - q}\right] = \frac{1}{p}.$$

(The interchange of differentiation and summation is justified here since the series converges for $|q| < 1$.) A similar computation shows that $V(X) = q/p^2$. (We shall derive both results again in Chapter 10, using a different approach.) Summarizing the above, we have the following theorem.

Theorem 8.3. If X has a geometric distribution as given by Eq. (8.5),

$$E(X) = 1/p \qquad \text{and} \qquad V(X) = q/p^2.$$

Note: The fact that $E(X)$ is the reciprocal of p is intuitively appealing, since it says that small values of $p = P(A)$ require many repetitions in order to have A occur.

EXAMPLE 8.7. Suppose that the cost of performing an experiment is $1000. If the experiment fails, an additional cost of $300 occurs because of certain changes that have to be made before the next trial is attempted. If the probability of success on any given trial is 0.2, if the individual trials are independent, and if the experiments are continued until the first successful result is achieved, what is the expected cost of the entire procedure?

If C is the cost, and X is the number of trials required to achieve success, we have $C = 1000X + 300(X - 1) = 1300X - 300$. Hence

$$E(C) = 1300E(X) - 300 = 1300 \frac{1}{0.2} - 300 = \$6200.$$

EXAMPLE 8.8. In a certain locality the probability that a thunderstorm will occur on any given day during the summer (say July and August) equals 0.1. Assuming independence from day to day, what is the probability that the first thunderstorm of the summer season occurs on August 3?

We let X be the number of days (starting with July 1) until the first thunderstorm and we require $P(X = 34)$ which equals $(0.9)^{33}(0.1) = 0.003$.

EXAMPLE 8.9. If the probability that a certain test yields a "positive" reaction equals 0.4, what is the probability that fewer than 5 "negative" reactions occur before the first positive one? Letting Y be the number of negative reactions before the first positive one, we have

$$P(Y = k) = (0.6)^k(0.4), \qquad k = 0, 1, 2, \ldots$$

Hence

$$P(Y < 5) = \sum_{k=0}^{4} (0.6)^k(0.4) = 0.92.$$

Note: If X has a geometric distribution as described in Eq. (8.5) and if we let $Z = X - 1$, we may interpret Z as the number of failures preceding the first success. We have $P(Z = k) = q^k p, \, k = 0, 1, 2, \ldots$, where $p = P$ (success) and $q = P$ (failure).

The geometric distribution has an interesting property which is summarized in the following theorem.

Theorem 8.4. Suppose that X has a geometric distribution given by Eq. (8.5). Then for any two positive integers s and t,

$$P(X \geq s + t \mid X > s) = P(X > t). \tag{8.6}$$

Proof: See Problem 8.18.

Notes: (a) The above theorem states that the geometric distribution has "no memory" in the following sense. Suppose that the event A has *not* occurred during the first s repetitions of $\mathcal{E}$. Then the probability that it will not occur during the *next* t repetitions is the same as the probability that it will not occur during the *first* t repetitions. In other words, the information of no successes is "forgotten" so far as subsequent calculations are concerned.

(b) The converse of the above theorem is true also: If Eq. (8.6) holds for a random variable assuming only positive integer values, then the random variable *must* have a geometric distribution. (We shall not prove this here. A discussion may be found in Feller's *An Introduction to Probability Theory and Its Applications*, John Wiley and Sons, Inc., 2nd Edition, New York, 1957, p. 304.)

(c) We shall note in the next chapter that there is a continuous random variable with a distribution possessing the analogous property to Eq. (8.6), namely the exponential distribution.

EXAMPLE 8.10. Suppose that an item is inspected at the end of each day to see whether it is still functioning properly. Let $p = P$ [failure during any specified day]. Hence if X is the number of inspections required to obtain the first failure, X has a geometric probability distribution and we have $P(X = n) = (1 - p)^{n-1}p$. Equivalently, $(1 - p)^{n-1}p = P$ [item will be found to have failed at the nth inspection and not at the $(n - 1)$-inspection.]

The maximum value of this probability is obtained by solving

$$\frac{d}{dp} P(X = n) = 0.$$

This yields

$$p(n - 1)(1 - p)^{n-2}(-1) + (1 - p)^{n-1} = 0,$$

which is equivalent to

$$(1 - p)^{n-2}[(1 - p) - (n - 1)p] = 0,$$

from which we obtain $p = 1/n$.

8.5 The Pascal Distribution

An obvious generalization to the geometric distribution arises if we ask the following question. Suppose that an experiment is continued until a particular event A occurs for the rth time. If

$$P(A) = p, \qquad P(\overline{A}) = q = 1 - p$$

on each repetition, we define the random variable Y as follows.

Y is the number of repetitions needed in order to have A occur exactly r times.

We require the probability distribution of Y. It should be clear that if $r = 1$, Y has the geometric distribution given by Eq. (8.5).

Now $Y = k$ if and only if A occurs on the kth repetition *and* A occurred precisely $(r - 1)$ times in the previous $(k - 1)$ repetitions. The probability of this event is simply $p\binom{k-1}{r-1}p^{r-1}q^{k-r}$, since what happens on the first $(k - 1)$ repetitions is independent of what happens on the kth repetition. Hence

$$P(Y = k) = \binom{k - 1}{r - 1}p^{r}q^{k-r}, \qquad k = r, r + 1, \ldots \qquad (8.7)$$

It is easily seen that for $r = 1$, the above reduces to Eq. (8.5). A random variable having a probability distribution given by Eq. (8.7) has a *Pascal* distribution.

Note: The preceding Pascal distribution is also commonly known as the *negative binomial* distribution. The reason for this is that in checking the condition

$$\sum_{k=r}^{\infty} P(Y = k) = 1$$

we obtain

$$\sum_{k=r}^{\infty} \binom{k - 1}{r - 1} p^{r}q^{k-r} = p^{r} \sum_{k=r}^{\infty} \binom{k - 1}{r - 1} q^{k-r} = p^{r}(1 - q)^{-r}$$

which obviously equals 1. The last equality results from the series expansion of

$$(1 - q)^{-r} = \sum_{n=0}^{\infty} \binom{-r}{n}(-q)^{n},$$

which equals

$$\sum_{k=r}^{\infty} \binom{k-1}{r-1} q^{k-r}$$

after some algebraic simplification and recalling the definition of the (generalized) binomial coefficient (see the note before Example 2.7). It is because of the negative exponent $(-r)$ in the above expression that the distribution is called the negative binomial distribution. To evaluate $E(Y)$ and $V(Y)$ we can either proceed directly, trying to evaluate the various summations, or we can proceed in the following manner.

Let

Z_1 = number of repetitions required up to and including the first occurrence of A.

Z_2 = number of repetitions required between first occurrence of A up to and including the second occurrence of A.

$\vdots$

Z_r = number of repetitions required between the $(r-1)$-occurrence up to and including the rth occurrence of A.

Thus we see that all the Z_i's are independent random variables, each having a geometric distribution. Also, $Y = Z_1 + \cdots + Z_r$. Hence using Theorem 8.3, we have the following theorem.

Theorem 8.5. If Y has a Pascal distribution given by Eq. (8.7), then

$$E(Y) = r/p, \qquad V(Y) = rq/p^2. \tag{8.8}$$

EXAMPLE 8.11. The probability that an experiment will succeed is 0.8. If the experiment is repeated until four successful outcomes have occurred, what is the expected number of repetitions required? From the above, we have E (number of repetitions) $= 4/0.8 = 5$.

8.6 Relationship between the Binomial and Pascal Distributions

Let X have a binomial distribution with parameters n and p. (That is, $X =$ number of successes in n Bernoulli trials with $P(\text{success}) = p$.) Let Y have a Pascal distribution with parameters r and p. (That is, $Y =$ number of Bernoulli trials required to obtain r successes with $P(\text{success}) = p$.) Then the following relationships hold:

(a) $P(Y \leq n) = P(X \geq r)$,
(b) $P(Y > n) = P(X < r)$.

Proof: (a) If there are r or more successes in the first n trials, then it required n or fewer trials to obtain the first r successes.

(b) If there are fewer than r successes on the first n trials, then it takes more than n trials to obtain r successes.

Notes: (a) The above properties make it possible to use the tabulated binomial distribution for evaluating probabilities associated with the Pascal distribution. For example, suppose we wish to evaluate the probability that more than 10 repetitions are necessary to obtain the third success when $p = P(\text{success}) = 0.2$. We have, using the above notation for X and Y,

$$P(Y > 10) = P(X < 3) = \sum_{k=0}^{2} \binom{10}{k} (0.2)^k (0.8)^{10-k} = 0.678$$

(from the table in the appendix).

(b) Let us briefly contrast the binomial and Pascal distributions. In each case we are concerned with repeated Bernoulli trials. The binomial distribution arises when we deal with a fixed number (say n) of such trials and are concerned with the number of successes which occur. The Pascal distribution is encountered when we pre-assign the number of successes to be obtained and then record the number of Bernoulli trials required. This is particularly relevant for a statistical problem which we shall discuss in more detail later (see Example 14.1).

8.7 The Hypergeometric Distribution

Suppose that we have a lot of N items, r of which are defective and $(N - r)$ of which are nondefective. Suppose that we choose, at random, n items from the lot ($n \leq N$), without replacement. Let X be the number of defectives found. Since $X = k$ if and only if we obtain precisely k defective items (from the r defectives in the lot) and precisely $(n - k)$ nondefectives [from the $(N - r)$ nondefectives in the lot], we have

$$P(X = k) = \frac{\binom{r}{k}\binom{N-r}{n-k}}{\binom{N}{n}}, \qquad k = 0, 1, 2, \ldots \tag{8.9}$$

A discrete random variable having probability distribution Eq. (8.9) is said to have a *hypergeometric distribution.*

Note: Since $\binom{a}{b} = 0$ whenever $b > a$, if a and b are nonnegative integers, we may define the above probabilities for all $k = 0, 1, 2, \ldots$ We obviously cannot obtain more than r defectives, but probability zero will be assigned to that event by Eq. (8.9).

EXAMPLE 8.12. Small electric motors are shipped in lots of 50. Before such a shipment is accepted, an inspector chooses 5 of these motors and inspects them. If none of these tested motors are defective, the lot is accepted. If one or more are found to be defective, the entire shipment is inspected. Suppose that there are, in fact, three defective motors in the lot. What is the probability that 100 percent inspection is required?

If we let X be the number of defective motors found, 100 percent inspection will be required if and only if $X \geq 1$. Hence

$$P(X \geq 1) = 1 - P(X = 0) = 1 - \frac{\binom{3}{0}\binom{47}{5}}{\binom{50}{5}} = 0.28.$$

Theorem 8.6. Let X have a hypergeometric distribution as given by Eq. (8.9). Let $p = r/N, q = 1 - p$. Then we have

(a) $E(X) = np$;

(b) $V(X) = npq \dfrac{N - n}{N - 1}$;

(c) $P(X = k) \simeq \dbinom{n}{k} p^k (1 - p)^{n-k}$,

for large N.

Proof: We shall leave the details of the proof to the reader. (See Problem 8.19.)

Note: Property (c) of Theorem 8.6 states that if the lot size N is sufficiently large, the distribution of X may be approximated by the binomial distribution. This is intuitively reasonable. For the binomial distribution is applicable when we sample *with* replacement (since in that case the probability of obtaining a defective item remains constant), while the hypergeometric distribution is applicable when we sample without replacement. If the lot size is large, it should not make a great deal of difference whether or not a particular item is returned to the lot before the next one is chosen. Property (c) of Theorem 8.6 is simply a mathematical statement of that fact. Also note that the expected value of a hypergeometric random variable X is the same as that of the corresponding binomially distributed random variable, while the variance of X is somewhat smaller than the corresponding one for the binomial case. The "correction term" $(N - n)/(N - 1)$ is approximately equal to 1, for large N.

We can illustrate the meaning of (c) with the following simple example. Suppose that we want to evaluate $P(X = 0)$.

For $n = 1$, we obtain from the hypergeometric distribution, $P(X = 0) = (N - r)/N = 1 - r/N = q$. From the binomial distribution we directly obtain $P(X = 0) = q$. Hence these answers are the same as indeed they should be for $n = 1$.

For $n = 2$, we obtain from the hypergeometric distribution

$$P(X = 0) = \frac{N - r}{N} \frac{N - r - 1}{N - 1} = \left(1 - \frac{r}{N}\right)\left(1 - \frac{r}{N - 1}\right).$$

From the binomial distribution we obtain $P(X = 0) = q^2$. It should be noted that $(1 - r/N) = q$, while $[1 - r/(N - 1)]$ is almost equal to q.

In general, the approximation of the hypergeometric distribution by the binomial distribution is very good if $n/N \leq 0.1$.

8.8 The Multinomial Distribution

Finally, we shall consider an important higher-dimensional discrete random variable, which may be thought of as a generalization of the binomial distribution. Consider an experiment $\mathcal{E}$, its sample space S, and a partition of S into k mutually exclusive events $A_1, \ldots, A_k$. (That is, when ε is performed one and only one of the events A_i occurs.) Consider n independent repetitions of ε. Let $p_i = P(A_i)$ and suppose that p_i remains constant during all repetitions. Clearly we have $\sum_{i=1}^{k} p_i = 1$. Define the random variables $X_1, \ldots, X_k$ as follows.

X_i is the number of times A_i occurs among the n repetitions of ε, $i = 1, \ldots, k$.

The X_i's are not independent random variables since $\sum_{i=1}^{k} X_i = n$. Hence as soon as the value of any $(k - 1)$ of these random variables is known, the value of the other one is determined. We have the following result.

Theorem 8.7. If X_i, $i = 1, 2, \ldots, k$, are as defined above, we have

$$P(X_1 = n_1, X_2 = n_2, \ldots, X_k = n_k) = \frac{n!}{n_1! n_2! \cdots n_k!} p_1^{n_1} \cdots p_k^{n_k}, \qquad (8.10)$$

where $\sum_{i=1}^{k} n_i = n$.

Proof: The argument is identical to the one used to establish the binomial probabilities. We must simply observe that the number of ways of arranging n objects, n_1 of which are of one kind, n_2 of which are of a second kind, $\ldots$, n_k of which are of a kth kind is given by

$$\frac{n!}{n_1! \cdots n_k!}.$$

Notes: (a) If $k = 2$ the above reduces to the binomial distribution. In this case we labeled the two possible events "success" and "failure."

(b) The above distribution is known as the *multinomial probability distribution*. Let us recall that the terms of the binomial distribution were obtainable from an expansion of the binomial expression $[p + (1 - p)]^n = \sum_{k=0}^{n} \binom{n}{k} p^k (1 - p)^{n-k}$. In an analogous way, the above probabilities may be obtained from an expansion of the multinomial expression $(p_1 + p_2 + \cdots + p_k)^n$.

Theorem 8.8. Suppose that $(X_1, \ldots, X_k)$ has a multinomial distribution given by Eq. (8.10). Then

$$E(X_i) = np_i \qquad \text{and} \qquad V(X_i) = np_i(1 - p_i), \quad i = 1, 2, \ldots, k.$$

Proof: This is an immediate consequence of the observation that each X_i as defined above has a binomial distribution, with probability of success (that is, the occurrence of A_i) equal to p_i.

EXAMPLE 8.13. A rod of specified length is manufactured. Assume that the actual length X (inches) is a random variable uniformly distributed over $[10, 12]$. Suppose that it is only of interest to know whether one of the following three events has occurred:

$$A_1 = \{X < 10.5\}, \quad A_2 = \{10.5 \le X \le 11.8\}, \quad \text{and} \quad A_3 = \{X > 11.8\}.$$

We have

$$p_1 = P(A_1) = 0.25, \qquad p_2 = P(A_2) = 0.65, \qquad \text{and} \qquad p_3 = P(A_3) = 0.1.$$

Thus if 10 such rods are manufactured, the probability of obtaining exactly 5 rods of length less than 10.5 inches and exactly 2 of length greater than 11.8 inches is given by

$$\frac{10!}{5!3!2!} (0.25)^5(0.65)^3(0.1)^2.$$

PROBLEMS

8.1.   If X has a Poisson distribution with parameter β, and if $P(X = 0) = 0.2$, evaluate $P(X > 2)$.

8.2.   Let X have a Poisson distribution with parameter λ. Find that value of k for which $P(X = k)$ is largest. [*Hint:* Compare $P(X = k)$ with $P(X = k - 1)$.]

8.3.   (This problem is taken from *Probability and Statistical Inference for Engineers* by Derman and Klein, Oxford University Press, London, 1959.) The number of oil tankers, say N, arriving at a certain refinery each day has a Poisson distribution with parameter $\lambda = 2$. Present port facilities can service three tankers a day. If more than three tankers arrive in a day, the tankers in excess of three must be sent to another port.

(a) On a given day, what is the probability of having to send tankers away?

(b) How much must present facilities be increased to permit handling all tankers on approximately 90 percent of the days?

(c) What is the expected number of tankers arriving per day?

(d) What is the most probable number of tankers arriving daily?

(e) What is the expected number of tankers serviced daily?

(f) What is the expected number of tankers turned away daily?

8.4.   Suppose that the probability that an item produced by a particular machine is defective equals 0.2. If 10 items produced from this machine are selected at random, what is the probability that not more than one defective is found? Use the binomial and Poisson distributions and compare the answers.

8.5.   An insurance company has discovered that only about 0.1 percent of the population is involved in a certain type of accident each year. If its 10,000 policy holders were randomly selected from the population, what is the probability that not more than 5 of its clients are involved in such an accident next year?

8.6.   Suppose that X has a Poisson distribution. If

$$P(X = 2) = \tfrac{2}{3}P(X = 1),$$

evaluate $P(X = 0)$ and $P(X = 3)$.

8.7. A film supplier produces 10 rolls of a specially sensitized film each year. If the film is not sold within the year, it must be discarded. Past experience indicates that D, the (small) demand for the film, is a Poisson-distributed random variable with parameter 8. If a profit of $7 is made on every roll which is sold, while a loss of $3 is incurred on every roll which must be discarded, compute the expected profit which the supplier may realize on the 10 rolls which he produces.

8.8. Particles are emitted from a radioactive source. Suppose that the number of such particles emitted during a one-hour period has a Poisson distribution with parameter λ. A counting device is used to record the number of such particles emitted. If more than 30 particles arrive during any one-hour period, the recording device is incapable of keeping track of the excess and simply records 30. If Y is the random variable defined as the number of particles *recorded* by the counting device, obtain the probability distribution of Y.

8.9. Suppose that particles are emitted from a radioactive source and that the number of particles emitted during a one-hour period has a Poisson distribution with parameter λ. Assume that the counting device recording these emissions occasionally fails to record an emitted particle. Specifically, suppose that any emitted particle has a probability p of being recorded.

(a) If Y is defined as the number of particles recorded, what is an expression for the probability distribution of Y?

(b) Evaluate $P(Y = 0)$ if $\lambda = 4$ and $p = 0.9$.

8.10. Suppose that a container contains 10,000 particles. The probability that such a particle escapes from the container equals 0.0004. What is the probability that more than 5 such escapes occur? (You may assume that the various escapes are independent of one another.)

8.11. Suppose that a book of 585 pages contains 43 typographical errors. If these errors are randomly distributed throughout the book, what is the probability that 10 pages, selected at random, will be free of errors? [*Hint:* Suppose that $X = $ number of errors per page has a Poisson distribution.]

8.12. A radioactive source is observed during 7 time intervals each of ten seconds in duration. The number of particles emitted during each period is counted. Suppose that the number of particles emitted, say X, during each observed period has a Poisson distribution with parameter 5.0. (That is, particles are emitted at the rate of 0.5 particles per second.)

(a) What is the probability that in each of the 7 time intervals, 4 or more particles are emitted?

(b) What is the probability that in at least 1 of the 7 time intervals, 4 or more particles are emitted?

8.13. It has been found that the number of transistor failures on an electronic computer in any one-hour period may be considered as a random variable having a Poisson distribution with parameter 0.1. (That is, on the average there is one transistor failure every 10 hours.) A certain computation requiring 20 hours of computing time is initiated.

(a) Find the probability that the above computation can be successfully completed without a breakdown. (Assume that the machine becomes inoperative only if 3 or more transistors fail.)

(b) Same as (a), except that the machine becomes inoperative if 2 or more transistors fail.

8.14.   In forming binary numbers with n digits, the probability that an incorrect digit will appear is, say 0.002. If the errors are independent, what is the probability of finding zero, one, or more than one incorrect digits in a 25-digit binary number? If the computer forms 10^6 such 25-digit numbers per second, what is the probability that an incorrect number is formed during any one-second period?

8.15.   Two independently operating launching procedures are used every week for launching rockets. Assume that each procedure is continued until *it* produces a successful launching. Suppose that using procedure I, $P(S)$, the probability of a successful launching, equals p_1, while for procedure II, $P(S) = p_2$. Assume furthermore, that one attempt is made every week with each of the two methods. Let X_1 and X_2 represent the number of weeks required to achieve a successful launching by means of I and II, respectively. (Hence X_1 and X_2 are independent random variables, each having a geometric distribution.) Let W be the minimum (X_1, X_2) and Z be the maximum (X_1, X_2). Thus W represents the number of weeks required to obtain *a* successful launching while Z represents the number of weeks needed to achieve successful launchings with both procedures. (Thus if procedure I results in $\overline{S}\,\overline{S}\,\overline{S}\,S$, while procedure II results in $\overline{S}\,\overline{S}\,S$, we have $W = 3, Z = 4$.)

(a) Obtain an expression for the probability distribution of W. [*Hint:* Express, in terms of X_1 and X_2, the event $\{W = k\}$.]
(b) Obtain an expression for the probability distribution of Z.
(c) Rewrite the above expressions if $p_1 = p_2$.

8.16.   Four components are assembled into a single apparatus. The components originate from independent sources and $p_i = P(i$th component is defective$)$, $i = 1, 2, 3, 4$.

(a) Obtain an expression for the probability that the entire apparatus is functioning.
(b) Obtain an expression for the probability that at least 3 components are functioning.
(c) If $p_1 = p_2 = 0.1$ and $p_3 = p_4 = 0.2$, evaluate the probability that exactly 2 components are functioning.

8.17.   A machinist keeps a large number of washers in a drawer. About 50 percent of these washers are $\frac{1}{4}$ inch in diameter, about 30 percent are $\frac{1}{8}$ inch in diameter, and the remaining 20 percent are $\frac{3}{8}$ inch in diameter. Suppose that 10 washers are chosen at random.

(a) What is the probability that there are exactly five $\frac{1}{4}$-inch washers, four $\frac{1}{8}$-inch washers, and one $\frac{3}{8}$-inch washer?
(b) What is the probability that only two kinds of washers are among the chosen ones?
(c) What is the probability that all three kinds of washers are among the chosen ones?
(d) What is the probability that there are three of one kind, three of another kind, and four of the third kind in a sample of 10?

8.18.   Prove Theorem 8.4.

8.19.   Prove Theorem 8.6.

8.20.   The number of particles emitted from a radioactive source during a specified period is a random variable with a Poisson distribution. If the probability of no emissions equals $\frac{1}{3}$, what is the probability that 2 or more emissions occur?

8.21.   Suppose that X_t, the number of particles emitted in t hours from a radioactive source, has a Poisson distribution with parameter $20t$. What is the probability that exactly 5 particles are emitted during a 15-minute period?

8.22.   The probability of a successful rocket launching equals 0.8. Suppose that launching attempts are made until 3 successful launchings have occurred. What is the probability that exactly 6 attempts will be necessary? What is the probability that fewer than 6 attempts will be required?

8.23.   In the situation described in Problem 8.22, suppose that launching attempts are made until three *consecutive* successful launchings occur. Answer the questions raised in the previous problem in this case.

8.24.   Consider again the situation described in Problem 8.22. Suppose that each launching attempt costs $5000. In addition, a launching failure results in an additional cost of $500. Evaluate the expected cost for the situation described.

8.25.   With X and Y defined as in Section 8.6, prove or disprove the following:

$$P(Y < n) = P(X > r).$$

9

Some Important Continuous Random Variables

9.1 Introduction

In this chapter we shall pursue the task we set for ourselves in Chapter 8, and study in considerable detail, a number of important continuous random variables and their characteristics. As we have pointed out before, in many problems it becomes mathematically simpler to consider an "idealized" range space for a random variable X, in which *all* possible real numbers (in some specified interval or set of intervals) may be considered as possible outcomes. In this manner we are led to continuous random variables. Many of the random variables we shall now introduce have important applications and we shall defer until a later chapter the discussion of some of these applications.

9.2 The Normal Distribution

One of the most important continuous random variables is the following.

Definition. The random variable X, assuming all real values $-\infty < x < \infty$, has a *normal (or Gaussian)* distribution if its pdf is of the form

$$f(x) = \frac{1}{\sqrt{2\pi}\,\sigma} \exp\left(-\frac{1}{2}\left[\frac{x-\mu}{\sigma}\right]^2\right), \qquad -\infty < x < \infty. \qquad (9.1)$$

The parameters μ and σ must satisfy the conditions $-\infty < \mu < \infty, \sigma > 0$. Since we shall have many occasions to refer to the above distribution we shall use the following notation: X has distribution $N(\mu, \sigma^2)$ if and only if its probability distribution is given by Eq. (9.1). [We shall frequently use the notation exp (t) to represent e^t.]

We shall delay until Chapter 12 discussing the reason for the great importance of this distribution. Let us simply state now that the *normal distribution serves as an excellent approximation to a large class of distributions* which have great practical importance. Furthermore, this distribution has a number of very desirable mathematical properties which make it possible to derive important theoretical results.

9.3 Properties of the Normal Distribution

(a) Let us state that f is a legitimate pdf. Obviously $f(x) \geq 0$. We must still check that $\int_{-\infty}^{+\infty} f(x)\,dx = 1$. We observe that by letting $t = (x - \mu)/\sigma$, we may write $\int_{-\infty}^{+\infty} f(x)\,dx$ as $(1/\sqrt{2\pi}) \int_{-\infty}^{+\infty} e^{-t^2/2}\,dt = I$.

The "trick" used to evaluate this integral (and it *is* a trick) is to consider, instead of I, the square of this integral, namely I^2. Thus

$$I^2 = \frac{1}{2\pi} \int_{-\infty}^{+\infty} e^{-t^2/2}\,dt \int_{-\infty}^{+\infty} e^{-s^2/2}\,ds$$

$$= \frac{1}{2\pi} \int_{-\infty}^{+\infty} \int_{-\infty}^{+\infty} e^{-(s^2+t^2)/2}\,ds\,dt.$$

Let us introduce polar coordinates to evaluate this double integral:

$$s = r\cos\alpha, \qquad t = r\sin\alpha.$$

Hence the element of area $ds\,dt$ becomes $r\,dr\,d\alpha$. As s and t vary between $-\infty$ and $+\infty$, r varies between 0 and ∞, while α varies between 0 and 2π. Thus

$$I^2 = \frac{1}{2\pi} \int_0^{2\pi} \int_0^\infty r e^{-r^2/2}\,dr\,d\alpha$$

$$= \frac{1}{2\pi} \int_0^{2\pi} -e^{-r^2/2}\Big|_0^\infty\,d\alpha$$

$$= \frac{1}{2\pi} \int_0^{2\pi} d\alpha = 1.$$

Hence $I = 1$ as was to be shown.

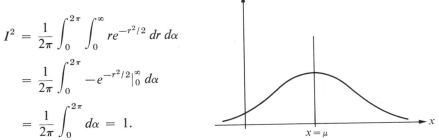

FIGURE 9.1

(b) Let us consider the appearance of the graph of f. It has the well-known bell shape indicated in Fig. 9.1. Since f depends on x only through the expression $(x - \mu)^2$, it is evident that the graph of f will be *symmetric* with respect to μ. For example, if $x = \mu + 2$, $(x - \mu)^2 = (\mu + 2 - \mu)^2 = 4$, while for $x = \mu - 2$, $(x - \mu)^2 = (\mu - 2 - \mu)^2 = 4$ also.

The parameter σ may also be interpreted geometrically. We note that at $x = \mu$ the graph of f is concave downward. As $x \to \pm\infty$, $f(x) \to 0$, asymptotically. Since $f(x) \geq 0$ for all x, this means that for large (positive or negative) values of x, the graph of f is concave upward. The point at which the concavity changes is called a point of inflection, and it is located by solving the equation $f''(x) = 0$. When we do this we find that the points of inflection occur at $x = \mu \pm \sigma$. That is, σ units to the right and to the left of μ the graph of f changes concavity. Thus if σ is relatively large, the graph of f tends to be "flat," while if σ is small the graph of f tends to be quite "peaked."

(c) In addition to the geometric interpretation of the parameters μ and σ, the following important probabilistic meaning may be associated with these quantities. Consider

$$E(X) = \frac{1}{\sqrt{2\pi}\,\sigma} \int_{-\infty}^{+\infty} x \exp\left(-\frac{1}{2}\left[\frac{x - \mu}{\sigma}\right]^2\right) dx.$$

Letting $z = (x - \mu)/\sigma$ and noting that $dx = \sigma\,dz$, we obtain

$$E(X) = \frac{1}{\sqrt{2\pi}} \int_{-\infty}^{+\infty} (\sigma z + \mu)e^{-z^2/2}\,dz$$

$$= \frac{1}{\sqrt{2\pi}}\,\sigma \int_{-\infty}^{+\infty} ze^{-z^2/2}\,dz + \mu\,\frac{1}{\sqrt{2\pi}} \int_{-\infty}^{+\infty} e^{-z^2/2}\,dz.$$

The first of the above integrals equals zero since the integrand, say $g(z)$, has the property that $g(z) = -g(-z)$, and hence g is an odd function. The second integral (without the factor μ) represents the total area under the normal pdf and hence equals unity. Thus $E(X) = \mu$.

Consider next

$$E(X^2) = \frac{1}{\sqrt{2\pi}\,\sigma} \int_{-\infty}^{+\infty} x^2 \exp\left(-\frac{1}{2}\left[\frac{x - \mu}{\sigma}\right]^2\right) dx.$$

Again letting $z = (x - \mu)/\sigma$, we obtain

$$E(X^2) = \frac{1}{\sqrt{2\pi}} \int_{-\infty}^{+\infty} (\sigma z + \mu)^2 e^{-z^2/2}\,dz$$

$$= \frac{1}{\sqrt{2\pi}} \int_{-\infty}^{+\infty} \sigma^2 z^2 e^{-z^2/2}\,dz + 2\mu\sigma\,\frac{1}{\sqrt{2\pi}} \int_{-\infty}^{+\infty} ze^{-z^2/2}\,dz$$

$$+ \mu^2\,\frac{1}{\sqrt{2\pi}} \int_{-\infty}^{+\infty} e^{-z^2/2}\,dz.$$

The second integral again equals zero by the argument used above. The last integral (without the factor μ^2) equals unity. To evaluate $(1/\sqrt{2\pi}) \int_{-\infty}^{+\infty} z^2 e^{-z^2/2}\,dz$, we integrate by parts letting $ze^{-z^2/2} = dv$ and $z = u$. Hence $v = -e^{-z^2/2}$ while $dz = du$. We obtain

$$\frac{1}{\sqrt{2\pi}} \int_{-\infty}^{+\infty} z^2 e^{-z^2/2}\,dz = \frac{-ze^{-z^2/2}}{\sqrt{2\pi}}\Big|_{-\infty}^{+\infty} + \frac{1}{\sqrt{2\pi}} \int_{-\infty}^{+\infty} e^{-z^2/2}\,dz = 0 + 1 = 1.$$

Therefore, $E(X^2) = \sigma^2 + \mu^2$, and hence $V(X) = E(X^2) - (E(X))^2 = \sigma^2$. *Thus we find that the two parameters μ and σ^2 which characterize the normal distribution are the expectation and variance of X, respectively.* To put it differently, if we know that X is normally distributed we know only that its probability distribution is of a certain type (or belongs to a certain family.) If in addition, we know $E(X)$ and $V(X)$, the distribution of X is completely specified. As we mentioned above, the graph of the pdf of a normally distributed random variable is symmetric about μ. The steepness of the graph is determined by σ^2 in the sense that if X has distribution $N(\mu, \sigma_1^2)$ and Y has distribution $N(\mu, \sigma_2^2)$, where $\sigma_1^2 > \sigma_2^2$, then their pdf's would have the *relative* shapes shown in Fig. 9.2.

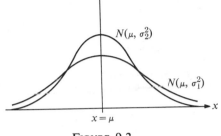

FIGURE 9.2

(d) If X has distribution $N(0, 1)$ we say that X has a *standardized normal distribution*. That is, the pdf of X may be written as

$$\varphi(x) = \frac{1}{\sqrt{2\pi}} e^{-x^2/2}. \qquad (9.2)$$

(We shall use the letter φ exclusively for the pdf of the above random variable X.) The importance of the standardized normal distribution is due to the fact that it is *tabulated*. Whenever X has distribution $N(\mu, \sigma^2)$ we can always obtain the standardized form by simply taking a *linear* function of X as the following theorem indicates.

Theorem 9.1. If X has the distribution $N(\mu, \sigma^2)$ and if $Y = aX + b$, then Y has the distribution $N(a\mu + b, a^2\sigma^2)$.

Proof: The fact that $E(Y) = a\mu + b$ and that $V(Y) = a^2\sigma^2$ follows immediately from the properties of expectation and variance discussed in Chapter 7. To show that in fact Y is normally distributed, we may apply Theorem 5.1, since $aX + b$ is either a decreasing or increasing function of X, depending on the sign of a. Hence if g is the pdf of Y, we have

$$g(y) = \frac{1}{\sqrt{2\pi}\,\sigma} \exp\left(-\frac{1}{2\sigma^2}\left[\frac{y-b}{a} - \mu\right]^2\right)\left|\frac{1}{a}\right|$$

$$= \frac{1}{\sqrt{2\pi}\,\sigma|a|} \exp\left(-\frac{1}{2\sigma^2 a^2}[y - (a\mu + b)]^2\right),$$

which represents the pdf of a random variable with distribution $N(a\mu + b, a^2\sigma^2)$.

Corollary. If X has distribution $N(\mu, \sigma^2)$ and if $Y = (X - \mu)/\sigma$, then Y has distribution $N(0, 1)$.

Proof: It is evident that Y is a linear function of X, and hence Theorem 9.1 applies.

Note: The importance of this corollary is that by *changing the units* in which the variable is measured we can obtain the standardized distribution (see d). By doing this we obtain a distribution with no unspecified parameters, a most desirable situation from the point of view of tabulating the distribution (see the next section).

9.4 Tabulation of the Normal Distribution

Suppose that X has distribution $N(0, 1)$. Then

$$P(a \le X \le b) = \frac{1}{\sqrt{2\pi}} \int_a^b e^{-x^2/2} \, dx.$$

This integral cannot be evaluated by ordinary means. (The difficulty stems from the fact that we cannot apply the Fundamental Theorem of the Calculus since we cannot find a function whose derivative equals $e^{-x^2/2}$.) However, methods of numerical integration can be used to evaluate integrals of the above form, and in fact $P(X \le s)$ has been *tabulated*.

The cdf of the standardized normal distribution will be consistently denoted by Φ. That is,

$$\Phi(s) = \frac{1}{\sqrt{2\pi}} \int_{-\infty}^s e^{-x^2/2} \, dx. \tag{9.3}$$

(See Fig. 9.3.) The function Φ has been extensively tabulated, and an excerpt of such a table is given in the Appendix. We may now use the tabulation of the function Φ in order to evaluate $P(a \le X \le b)$, where X has the standardized $N(0, 1)$ distribution since

$P(a \le X \le b) = \Phi(b) - \Phi(a).$

The particular usefulness of the above tabulation is due to the fact that if X has *any* normal distribution $N(\mu, \sigma^2)$, the tabulated function Φ may be used to evaluate probabilities associated with X.

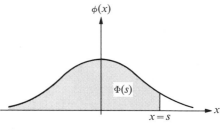

FIGURE 9.3

We simply use Theorem 9.1 to note that if X has distribution $N(\mu, \sigma^2)$, then $Y = (X - \mu)/\sigma$ has distribution $N(0, 1)$. Hence

$$P(a \le X \le b) = P\left(\frac{a - \mu}{\sigma} \le Y \le \frac{b - \mu}{\sigma}\right)$$

$$= \Phi\left(\frac{b - \mu}{\sigma}\right) - \Phi\left(\frac{a - \mu}{\sigma}\right). \tag{9.4}$$

It is also evident from the definition of Φ (see Fig. 9.3) that

$$\Phi(-x) = 1 - \Phi(x). \tag{9.5}$$

This relationship is particularly useful since in most tables the function Φ is tabulated only for positive values of x.

Finally let us compute $P(\mu - k\sigma \leq X \leq \mu + k\sigma)$, where X has distribution $N(\mu, \sigma^2)$. The above probability may be expressed in terms of the function Φ by writing

$$P(\mu - k\sigma \leq X \leq \mu + k\sigma) = P\left(-k \leq \frac{X - \mu}{\sigma} \leq k\right)$$

$$= \Phi(k) - \Phi(-k).$$

Using Eq. (9.5), we have for $k > 0$,

$$P(\mu - k\sigma \leq X \leq \mu + k\sigma) = 2\Phi(k) - 1. \tag{9.6}$$

Note that the above probability is independent of μ and σ. In words: The probability that a random variable with distribution $N(\mu, \sigma^2)$ takes values within k standard deviations of the expected value depends only on k and is given by Eq. (9.6).

Note: We shall have many occasions to refer to "tabulated functions." In a sense when an expression can be written in terms of tabulated functions, the problem is "solved." (With the availability of modern computing facilities, many functions which are not tabulated may be readily evaluated. Although we do not expect that everyone has easy access to a computer, it does not seem too unreasonable to suppose that certain common tables are available.) Thus we should feel as much at ease with the function $\Phi(x) = (1/\sqrt{2\pi}) \int_{-\infty}^{x} e^{-s^2/2} ds$ as with the function $f(x) = \sqrt{x}$. Both these functions are tabulated, and in either case we might experience some difficulty in evaluating the function directly for $x = 0.43$, for example. In the Appendix various tables are listed of some of the most important functions we shall encounter in our work. Occasional references will be given to other tables not listed in this text.

EXAMPLE 9.1. Suppose that X has distribution $N(3, 4)$. We want to find a number c such that

$$P(X > c) = 2P(X \leq c).$$

We note that $(X - 3)/2$ has distribution $N(0, 1)$. Hence

$$P(X > c) = P\left(\frac{X - 3}{2} > \frac{c - 3}{2}\right) = 1 - \Phi\left(\frac{c - 3}{2}\right).$$

Also,

$$P(X \leq c) = P\left(\frac{X - 3}{2} \leq \frac{c - 3}{2}\right) = \Phi\left(\frac{c - 3}{2}\right).$$

The above condition may therefore be written as $1 - \Phi[(c - 3)/2] = 2\Phi[(c - 3)/2]$. This becomes $\Phi[(c - 3)/2] = \frac{1}{3}$. Hence (from the tables of the normal distribution) we find that $(c - 3)/2 = -0.43$, yielding $c = 2.14$.

EXAMPLE 9.2. Suppose that the breaking strength of cotton fabric (in pounds), say X, is normally distributed with $E(X) = 165$ and $V(X) = 9$. Assume furthermore that a sample of this fabric is considered to be defective if $X < 162$. What is the probability that a fabric chosen at random will be defective?

We must compute $P(X < 162)$. However

$$P(X < 162) = P\left(\frac{X - 165}{3} < \frac{162 - 165}{3}\right)$$

$$= \Phi(-1) = 1 - \Phi(1) = 0.159.$$

Note: An immediate objection to the use of the normal distribution may be raised here. For it is obvious that X, the strength of cotton fabric, cannot assume negative values, while a normally distributed random variable may assume all positive and negative values. However, the above model (apparently not valid in view of the objection just raised) assigns negligible probability to the event $\{X < 0\}$. That is,

$$P(X < 0) = P\left(\frac{X - 165}{3} < \frac{0 - 165}{3}\right) = \Phi(-55) \simeq 0.$$

The point raised here will occur frequently: A certain random variable X which we know cannot assume negative values (say) will be assumed to have a normal distribution, thus taking on (theoretically, at least) both positive and negative values. So long as the parameters μ and σ^2 are chosen so that $P(X < 0)$ is essentially zero, such a representation is perfectly valid.

The problem of finding the pdf of a function of a random variable, say $Y = H(X)$, as discussed in Chapter 5, occurs in the present context in which the random variable X is normally distributed.

EXAMPLE 9.3. Suppose that the radius R of a ball bearing is normally distributed with expected value 1 and variance 0.04. Find the pdf of the volume of the ball bearing.

The pdf of the random variable R is given by

$$f(r) = \frac{1}{\sqrt{2\pi}\,(0.2)} \exp\left(-\frac{1}{2}\left[\frac{r - 1}{0.2}\right]^2\right).$$

Since V is a monotonically increasing function of R, we may directly apply Theorem 5.1 for the pdf of $V = \frac{4}{3}\pi R^3$, and obtain $g(v) = f(r)(dr/dv)$, where r is everywhere expressed in terms of v. From the above relationship, we obtain $r = \sqrt[3]{3v/4\pi}$. Hence $dr/dv = (1/4\pi)(3v/4\pi)^{-2/3}$. By substituting these expressions into the above equation, we obtain the desired pdf of V.

EXAMPLE 9.4. Suppose that X, the inside diameter (millimeters) of a nozzle, is a normally distributed random variable with expectation μ and variance 1. If X does not meet certain specifications, a loss is assessed to the manufacturer. Specifically, suppose that the profit T (per nozzle) is the following function of X:

$$
\begin{aligned}
T &= C_1 \text{ (dollars)} && \text{if } 10 \le X \le 12, \\
&= -C_2 && \text{if } X < 10, \\
&= -C_3 && \text{if } X > 12.
\end{aligned}
$$

Hence the expected profit (per nozzle) may be written as

$$
\begin{aligned}
E(T) &= C_1[\Phi(12 - \mu) - \Phi(10 - \mu)] \\
&\quad - C_2[\Phi(10 - \mu)] - C_3[1 - \Phi(12 - \mu)] \\
&= (C_1 + C_3)\Phi(12 - \mu) - (C_1 + C_2)\Phi(10 - \mu) - C_3.
\end{aligned}
$$

Suppose that the manufacturing process can be adjusted so that different values of μ may be achieved. For what value of μ is the expected profit maximum? We must compute $dE(T)/(d\mu)$ and set it equal to zero. Denoting, as usual, the pdf of the $N(0, 1)$ distribution by φ, we have

$$
\frac{dE(T)}{d\mu} = (C_1 + C_3)[-\varphi(12 - \mu)] - (C_1 + C_2)[-\varphi(10 - \mu)].
$$

Hence

$$
-(C_1 + C_3)\frac{1}{\sqrt{2\pi}} \exp\left(-\tfrac{1}{2}(12 - \mu)^2\right)
$$

$$
+ (C_1 + C_2)\frac{1}{\sqrt{2\pi}} \exp\left(-\tfrac{1}{2}(10 - \mu)^2\right) = 0.
$$

Or

$$
e^{22-2\mu} = \frac{C_1 + C_3}{C_1 + C_2}.
$$

Thus

$$
\mu = 11 - \tfrac{1}{2}\ln\left(\frac{C_1 + C_3}{C_1 + C_2}\right).
$$

[It is an easy matter for the reader to check that the above yields a maximum value for $E(T)$.]

Notes: (a) If $C_2 = C_3$, that is, if too large or too small a diameter X is an equally serious defect, then the value of μ for which the maximum value of $E(T)$ is attained is $\mu = 11$. If $C_2 > C_3$, the value of μ is > 11, while if $C_2 < C_3$, the value of μ is < 11. As $\mu \to +\infty$, $E(T) \to -C_3$, while if $\mu \to -\infty$, $E(T) \to -C_2$.

(b) Consider the following cost values: $C_1 = \$10$, $C_2 = \$3$, and $C_3 = \$2$. Hence the value of μ for which $E(T)$ is maximized equals $\mu = 11 - \tfrac{1}{2}\ln\left[\tfrac{12}{13}\right] = \$11.04$. Thus the maximum value attained by $E(T)$ equals $\$6.04$ per nozzle.

9.5 The Exponential Distribution

Definition. A continuous random variable X assuming all nonnegative values is said to have an *exponential distribution* with parameter $\alpha > 0$ if its pdf is given by

$$f(x) = \alpha e^{-\alpha x}, \qquad x > 0$$
$$\quad = 0, \qquad\quad \text{elsewhere.} \qquad (9.7)$$

(See Fig. 9.4.) [A straightforward integration reveals that

$$\int_0^\infty f(x)\, dx = 1$$

and hence Eq. (9.7) does represent a pdf.]

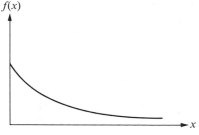

FIGURE 9.4

The exponential distribution plays an important role in describing a large class of phenomena, particularly in the area of reliability theory. We shall devote Chapter 11 to some of these applications. For the moment, let us simply investigate some of the properties of the exponential distribution.

9.6 Properties of the Exponential Distribution

(a) The cdf F of the exponential distribution is given by

$$F(x) = P(X \le x) = \int_0^x \alpha e^{-\alpha t}\, dt = 1 - e^{-\alpha x}, \qquad x \ge 0 \qquad (9.8)$$
$$= 0, \qquad \text{elsewhere.}$$

[Hence $P(X > x) = e^{-\alpha x}$.]

(b) The expected value of X is obtained as follows:

$$E(X) = \int_0^\infty x\alpha e^{-\alpha x}\, dx.$$

Integrating by parts and letting $\alpha e^{-\alpha x}\, dx = dv$, $x = u$, we obtain $v = -e^{-\alpha x}$, $du = dx$. Thus,

$$E(X) = [-xe^{-\alpha x}]\big|_0^\infty + \int_0^\infty e^{-\alpha x}\, dx = \frac{1}{\alpha}. \qquad (9.9)$$

Thus the expected value equals the reciprocal of the parameter α. [By simply relabeling the parameter $\alpha = 1/\beta$, we could have written the pdf of X as $f(x) = (1/\beta)e^{-x/\beta}$. In this form, the parameter β equals the expected value of X. However, we shall continue to use the form of Eq. (9.7).]

(c) The variance of X may be obtained by a similar integration. We find that $E(X^2) = 2/\alpha^2$ and therefore

$$V(X) = E(X^2) - [E(X)]^2 = \frac{1}{\alpha^2}. \qquad (9.10)$$

(d) The exponential distribution has the following interesting property, analogous to Eq. (8.6) described for the geometric distribution. Consider for any $s, t > 0$, $P(X > s + t \mid X > s)$. We have

$$P(X > s + t \mid X > s) = \frac{P(X > s + t)}{P(X > s)} = \frac{e^{-\alpha(s+t)}}{e^{-\alpha s}} = e^{-\alpha t}.$$

Hence

$$P(X > s + t \mid X > s) = P(X > t). \tag{9.11}$$

Thus we have shown that the exponential distribution also has the property of having "no memory" as did the geometric distribution. (See Note following Theorem 8.4.) We shall make considerable use of this property in applying the exponential distribution to fatigue models in Chapter 11.

Note: As was true in the case of the geometric distribution, the *converse* of Property (d) also holds. The only continuous random variable X assuming nonnegative values for which $P(X > s + t \mid X > s) = P(X > t)$ for all $s, t > 0$, is an exponentially distributed random variable. [Although we shall not prove this here, it might be pointed out that the crux of the argument involves the fact that the only continuous function G having the property that $G(x + y) = G(x)G(y)$ for all $x, y > 0$, is $G(x) = e^{-kx}$. It is easily seen that if we define $G(x) = 1 - F(x)$, where F is the cdf of X, then G will satisfy this condition.]

EXAMPLE 9.5. Suppose that a fuse has a life length X which may be considered as a continuous random variable with an exponential distribution. There are two processes by which the fuse may be manufactured. Process I yields an expected life length of 100 hours (that is, the parameter equals 100^{-1}), while process II yields an expected life length of 150 hours (that is, the parameter equals 150^{-1}). Suppose that process II is twice as costly (per fuse) as process I, which costs C dollars per fuse. Assume, furthermore, that if a fuse lasts less than 200 hours, a loss of K dollars is assessed against the manufacturer. Which process should be used? Let us compute the *expected* cost for each process. For process I, we have

$$C_{\mathrm{I}} = \text{cost (per fuse)} = C \quad \text{if} \quad X > 200$$
$$= C + K \quad \text{if} \quad X \le 200.$$

Therefore,

$$E(C_{\mathrm{I}}) = CP(X > 200) + (C + K)P(X \le 200)$$
$$= Ce^{-(1/100)200} + (C + K)(1 - e^{-(1/100)200})$$
$$= Ce^{-2} + (C + K)(1 - e^{-2}) = K(1 - e^{-2}) + C.$$

By a similar computation we find that

$$E(C_{\mathrm{II}}) = K(1 - e^{-4/3}) + 2C.$$

Thus

$$E(C_{\mathrm{II}}) - E(C_{\mathrm{I}}) = C + K(e^{-2} - e^{-4/3}) = C - 0.13K.$$

Hence we prefer process I, provided that $C > 0.13K$.

EXAMPLE 9.6. Suppose that X has an exponential distribution with parameter α. Then $E(X) = 1/\alpha$. Let us compute the probability that X exceeds its expected value (Fig. 9.5). We have

$$P\left(X > \frac{1}{\alpha}\right) = e^{-\alpha(1/\alpha)}$$

$$= e^{-1} < \tfrac{1}{2}.$$

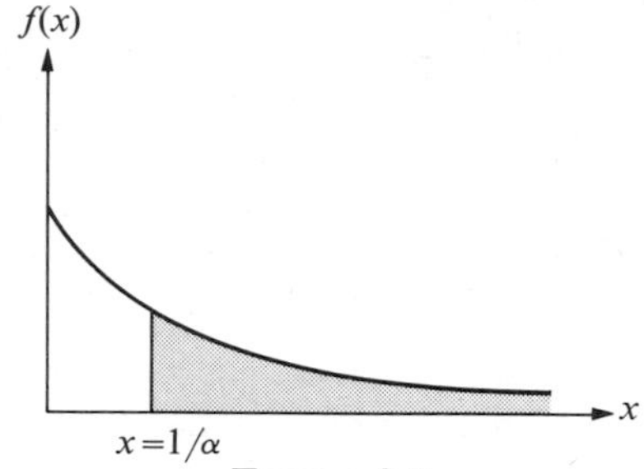

$x = 1/\alpha$

FIGURE 9.5

EXAMPLE 9.7. Suppose that T, the time to failure of a component, is exponentially distributed. Hence $f(t) = \alpha e^{-\alpha t}$. If n such components are installed, what is the probability that one-half or more of these components are still functioning at the end of t hours? The required probability is

$$\sum_{k=n/2}^{n} \binom{n}{k} (1 - e^{-\alpha t})^{n-k} (e^{-\alpha t k}) \qquad \text{if } n \text{ is even;}$$

$$\sum_{k=(n+1)/2}^{n} \binom{n}{k} (1 - e^{-\alpha t})^{n-k} (e^{-\alpha t k}) \qquad \text{if } n \text{ is odd.}$$

EXAMPLE 9.8. Suppose that the life length in hours, say T, of a certain electronic tube is a random variable with exponential distribution with parameter β. That is, the pdf is given by $f(t) = \beta e^{-\beta t}$, $t > 0$. A machine using this tube costs C_1 dollars/hour to run. While the machine is functioning, a profit of C_2 dollars/hour is realized. An operator must be hired for a *prearranged* number of hours, say H, and he gets paid C_3 dollars/hour. For what value of H is the *expected profit* greatest?

Let us first get an expression for the profit, say R. We have

$$R = C_2 H - C_1 H - C_3 H \qquad \text{if } T > H$$
$$= C_2 T - C_1 T - C_3 H \qquad \text{if } T \leq H.$$

Note that R is a random variable since it is a function of T. Hence

$$E(R) = H(C_2 - C_1 - C_3)P(T > H) - C_3 H P(T \leq H)$$
$$+ (C_2 - C_1) \int_0^H t\beta e^{-\beta t}\, dt$$
$$= H(C_2 - C_1 - C_3)e^{-\beta H} - C_3 H(1 - e^{-\beta H})$$
$$+ (C_2 - C_1)[\beta^{-1} - e^{-\beta H}(\beta^{-1} + H)]$$
$$= (C_2 - C_1)[He^{-\beta H} + \beta^{-1} - e^{-\beta H}(\beta^{-1} + H)] - C_3 H.$$

To obtain the maximum value of $E(R)$ we differentiate it with respect to H and set the derivative equal to zero. We have

$$\frac{dE(R)}{dH} = (C_2 - C_1)[H(-\beta)e^{-\beta H} + e^{-\beta H} - e^{-\beta H} + (\beta^{-1} + H)(\beta)e^{-\beta H}] - C_3$$

$$= (C_2 - C_1)e^{-\beta H} - C_3.$$

Hence $dE(R)/dH = 0$ implies that

$$H = -\left(\frac{1}{\beta}\right)\ln\left[\frac{C_3}{C_2 - C_1}\right].$$

[In order for the above solution to be meaningful, we must have $H > 0$ which occurs if and only if $0 < C_3/(C_2 - C_1) < 1$, which in turn is equivalent to $C_2 - C_1 > 0$ and $C_2 - C_1 - C_3 > 0$. However, the last condition simply requires that the cost figures be of such a magnitude that a profit may be realized.] Suppose in particular that $\beta = 0.01$, $C_1 = \$3$, $C_2 = \$10$, and $C_3 = \$4$. Then $H = -100 \ln [\frac{4}{7}] = 55.9$ hours $\simeq 56$ hours. Thus, the operator should be hired for 56 hours in order to achieve the maximum profit. (For a slight modification of the above example, see Problem 9.18.)

9.7 The Gamma Distribution

Let us first introduce a function which is most important not only in probability theory but in many areas of mathematics.

Definition. The *Gamma function*, denoted by Γ, is defined as follows:

$$\Gamma(p) = \int_0^\infty x^{p-1}e^{-x}\,dx, \qquad \text{defined for } p > 0. \qquad (9.12)$$

[It can be shown that the above improper integral exists (converges) whenever $p > 0$.] If we integrate the above by parts, letting $e^{-x}\,dx = dv$ and $x^{p-1} = u$, we obtain

$$\Gamma(p) = -e^{-x}x^{p-1}\big|_0^\infty - \int_0^\infty [-e^{-x}(p - 1)x^{p-2}\,dx]$$

$$= 0 + (p - 1)\int_0^\infty e^{-x}x^{p-2}\,dx$$

$$= (p - 1)\Gamma(p - 1). \qquad (9.13)$$

Thus we have shown that the Gamma function obeys an interesting recursion relationship. Suppose that p is a *positive integer*, say $p = n$. Then applying Eq. (9.13) repeatedly, we obtain

$$\Gamma(n) = (n - 1)\Gamma(n - 1)$$

$$= (n - 1)(n - 2)\Gamma(n - 2) = \cdots$$

$$= (n - 1)(n - 2) \cdots \Gamma(1).$$

However, $\Gamma(1) = \int_0^\infty e^{-x}\,dx = 1$, and hence we have

$$\Gamma(n) = (n - 1)! \qquad (9.14)$$

(if n is a positive integer). (Thus we may consider the Gamma function to be a generalization of the factorial function.) It is also easy to verify that

$$\Gamma(\tfrac{1}{2}) = \int_0^\infty x^{-1/2} e^{-x} \, dx = \sqrt{\pi}. \tag{9.15}$$

(See Problem 9.19.)

With the aid of the Gamma function we can now introduce the Gamma probability distribution.

Definition. Let X be a continuous random variable assuming only nonnegative values. We say that X has a *Gamma probability distribution* if its pdf is given by

$$f(x) = \frac{\alpha}{\Gamma(r)} (\alpha x)^{r-1} e^{-\alpha x}, \qquad x > 0$$

$$= 0, \qquad \text{elsewhere.} \qquad (9.16)$$

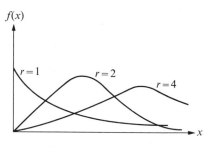

FIGURE 9.6

This distribution depends on *two parameters, r and α*, of which we require $r > 0$, $\alpha > 0$. [Because of the definition of the Gamma function, it is easy to see that $\int_{-\infty}^{+\infty} f(x)\, dx = 1$.] Figure 9.6 shows graphs of the pdf Eq. (9.16) for various values of r and $\alpha = 1$.

9.8 Properties of the Gamma Distribution

(a) If $r = 1$, Eq. (9.16) becomes $f(x) = \alpha e^{-\alpha x}$. Hence the *exponential distribution* is a *special case of the Gamma distribution*. (If r is a positive integer > 1, the Gamma distribution is also related to the exponential distribution but in a slightly different way. We shall refer to this in Chapter 10.)

(b) In most of our applications, the parameter r will be a positive *integer*. In this case, an interesting relationship between the cdf of the Gamma distribution and the Poisson distribution exists which we shall now develop.

Consider the integral $I = \int_a^\infty (e^{-y} y^r / r!) \, dy$, where r is a positive integer and $a > 0$. Then $r! I = \int_a^\infty e^{-y} y^r \, dy$. Integrating by parts, letting $u = y^r$ and $dv = e^{-y}\, dy$, will yield $du = ry^{r-1}\, dy$ and $v = -e^{-y}$. Hence $r! I = e^{-a} a^r + r \int_a^\infty e^{-y} y^{r-1} \, dy$. The integral in this expression is exactly of the same form as the original integral with r replaced by $(r - 1)$. Thus continuing to integrate by parts we obtain, since r is a positive integer,

$$r! I = e^{-a} [a^r + r a^{r-1} + r(r - 1) a^{r-2} + \cdots + r!].$$

Therefore

$$I = e^{-a}[1 + a + a^2/2! + \cdots + a^r/r!]$$

$$= \sum_{k=0}^{r} P(Y = k),$$

where Y has a Poisson distribution with parameter a.

We now consider the cdf of the random variable whose pdf is given by Eq. (9.16). Since r is a positive integer, Eq. (9.16) may be written as

$$f(x) = \frac{\alpha}{(r - 1)!} (\alpha x)^{r-1} e^{-\alpha x}, \qquad 0 < x$$

and consequently the cdf of X becomes

$$F(x) = 1 - P(X > x)$$

$$= 1 - \int_{x}^{\infty} \frac{\alpha}{(r - 1)!} (\alpha s)^{r-1} e^{-\alpha s} \, ds, \qquad x > 0.$$

Letting $(\alpha s) = u$, we find that this becomes

$$F(x) = 1 - \int_{\alpha x}^{\infty} \frac{u^{r-1} e^{-u}}{(r - 1)!} \, du, \qquad x > 0.$$

This integral is precisely of the form considered above, namely I (with $a = \alpha x$), and thus

$$F(x) = 1 - \sum_{k=0}^{r-1} e^{-\alpha x} (\alpha x)^k / k!, \qquad x > 0. \tag{9.17}$$

Hence, *the cdf of the Gamma distribution may be expressed in terms of the tabulated cdf of the Poisson distribution.* (We recall that this is valid if the parameter r is a positive integer.)

Note: The result stated in Eq. (9.17), relating the cdf of the Poisson distribution to the cdf of the Gamma distribution, is not as surprising as it might first appear, as the following discussion will indicate.

First of all, recall the relationship between the binomial and Pascal distributions (see Note (b), Section 8.6). A similar relationship exists between the Poisson and Gamma distribution except that the latter is a continuous distribution. When we deal with a Poisson distribution we are essentially concerned about the number of occurrences of some event during a fixed time period. And, as will be indicated, the Gamma distribution arises when we ask for distribution of the *time* required to obtain a specified number of occurrences of the event.

Specifically, suppose X = number of occurrences of the event A during $(0, t]$. Then, under suitable conditions (e.g., satisfying assumptions A_1 through A_5 in Section 8.3) X has a Poisson distribution with parameter αt, where α is the expected number of occurrences of A during a unit time interval. Let T = time required to observe r occurrences of A. We have:

$$H(t) = P(T \leq t) = 1 - P(T > t)$$
$$= 1 - P(\text{fewer than } r \text{ occurrences of } A \text{ occur in } (0, t])$$
$$= 1 - P(X < r)$$
$$= 1 - \sum_{k=0}^{r-1} \frac{e^{-\alpha t}(\alpha t)^k}{k!}.$$

Comparing this with Eq. (9.17) establishes the desired relationship.

(c) If X has a Gamma distribution given by Eq. (9.16), we have

$$E(X) = r/\alpha, \qquad V(X) = r/\alpha^2. \tag{9.18}$$

Proof: See Problem 9.20.

9.9 The Chi-Square Distribution

A special, very important, case of the Gamma distribution Eq. (9.16) is obtained if we let $\alpha = \frac{1}{2}$ and $r = n/2$, where n is a positive integer. We obtain a one-parameter family of distributions with pdf

$$f(z) = \frac{1}{2^{n/2}\Gamma(n/2)} z^{n/2-1} e^{-z/2}, \qquad z > 0. \tag{9.19}$$

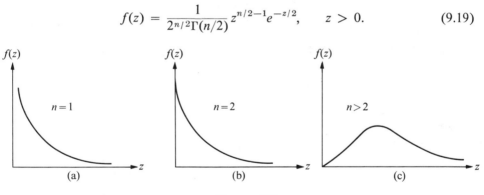

(a)   (b)   (c)

FIGURE 9.7

A random variable Z having pdf given by Eq. (9.19) is said to have a *chi-square distribution with n degrees of freedom* (denoted by χ_n^2). In Fig. 9.7, the pdf for $n = 1, 2$ and $n > 2$ is shown. It is an immediate consequence of Eq. (9.18) that if Z has pdf Eq. (9.19), we have

$$E(Z) = n, \quad V(Z) = 2n. \tag{9.20}$$

The chi-square distribution has many important applications in statistical inference, some of which we shall refer to later. Because of its importance, the chi-square distribution is tabulated for various values of the parameter n. (See Appendix.) Thus we may find in the table that value, denoted by χ_α^2, satisfying $P(Z \leq \chi_\alpha^2) = \alpha$, $0 < \alpha < 1$ (Fig. 9.8). Example 9.9 deals with a special case of

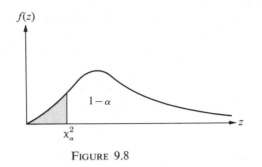

FIGURE 9.8

a general characterization of the chi-square distribution which we shall study in a later chapter.

EXAMPLE 9.9. Suppose that the velocity V of an object has distribution $N(0, 1)$ Let $K = mV^2/2$ be the kinetic energy of the object. To find the pdf of K, let us first find the pdf of $S = V^2$. Applying Theorem 5.2 directly we have

$$g(s) = \frac{1}{2\sqrt{s}}[\varphi(\sqrt{s}) + \varphi(-\sqrt{s})]$$

$$= s^{-1/2} \frac{1}{\sqrt{2\pi}} e^{-s/2}.$$

If we compare this with Eq. (9.19) and recall that $\Gamma(\frac{1}{2}) = \sqrt{\pi}$, we note that S has a χ_1^2-distribution. Thus we find that *the square of a random variable with distribution $N(0, 1)$ has a χ_1^2-distribution*. (It is this result which we shall generalize later.)

We can now obtain the pdf h of the kinetic energy K. Since K is a monotone function of V^2 whose pdf is given by g above, we have directly

$$h(k) = \frac{2}{m} g\left(\frac{2}{m}k\right) = \frac{2}{m} \frac{1}{\sqrt{2\pi}} \left(\frac{2}{m}k\right)^{-1/2} e^{-k/m}, \qquad k > 0.$$

In order to evaluate $P(K \le 5)$ for example, we need not use the pdf of K but may simply use the tabulated chi-square distribution as follows.

$$P(K \le 5) = P((m/2)V^2 \le 5) = P(V^2 \le 10/m).$$

This latter probability can be obtained directly from the tables of the chi-square distribution (if m is known) since V^2 has a χ_1^2-distribution. Since $E(V^2) = 1$ and variance $(V^2) = 2$ [see Eq. (9.20)], we find directly

$$E(K) = m/2 \qquad \text{and} \qquad V(K) = m^2/2.$$

Note: The tabulation of the chi-square distribution as given in the Appendix only tabulates those values for which n, the number of degrees of freedom, is less than or equal to 45. The reason for this is that if n is large, we may approximate the chi-square distribution with the normal distribution, as indicated by the following theorem.

Theorem 9.2. Suppose that the random variable Y has distribution χ_n^2. Then for sufficiently large n the random variable $\sqrt{2Y}$ has approximately the distribution $N(\sqrt{2n - 1}, 1)$. (The proof is not given here.)

This theorem may be used as follows. Suppose that we require $P(Y \leq t)$, where Y has distribution χ_n^2, and n is so large that the above probability cannot be directly obtained from the table of the chi-square distribution. Using Theorem 9.2 we may write,

$$
\begin{aligned}
P(Y \leq t) &= P(\sqrt{2Y} \leq \sqrt{2t}) \\
&= P(\sqrt{2Y} - \sqrt{2n - 1} \leq \sqrt{2t} - \sqrt{2n - 1}) \\
&\simeq \Phi(\sqrt{2t} - \sqrt{2n - 1}).
\end{aligned}
$$

The value of Φ may be obtained from the tables of the normal distribution.

9.10 Comparisons among Various Distributions

We have by now introduced a number of important probability distributions, both discrete and continuous: the binomial, Pascal, and Poisson among the discrete ones, and the exponential, geometric, and gamma among the continuous ones. We shall not restate the various assumptions which led to these distributions. Our principal concern here is to point out certain similarities (and differences) among the random variables having these distributions.

1. Assume that independent Bernoulli trials are being performed.

   (a) random variable: number of occurrences of event A in a *fixed* number of trials
   distribution: binomial

   (b) random variable: number of Bernoulli trials required to obtain first occurrence of A
   distribution: geometric

   (c) random variable: number of Bernoulli trials required to obtain rth occurrence of A
   distribution: Pascal

2. Assume a Poisson process (see note (c) preceeding Example 8.5).

   (d) random variable: number of occurrences of event A during a *fixed* time interval
   distribution: Poisson

   (e) random variable: time required until first occurrence of A
   distribution: exponential

(f) random variable: time required until rth occurrence of A
distribution: gamma

Note: Observe the similarity between (a) and (d), (b) and (e), and finally (c) and (f).

9.11 The Bivariate Normal Distribution

All the continuous random variables we have discussed have been one-dimensional random variables. As we mentioned in Chapter 6, higher-dimensional random variables play an important role in describing experimental outcomes. One of the most important continuous two-dimensional random variables, a direct generalization of the one-dimensional normal distribution, is defined as follows.

Definition.   Let (X, Y) be a two-dimensional, continuous random variable assuming all values in the euclidean plane. We say that (X, Y) has a *bivariate normal distribution* if its joint pdf is given by the following expression.

$$f(x, y) = \frac{1}{2\pi\sigma_x\sigma_y\sqrt{1 - \rho^2}}$$

$$\times \exp\left\{-\frac{1}{2(1 - \rho^2)}\left[\left(\frac{x - \mu_x}{\sigma_x}\right)^2 - 2\rho\frac{(x - \mu_x)(y - \mu_y)}{\sigma_x\sigma_y} + \left(\frac{y - \mu_y}{\sigma_y}\right)^2\right]\right\},$$

$$-\infty < x < \infty, \quad -\infty < y < \infty. \quad (9.21)$$

The above pdf depends on 5 parameters. For f to define a legitimate pdf [that is, $f(x, y) \geq 0$, $\int_{-\infty}^{+\infty} \int_{-\infty}^{+\infty} f(x, y)\, dx\, dy = 1$], we must place the following restrictions on the parameters: $-\infty < \mu_x < \infty$; $-\infty < \mu_y < \infty$; $\sigma_x > 0$; $\sigma_y > 0$; $-1 < \rho < 1$. The following properties of the bivariate normal distribution may easily be checked.

Theorem 9.3. Suppose that (X, Y) has pdf as given by Eq. (9.21). Then

(a) the marginal distributions of X and of Y are $N(\mu_x, \sigma_x^2)$ and $N(\mu_y, \sigma_y^2)$, respectively;

(b) the parameter ρ appearing above is the correlation coefficient between X and Y;

(c) the conditional distributions of X (given that $Y = y$) and of Y (given that $X = x$) are respectively

$$N\left[\mu_x + \rho\frac{\sigma_x}{\sigma_y}(y - \mu_y), \sigma_x^2(1 - \rho^2)\right], \quad N\left[\mu_y + \rho\frac{\sigma_y}{\sigma_x}(x - \mu_x), \sigma_y^2(1 - \rho^2)\right].$$

Proof: See Problem 9.21.

Notes: (a) The converse of (a) of Theorem 9.3 is not true. It is possible to have a joint pdf which is not bivariate normal and yet the marginal pdf's of X and of Y are one-dimensional normal.

(b) We observe from Eq. (9.21) that if $\rho = 0$, the joint pdf of (X, Y) may be factored and hence X and Y are independent. Thus we find that in the case of a bivariate normal distribution, zero correlation and independence *are* equivalent.

(c) Statement (c) of the above theorem shows that both regression functions of the mean are *linear*. It also shows that the variance of the conditional distribution is reduced in the same proportion as $(1 - \rho^2)$. That is, if ρ is close to zero, the conditional variance is essentially the same as the unconditional variance, while if ρ is close to ± 1, the conditional variance is close to zero.

The bivariate normal pdf has a number of interesting properties.   We shall state some of these as a theorem, leaving the proof to the reader.

Theorem 9.4. Consider the surface $z = f(x, y)$, where f is the bivariate normal pdf given by Eq. (9.3).

(a) $z = c$ (const) cuts the surface in an *ellipse*. (These are sometimes called contours of constant probability density.)

(b) If $\rho = 0$ and $\sigma_x = \sigma_y$, the above ellipse becomes a circle. (What happens to the above ellipse as $\rho \to \pm 1$?)

Proof: See Problem 9.22.

Note: Because of the importance of the bivariate normal distribution, various probabilities associated with it have been tabulated. (See D. B. Owen, *Handbook of Statistical Tables*, Addison-Wesley Publishing Company, Inc., Reading, Mass., 1962.)

9.12 Truncated Distributions

EXAMPLE 9.10.   Suppose that a certain type of bolt is manufactured and its length, say Y, is a random variable with distribution $N(2.2, 0.01)$. From a large lot of such bolts a new lot is obtained by discarding all those bolts for which $Y > 2$. Hence if X is the random variable representing the length of the bolts in the new lot, and if F is its cdf we have

$$F(x) = P(X \le x)$$
$$= P(Y \le x \mid Y \le 2) = 1 \qquad \text{if} \quad x > 2,$$
$$= P(Y \le x)/P(Y \le 2) \qquad \text{if} \quad x \le 2.$$

(See Fig. 9.9.) Thus f, the pdf of X is given by

$$f(x) = F'(x) = 0 \qquad \text{if} \quad x > 2,$$
$$= \frac{\dfrac{1}{\sqrt{2\pi}\,(0.1)} \exp\left(-\dfrac{1}{2}\left[\dfrac{x - 2.2}{0.1}\right]^2\right)}{\Phi(-2)} \qquad \text{if } x \le 2$$

FIGURE 9.9

since

$$P(Y \leq 2) = P\left(\frac{Y - 2.2}{0.1} \leq \frac{2 - 2.2}{0.1}\right) = \Phi(-2).$$

[Φ, as usual, is the cdf of the distribution $N(0, 1)$.]

The above is an illustration of a *truncated normal distribution* (specifically, truncated to the right at $X = 2$). This example may be generalized as follows.

Definition. We say that the random variable X has a normal distribution *truncated to the right* at $X = \tau$ if its pdf f is of the form

$$f(x) = 0 \qquad \text{if} \quad x > \tau,$$

$$= K \frac{1}{\sqrt{2\pi}\,\sigma} \exp\left(-\frac{1}{2}\left[\frac{x - \mu}{\sigma}\right]^2\right) \qquad \text{if} \quad x \leq \tau. \qquad (9.22)$$

We note that K is determined from the condition $\int_{-\infty}^{+\infty} f(x)\,dx = 1$ and hence

$$K = \frac{1}{\Phi[(\tau - \mu)/\sigma]} = \frac{1}{P(Z \leq \tau)}$$

where Z has distribution $N(\mu, \sigma^2)$. Analogously to the above we have the following definition.

Definition. We say that the random variable X has a normal distribution *truncated to the left* at $X = \gamma$, if its pdf f is of the form

$$f(x) = 0 \qquad \text{if} \quad x < \gamma,$$

$$= \frac{K}{\sqrt{2\pi}\,\sigma} \exp\left(-\frac{1}{2}\left[\frac{x - \mu}{\sigma}\right]^2\right) \qquad \text{if} \quad x \geq \gamma. \qquad (9.23)$$

Again, K is determined from the condition $\int_{-\infty}^{+\infty} f(x)\,dx = 1$ and thus

$$K = \left[1 - \Phi\left(\frac{\gamma - \mu}{\sigma}\right)\right]^{-1}.$$

The concepts introduced above for the normal distribution may be extended in an obvious way to other distributions. For example, an exponentially distributed random variable X, truncated to the left at $X = \gamma$, would have the following pdf:

$$f(x) = 0 \qquad \text{if} \quad x < \gamma,$$

$$= C\alpha e^{-\alpha x} \qquad \text{if} \quad x \geq \gamma. \qquad (9.24)$$

Again, C is determined from the condition $\int_{-\infty}^{+\infty} f(x)\,dx = 1$ and hence

$$C = e^{\alpha\gamma}.$$

We can consider also a truncated random variable in the discrete case. For instance, if a Poisson-distributed random variable X (with parameter λ) is truncated to the right at $X = k + 1$, it means that X has the following distribution:

$$P(X = i) = 0 \qquad \text{if} \quad i \geq k + 1,$$

$$= C \frac{\lambda^i}{i!} e^{-\lambda} \qquad \text{if} \quad i = 0, 1, \dots, k. \qquad (9.25)$$

We determine C from the condition $\sum_{i=0}^{\infty} P(X = i) = 1$ and find

$$C = \frac{1}{\sum_{j=0}^{k} (\lambda^j / j!) e^{-\lambda}}.$$

Thus

$$P(X = i) = \frac{\lambda^i}{i!} \frac{1}{\sum_{j=0}^{k} (\lambda^j / j!)}, \qquad i = 0, 1, \dots, k \quad \text{and} \quad 0 \text{ elsewhere.}$$

Truncated distributions may arise in many important applications.   We shall consider a few examples below.

EXAMPLE 9.11. Suppose that X represents the life length of a component. If X is normally distributed with

$$E(X) = 4 \qquad \text{and} \qquad V(X) = 4,$$

we find that

$$P(X < 0) = \Phi(-2) = 0.023.$$

Thus this model is not very meaningful since it assigns probability 0.023 to an event which we know *cannot* occur. We might consider, instead, the above random variable X truncated to the left at $X = 0$. Hence we shall suppose that the pdf of the random variable X is given by

$$f(x) = 0 \qquad \text{if} \quad x \leq 0,$$

$$= \frac{1}{\sqrt{(2\pi)}\,(2)} \exp\left[-\frac{1}{2} \left(\frac{x - 4}{2} \right)^2 \right] \frac{1}{\Phi(2)} \qquad \text{if} \quad x > 0.$$

Note: We have indicated that we often use the normal distribution to represent a random variable X about which we *know* that it cannot assume negative values. (For instance, time to failure, the length of a rod, etc.) For certain parameter values $\mu = E(X)$ and $\sigma^2 = V(X)$ the value of $P(X < 0)$ will be negligible. However, if this is not the case (as in Example 9.11) we should consider using the normal distribution truncated to the left at $X = 0$.

EXAMPLE 9.12. Suppose that a system is made up of n components which function independently, each having the same probability p of functioning properly. Whenever the system malfunctions, it is inspected in order to discover which and how many components are faulty. Let the random variable X be defined as the number of components that are found to be faulty in a system which has broken down. If we suppose that the system fails if and only if at least one component fails, then X has a *binomial distribution truncated to the left* at $X = 0$. For the very fact that the system *has* failed precludes the possibility that $X = 0$. Specifically we have

$$P(X = k) = \frac{\binom{n}{k}(1 - p)^k p^{n-k}}{P \text{ (system fails)}}, \qquad k = 1, 2, \ldots, n.$$

Since P (system fails) $= 1 - p^n$, we may write

$$P(X = k) = \frac{\binom{n}{k}(1 - p)^k p^{n-k}}{1 - p^n}, \qquad k = 1, 2, \ldots, n.$$

EXAMPLE 9.13. Suppose that particles are emitted from a radioactive source according to the Poisson distribution with parameter λ. A counting device, recording these emissions only functions if fewer than three particles arrive. (That is, if more than three particles arrive during a specified time period, the device ceases to function because of some "locking" that takes place.) Hence if Y is the number of particles recorded during the specified time interval, Y has possible values 0, 1, and 2. Thus

$$P(Y = k) = \frac{e^{-\lambda}}{k!} \frac{\lambda^k}{e^{-\lambda}[1 + \lambda + (\lambda^2/2)]}, \qquad k = 0, 1, 2,$$

$$= 0, \qquad \text{elsewhere.}$$

Since the truncated normal distribution is particularly important, let us consider the following problem associated with this distribution.

Suppose that X is a normally distributed random variable truncated to the right at $X = \tau$. Hence the pdf f is of the form

$$f(x) = 0 \qquad \text{if} \quad x \geq \tau,$$

$$= \frac{1}{\sqrt{2\pi}\,\sigma} \exp\left[-\frac{1}{2}\left(\frac{x - \mu}{\sigma}\right)^2\right] \frac{1}{\Phi[(\tau - \mu)/\sigma]} \qquad \text{if} \quad x \leq \tau.$$

We therefore have

$$E(X) = \int_{-\infty}^{+\infty} x\, f(x)\, dx = \frac{1}{\Phi[(\tau - \mu)/\sigma]} \int_{-\infty}^{\tau} \frac{x}{\sqrt{2\pi}\,\sigma} \exp\left[-\frac{1}{2}\left(\frac{x - \mu}{\sigma}\right)^2\right] dx$$

$$= \frac{1}{\Phi[(\tau - \mu)/\sigma]} \frac{1}{\sqrt{2\pi}} \int_{-\infty}^{(\tau-\mu)/\sigma} (s\sigma + \mu)e^{-s^2/2}\, ds$$

$$= \frac{1}{\Phi[(\tau - \mu)/\sigma]} \left[\mu\Phi\left(\frac{\tau - \mu}{\sigma}\right) + \sigma\frac{1}{\sqrt{2\pi}} \int_{-\infty}^{(\tau-\mu)/\sigma} s e^{-s^2/2}\, ds\right]$$

$$= \mu + \frac{\sigma}{\Phi[(\tau - \mu)/\sigma]} \frac{1}{\sqrt{2\pi}} e^{-s^2/2}(-1)\Big|_{-\infty}^{(\tau-\mu)/\sigma}$$

$$= \mu - \frac{\sigma}{\Phi[(\tau - \mu)/\sigma]} \frac{1}{\sqrt{2\pi}} \exp\left[-\frac{1}{2}\left(\frac{\tau - \mu}{\sigma}\right)^2\right].$$

Note that the expression obtained for $E(X)$ is expressed in terms of tabulated functions. The function Φ is of course the usual cdf of the distribution $N(0, 1)$, while $(1/\sqrt{2\pi})e^{-x^2/2}$ is the ordinate of the pdf of the $N(0, 1)$ distribution and is also tabulated. In fact the quotient

$$\frac{(1/\sqrt{2\pi})e^{-x^2/2}}{\Phi(x)}$$

is tabulated. (See D. B. Owen, *Handbook of Statistical Tables*, Addison-Wesley Publishing Company, Inc., Reading, Mass., 1962.)

Using the above result, we may now ask the following question: For given μ and σ where should the truncation occur (that is, what should the value of τ be) so that the expected value *after* truncation has some preassigned value, say A? We can answer this question with the aid of the tabulated normal distribution. Suppose that $\mu = 10, \sigma = 1$, and we require that $A = 9.5$. Hence we must solve

$$9.5 = 10 - \frac{1}{\Phi(\tau - 10)} \frac{1}{\sqrt{2\pi}} e^{-(\tau-10)^2/2}.$$

This becomes

$$\frac{1}{2} = \frac{(1/\sqrt{2\pi})e^{-(\tau-10)^2/2}}{\Phi(\tau - 10)}.$$

Using the tables referred to above, we find that $\tau - 10 = 0.52$. and hence $\tau = 10.52$.

Note: The problem raised above may be solved only for certain values of μ, σ, and A. That is, for given μ and σ, it may not be possible to obtain a specified value of A. Consider the equation which must be solved:

$$\mu - A = \frac{\sigma}{\Phi[(\tau - \mu)/\sigma]} \frac{1}{\sqrt{2\pi}} \exp\left[-\frac{1}{2}\left(\frac{\tau - \mu}{\sigma}\right)^2\right].$$

The right-hand side of this equation is obviously positive. Hence we must have $(\mu - A) > 0$ in order for the above problem to have a solution. This condition is not very surprising since it says simply that the expected value (after truncation on the *right*) must be less than the original expected value.

PROBLEMS

9.1. Suppose that X has distribution $N(2, 0.16)$. Using the table of the normal distribution, evaluate the following probabilities.

(a) $P(X \geq 2.3)$ (b) $P(1.8 \leq X \leq 2.1)$

9.2. The diameter of an electric cable is normally distributed with mean 0.8 and variance 0.0004. What is the probability that the diameter will exceed 0.81 inch?

9.3.   Suppose that the cable in Problem 9.2 is considered defective if the diameter differs from its mean by more than 0.025. What is the probability of obtaining a defective cable?

9.4.   The errors in a certain length-measuring device are known to be normally distributed with expected value zero and standard deviation 1 inch. What is the probability that the error in measurement will be greater than 1 inch? 2 inches? 3 inches?

9.5.   Suppose that the life lengths of two electronic devices, say D_1 and D_2, have distributions $N(40, 36)$ and $N(45, 9)$, respectively. If the electronic device is to be used for a 45-hour period, which device is to be preferred? If it is to be used for a 48-hour period, which device is to be preferred?

9.6.   We may be interested only in the magnitude of X, say $Y = |X|$. If X has distribution $N(0, 1)$, determine the pdf of Y, and evaluate $E(Y)$ and $V(Y)$.

9.7.   Suppose that we are measuring the position of an object in the plane. Let X and Y be the errors of measurement of the x- and y-coordinates, respectively. Assume that X and Y are independently and identically distributed, each with distribution $N(0, \sigma^2)$. Find the pdf of $R = \sqrt{X^2 + Y^2}$. (The distribution of R is known as the *Rayleigh distribution*.) [*Hint*: Let $X = R \cos \psi$ and $Y = R \sin \psi$. Obtain the joint pdf of (R, ψ) and then obtain the marginal pdf of R.]

9.8.   Find the pdf of the random variable $Q = X/Y$, where X and Y are distributed as in Problem 9.7. (The distribution of Q is known as the *Cauchy* distribution.) Can you compute $E(Q)$?

9.9.   A distribution closely related to the normal distribution is the *lognormal distribution*. Suppose that X is normally distributed with mean μ and variance σ^2. Let $Y = e^X$. Then Y has the lognormal distribution. (That is, Y is lognormal if and only if ln Y is normal.) Find the pdf of Y. *Note:* The following random variables may be represented by the above distribution: the diameter of small particles after a crushing process, the size of an organism subject to a number of small impulses, and the life length of certain items.

9.10.   Suppose that X has distribution $N(\mu, \sigma^2)$. Determine c (as a function of μ and σ) such that $P(X \leq c) = 2P(X > c)$.

9.11.   Suppose that temperature (measured in degrees centigrade) is normally distributed with expectation $50°$ and variance 4. What is the probability that the temperature T will be between $48°$ and $53°$ centigrade?

9.12.   The outside diameter of a shaft, say D, is specified to be 4 inches. Consider D to be a normally distributed random variable with mean 4 inches and variance 0.01 inch^2. If the actual diameter differs from the specified value by more than 0.05 inch but less than 0.08 inch, the loss to the manufacturer is $0.50. If the actual diameter differs from the specified diameter by more than 0.08 inch, the loss is $1.00. The loss, L, may be considered as a random variable. Find the probability distribution of L and evaluate $E(L)$.

9.13.   Compare the *upper bound* on the probability $P[|X - E(X)| \geq 2\sqrt{V(X)}]$ obtained from Chebyshev's inequality with the exact probability in each of the following cases.

(a) X has distribution $N(\mu, \sigma^2)$.

(b) X has Poisson distribution with parameter λ.

(c) X has exponential distribution with parameter α.

9.14. Suppose that X is a random variable for which $E(X) = \mu$ and $V(X) = \sigma^2$. Suppose that Y is uniformly distributed over the interval (a, b). Determine a and b so that $E(X) = E(Y)$ and $V(X) = V(Y)$.

9.15. Suppose that X, the breaking strength of rope (in pounds), has distribution $N(100, 16)$. Each 100-foot coil of rope brings a profit of $25, provided $X > 95$. If $X \leq 95$, the rope may be used for a different purpose and a profit of $10 per coil is realized. Find the expected profit per coil.

9.16. Let X_1 and X_2 be independent random variables each having distribution $N(\mu, \sigma^2)$. Let $Z(t) = X_1 \cos \omega t + X_2 \sin \omega t$. This random variable is of interest in the study of random signals. Let $V(t) = dZ(t)/dt$. (ω is assumed to be constant.)

(a) What is the probability distribution of $Z(t)$ and $V(t)$ for any fixed t?

(b) Show that $Z(t)$ and $V(t)$ are uncorrelated. [*Note:* One can actually show that $Z(t)$ and $V(t)$ are independent but this is somewhat more difficult to do.]

9.17. A rocket fuel is to contain a certain percent (say X) of a particular compound. The specifications call for X to be between 30 and 35 percent. The manufacturer will make a net profit on the fuel (per gallon) which is the following function of X:

$$T(X) = \$0.10 \text{ per gallon} \quad \text{if } 30 < X < 35,$$
$$= \$0.05 \text{ per gallon} \quad \text{if } 35 \leq X < 40 \text{ or } 25 < X \leq 30,$$
$$= -\$0.10 \text{ per gallon otherwise.}$$

(a) If X has distribution $N(33, 9)$, evaluate $E(T)$.

(b) Suppose that the manufacturer wants to increase his expected profit, $E(T)$, by 50 percent. He intends to do this by increasing his profit (per gallon) on those batches of fuel meeting the specifications, $30 < X < 35$. What must his new net profit be?

9.18. Consider Example 9.8. Suppose that the operator is paid C_3 dollars/hour while the machine is operating and C_4 dollars/hour ($C_4 < C_3$) for the remaining time he has been hired after the machine has failed. Again determine for what value of H (the number of hours the operator is being hired), the expected profit is maximized.

9.19. Show that $\Gamma(\tfrac{1}{2}) = \sqrt{\pi}$. (See 9.15.) [*Hint:* Make the change of variable $x = u^2/2$ in the integral $\Gamma(\tfrac{1}{2}) = \int_0^\infty x^{-1/2}e^{-x}\,dx$.]

9.20. Verify the expressions for $E(X)$ and $V(X)$, where X has a Gamma distribution [see Eq. (9.18)].

9.21. Prove Theorem 9.3.

9.22. Prove Theorem 9.4.

9.23. Suppose that the random variable X has a chi-square distribution with 10 degrees of freedom. If we are asked to find two numbers a and b such that $P(a < x < b) = 0.85$, say, we should realize that there are many pairs of this kind.

(a) Find two different sets of values (a, b) satisfying the above condition.

(b) Suppose that in addition to the above, we require that

$$P(X < a) = P(X > b).$$

How many sets of values are there?

9.24.   Suppose that V, the velocity (cm/sec) of an object having a mass of 1 kg, is a random variable having distribution $N(0, 25)$. Let $K = 1000V^2/2 = 500V^2$ represent the kinetic energy (KE) of the object. Evaluate $P(K < 200)$, $P(K > 800)$.

9.25.   Suppose that X has distribution $N(\mu, \sigma^2)$. Using Theorem 7.7, obtain an approximation expression for $E(Y)$ and $V(Y)$ if $Y = \ln X$.

9.26.   Suppose that X has a normal distribution truncated to the right as given by Eq. (9.22). Find an expression for $E(X)$ in terms of tabulated functions.

9.27.   Suppose that X has an exponential distribution truncated to the left as given by Eq. (9.24). Obtain $E(X)$.

9.28.   (a) Find the probability distribution of a binomially distributed random variable (based on n repetitions of an experiment) truncated to the right at $X - n$; that is, $X = n$ cannot be observed.

(b) Find the expected value and variance of the random variable described in (a).

9.29.   Suppose that a normally distributed random variable with expected value μ and variance σ^2 is truncated to the left at $X = \tau$ and to the right at $X = \gamma$. Find the pdf of this "doubly truncated" random variable.

9.30.   Suppose that X, the length of a rod, has distribution $N(10, 2)$. Instead of measuring the value of X, it is only specified whether certain requirements are met. Specifically, each manufactured rod is classified as follows: $X < 8$, $8 \le X < 12$, and $X \ge 12$. If 15 such rods are manufactured, what is the probability that an equal number of rods fall into each of the above categories?

9.31.   The annual rainfall at a certain locality is known to be a normally distributed random variable with mean value equal to 29.5 inches and standard deviation 2.5 inches. How many inches of rain (annually) is exceeded about 5 percent of the time?

9.32.   Suppose that X has distribution $N(0, 25)$. Evaluate $P(1 < X^2 < 4)$.

9.33.   Let X_t be the number of particles emitted in t hours from a radioactive source and suppose that X_t has a Poisson distribution with parameter βt. Let T equal the number of hours until the first emission. Show that T has an exponential distribution with parameter β. [*Hint:* Find the equivalent event (in terms of X_t) to the event $T > t$.]

9.34.   Suppose that X_t is defined as in Problem 9.33 with $\beta = 30$. What is the probability that the time between successive emissions will be > 5 minutes? > 10 minutes? < 30 seconds?

9.35.   In some tables for the normal distribution, $H(x) = (1/\sqrt{2\pi})\int_0^x e^{-t^2/2}\, dt$ is tabulated for positive values of x (instead of $\Phi(x)$ as given in the Appendix). If the random variable X has distribution $N(1, 4)$ express each of the following probabilities in terms of *tabulated* values of the function H.

(a) $P[|X| > 2]$ (b) $P[X < 0]$

9.36.   Suppose that a satellite telemetering device receives two kinds of signals which may be recorded as real numbers, say X and Y. Assume that X and Y are independent, continuous random variables with pdf's f and g, respectively. Suppose that during any specified period of time only one of these signals may be received and hence transmitted back to earth, namely that signal which arrives first. Assume furthermore that the signal giving rise to the value of X arrives first with probability p and hence the signal giving rise to Y arrives first with probability $1 - p$. Let Z denote the random variable whose value is actually received and transmitted.

(a) Express the pdf of Z in terms of f and g.

(b) Express $E(Z)$ in terms of $E(X)$ and $E(Y)$.

(c) Express $V(Z)$ in terms of $V(X)$ and $V(Y)$.

(d) Suppose that X has distribution $N(2, 4)$ and that Y has distribution $N(3, 3)$. If $p = \frac{2}{3}$, evaluate $P(Z > 2)$.

(e) Suppose that X and Y have distributions $N(\mu_1, \sigma_1^2)$ and $N(\mu_2, \sigma_2^2)$, respectively. Show that if $\mu_1 = \mu_2$, the distribution of Z is "uni-modal," that is, the pdf of Z has a unique relative maximum.

9.37.   Assume that the number of accidents in a factory may be represented by a Poisson process averaging 2 accidents per week. What is the probability that (a) the time from one accident to the next will be more than 3 days, (b) the time from one accident to the third accident will be more than a week? [*Hint:* In (a), let $T =$ time (in days) and compute $P(T > 3)$.]

9.38.   On the average a production process produces one defective item among every 300 manufactured. What is the probability that the *third* defective item will appear:

(a) before 1000 pieces have been produced?

(b) as the 1000th piece is produced?

(c) after the 1000th piece has been produced?

[*Hint:* Assume a Poisson process.]

10

The Moment-Generating Function

10.1 Introduction

In this chapter we shall introduce an important mathematical concept which has many applications to the probabilistic models we are considering. In order to present a rigorous development of this topic, mathematics of a considerably higher level than we are assuming here would be required. However, if we are willing to avoid certain mathematical difficulties that arise and if we are willing to accept that certain operations are valid, then we can obtain a sufficient understanding of the main ideas involved in order to use them intelligently.

In order to motivate what follows, let us recall our earliest encounter with the logarithm. It was introduced purely as a computational aid. With each positive real number x, we associated another number, denoted by $\log x$. (The value of this number could be obtained from suitable tables.) In order to compute xy, for example, we obtain the values of $\log x$ and of $\log y$ and then evaluate $\log x + \log y$, which represents $\log xy$. From the knowledge of $\log xy$ we were then able to obtain the value of xy (again with the aid of tables). In a similar way we may simplify the computation of other arithmetic calculations with the aid of the logarithm. The above approach is useful for the following reasons.

(a) To each positive number x there corresponds exactly one number, $\log x$, and this number is easily obtained from tables.

(b) To each value of $\log x$ there corresponds exactly one value of x, and this value is again available from tables. (That is, the relationship between x and $\log x$ is one to one.)

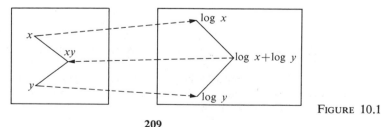

FIGURE 10.1

(c) Certain arithmetic operations involving the numbers x and y, such as multiplication and division, may be replaced by simpler operations, such as addition and subtraction, by means of the "transformed" numbers $\log x$ and $\log y$ (see the schematic in Fig. 10.1).

Instead of performing the arithmetic directly with the numbers x and y, we first obtain the numbers $\log x$ and $\log y$, do our arithmetic with these numbers, and then transform back.

10.2 The Moment-Generating Function

Let us now consider a more complicated situation. Suppose that X is a random variable; that is, X is a function from the sample space to the real numbers. In computing various characteristics of the random variable X, such as $E(X)$ or $V(X)$, we work directly with the probability distribution of X. [The probability distribution is given by a *function*: either the pdf in the continuous case, or the point probabilities $p(x_i) = P(X = x_i)$ in the discrete case. The latter may also be considered as a function assuming nonzero values only if $X = x_i, i = 1, 2, \ldots$] Possibly, we can introduce some other *function* and make our required computation in terms of this new function (just as above we associated with each *number*, some new *number*). This is, in fact, precisely what we shall do. Let us first make a formal definition.

Definition. Let X be a discrete random variable with probability distribution $p(x_i) = P(X = x_i)$, $i = 1, 2, \ldots$ The function M_X, called the *moment-generating function of X*, is defined by

$$M_X(t) = \sum_{j=1}^{\infty} e^{tx_j} p(x_j). \tag{10.1}$$

If X is a continuous random variable with pdf f, we define the moment-generating function by

$$M_X(t) = \int_{-\infty}^{+\infty} e^{tx} f(x)\, dx. \tag{10.2}$$

Notes: (a) In either the discrete or the continuous case, $M_X(t)$ is simply the expected value of e^{tX}. Hence we may combine the above expressions and write

$$M_X(t) = E(e^{tX}). \tag{10.3}$$

(b) $M_X(t)$ is the value which the function M_X assumes for the (real) variable t. The notation, indicating the dependence on X, is used because we may wish to consider two random variables, say X and Y, and then investigate the moment-generating function of each, say M_X and M_Y.

(c) We shall use the abbreviation mgf for the moment-generating function.

(d) The mgf as defined above is written as an infinite series or (improper) integral, depending on whether the random variable is discrete or continuous. Such a series (or integral) may not always exist (that is, converge to a finite value) for all values of t. Hence it may happen that the mgf is not defined for all values of t. However, we shall not concern ourselves with this potential difficulty. Whenever we make use of the mgf, we shall always assume it exists. (*At $t = 0$, the mgf always exists and equals 1.*)

(e) There is another function closely related to the mgf which is often used in its place. It is called the *characteristic function*, denoted by C_X, and is defined by $C_X(t) = E(e^{itX})$, where $i = \sqrt{-1}$, the imaginary unit. For theoretical reasons, there is considerable advantage in using $C_X(t)$ instead of $M_X(t)$. [For one thing, $C_X(t)$ always exists for *all* values of t.] However, in order to avoid computations with complex numbers we shall restrict our discussion to the moment-generating function.

(f) We shall delay until Section 10.4 the discussion of the reason for calling M_X the moment-generating function.

10.3 Examples of Moment-Generating Functions

Before considering some of the important applications of the mgf to the theory of probability, let us evaluate a few of these functions.

EXAMPLE 10.1. Suppose that X is *uniformly distributed* over the interval $[a, b]$. Therefore the mgf is given by

$$M_X(t) = \int_a^b \frac{e^{tx}}{b - a} \, dx$$

$$= \frac{1}{(b - a)t} [e^{bt} - e^{at}], \quad t \neq 0. \tag{10.4}$$

EXAMPLE 10.2. Suppose that X is *binomially* distributed with parameters n and p. Then

$$M_X(t) = \sum_{k=0}^{n} e^{tk} \binom{n}{k} p^k (1 - p)^{n-k}$$

$$= \sum_{k=0}^{n} \binom{n}{k} (pe^t)^k (1 - p)^{n-k}$$

$$= [pe^t + (1 - p)]^n. \tag{10.5}$$

(The last equality follows from a direct application of the binomial theorem.)

EXAMPLE 10.3. Suppose that X has a *Poisson* distribution with parameter λ. Thus

$$M_X(t) = \sum_{k=0}^{\infty} e^{tk} \frac{e^{-\lambda} \lambda^k}{k!} = e^{-\lambda} \sum_{k=0}^{\infty} \frac{(\lambda e^t)^k}{k!}$$

$$= e^{-\lambda} e^{\lambda e^t} = e^{\lambda(e^t - 1)}. \tag{10.6}$$

(The third equality follows from the expansion of e^y into $\sum_{n=0}^{\infty} (y^n/n!)$. We used this with $y = \lambda e^t$.)

EXAMPLE 10.4. Suppose that X has an *exponential distribution* with parameter α. Therefore

$$M_X(t) = \int_0^\infty e^{tx} \alpha e^{-\alpha x} \, dx = \alpha \int_0^\infty e^{x(t-\alpha)} \, dx.$$

(This integral converges only if $t < \alpha$. Hence the mgf exists only for those values of t. Assuming that this condition is satisfied, we shall proceed.) Thus

$$M_X(t) = \frac{\alpha}{t-\alpha} e^{x(t-\alpha)} \Big|_0^\infty$$

$$= \frac{\alpha}{\alpha - t}, \qquad t < \alpha. \tag{10.7}$$

Note: Since the mgf is simply an expected value X, we can obtain the mgf of a function of a random variable without first obtaining its probability distribution (see Theorem 7.3). For example, if X has distribution $N(0, 1)$ and we want to find the mgf of $Y = X^2$, we can proceed without first obtaining the pdf of Y. We may simply write

$$M_Y(t) = E(e^{tY}) = E(e^{tX^2}) = \frac{1}{\sqrt{2\pi}} \int_{-\infty}^{+\infty} \exp(tx^2 - x^2/2)dx = (1 - 2t)^{-1/2}$$

after a straightforward integration.

EXAMPLE 10.5. Suppose that X has distribution $N(\mu, \sigma^2)$. Hence

$$M_X(t) = \frac{1}{\sqrt{2\pi}\,\sigma} \int_{-\infty}^{+\infty} e^{tx} \exp\left(-\frac{1}{2}\left[\frac{x-\mu}{\sigma}\right]^2\right) dx.$$

Let $(x - \mu)/\sigma = s$; thus $x = \sigma s + \mu$ and $dx = \sigma \, ds$. Therefore

$$M_X(t) = \frac{1}{\sqrt{2\pi}} \int_{-\infty}^{+\infty} \exp[t(\sigma s + \mu)]e^{-s^2/2} \, ds$$

$$= e^{t\mu} \frac{1}{\sqrt{2\pi}} \int_{-\infty}^{+\infty} \exp\left(-\tfrac{1}{2}[s^2 - 2\sigma ts]\right) ds$$

$$= e^{t\mu} \frac{1}{\sqrt{2\pi}} \int_{-\infty}^{+\infty} \exp\left\{-\tfrac{1}{2}[(s - \sigma t)^2 - \sigma^2 t^2]\right\} ds$$

$$= e^{t\mu + \sigma^2 t^2/2} \frac{1}{\sqrt{2\pi}} \int_{-\infty}^{+\infty} \exp\left(-\tfrac{1}{2}[s - \sigma t]^2\right) ds.$$

Let $s - \sigma t = v$; then $ds = dv$ and we obtain

$$M_X(t) = e^{t\mu + \sigma^2 t^2/2} \frac{1}{\sqrt{2\pi}} \int_{-\infty}^{+\infty} e^{-v^2/2} \, dv$$

$$= e^{(t\mu + \sigma^2 t^2/2)}. \tag{10.8}$$

EXAMPLE 10.6. Let X have a *Gamma distribution* with parameters α and r (see Eq. 9.16). Then

$$M_X(t) = \frac{\alpha}{\Gamma(r)} \int_0^\infty e^{tx} (\alpha x)^{r-1} e^{-\alpha x} \, dx$$

$$= \frac{\alpha^r}{\Gamma(r)} \int_0^\infty x^{r-1} e^{-x(\alpha - t)} \, dx.$$

(This integral converges provided $\alpha > t$.) Let $x(\alpha - t) = u$; thus

$$dx = (du)/(\alpha - t),$$

and we obtain

$$M_X(t) = \frac{\alpha^r}{(\alpha - t)\Gamma(r)} \int_0^\infty \left(\frac{u}{\alpha - t}\right)^{r-1} e^{-u} \, du$$

$$= \left(\frac{\alpha}{\alpha - t}\right)^r \frac{1}{\Gamma(r)} \int_0^\infty u^{r-1} e^{-u} \, du.$$

Since the integral equals $\Gamma(r)$, we have

$$M_X(t) = \left(\frac{\alpha}{\alpha - t}\right)^r. \tag{10.9}$$

Notes: (a) If $r = 1$, the Gamma function becomes the exponential distribution. We see that if $r = 1$, Eqs. (10.7) and (10.9) are the same.

(b) Since the *chi-square* distribution is obtained as a special case of the Gamma distribution by letting $\alpha = \frac{1}{2}$ and $r = n/2$ (n a positive integer), we have that if Z has distribution χ_n^2, then

$$M_Z(t) = (1 - 2t)^{-n/2}. \tag{10.10}$$

10.4 Properties of the Moment-Generating Function

We shall now indicate the reason for calling M_X the *moment-generating* function. We recall the Maclaurin series expansion of the function e^x:

$$e^x = 1 + x + \frac{x^2}{2!} + \frac{x^3}{3!} + \cdots + \frac{x^n}{n!} + \cdots$$

(It is known that this series converges for all values of x.) Thus

$$e^{tx} = 1 + tx + \frac{(tx)^2}{2!} + \cdots + \frac{(tx)^n}{n!} + \cdots$$

Now

$$M_X(t) = E(e^{tX}) = E\left(1 + tX + \frac{(tX)^2}{2!} + \cdots + \frac{(tX)^n}{n!} + \cdots\right).$$

We have shown that for a *finite* sum, the expected value of the sum equals the sum of the expected values. However, above we are dealing with an infinite series and hence cannot, immediately, apply such a result. It turns out, however, that under fairly general conditions this operation is still valid. We shall assume that the required conditions are satisfied and proceed accordingly.

We recall that t is a constant so far as taking the expectation is concerned and we may write

$$M_X(t) = 1 + tE(X) + \frac{t^2 E(X^2)}{2!} + \cdots + \frac{t^n E(X^n)}{n!} + \cdots$$

Since M_X is a function of the real variable t, we may consider taking the derivative of $M_X(t)$ with respect to t, that is, $[d/(dt)]M_X(t)$ or for short, $M'(t)$. Again we are faced with a mathematical difficulty. The derivative of a *finite* sum is always equal to the sum of the derivatives (assuming, of course, that all derivatives in question exist). However, for an infinite sum this is not always so. Certain conditions must be satisfied in order to justify this operation; we shall simply assume that these conditions hold and proceed. (In most problems we shall encounter, such an assumption is justified.) Thus,

$$M'(t) = E(X) + tE(X^2) + \frac{t^2 E(X^3)}{2!} + \cdots + \frac{t^{n-1} E(X^n)}{(n-1)!} + \cdots$$

Setting $t = 0$ we find that only the first term survives and we have

$$M'(0) = E(X).$$

Thus the first derivative of the mgf evaluated at $t = 0$ yields the expected value of the random variable. If we compute the second derivative of $M_X(t)$, again proceeding as above, we obtain

$$M''(t) = E(X^2) + tE(X^3) + \cdots + \frac{t^{n-2} E(X^n)}{(n-2)!} + \cdots,$$

and setting $t = 0$, we have

$$M''(0) = E(X^2).$$

Continuing in this manner, we obtain [assuming that $M^{(n)}(0)$ exists] the following theorem.

Theorem 10.1

$$M^{(n)}(0) = E(X^n) \tag{10.11}$$

That is, the nth derivative of $M_X(t)$ evaluated at $t = 0$ yields $E(X^n)$.

Notes: (a) The numbers $E(X^n)$, $n = 1, 2, \ldots$, are called the nth *moments* of the random variable X about zero. We have therefore shown that from a knowledge of the function M_X, the moments may be "generated." (Hence the name "moment-generating function.")
     (b) Let us recall the general Maclaurin series expansion of a function, say h.

$$h(t) = h(0) + h'(0)t + \frac{h''(0)t^2}{2!} + \cdots + \frac{h^{(n)}(0)t^n}{n!} + \cdots,$$

where $h^{(n)}(0)$ is the nth derivative of the function h evaluated at $t = 0$. Applying this result to the function M_X, we may write

$$M_X(t) = M_X(0) + M'_X(0)t + \cdots + \frac{M_X^{(n)}(0)t^n}{n!} + \cdots$$

$$= 1 + \mu_1 t + \mu_2 t^2/2! + \cdots + \frac{\mu_n t^n}{n!} + \cdots$$

where $\mu_i = E(X^i)$, $i = 1, 2, \ldots$ In particular,

$$V(X) = E(X^2) - \left(E(X)\right)^2 = M''(0) - [M'(0)]^2.$$

(c) The reader may wonder why the above methods should be fruitful at all. Would it not be simpler (and more straightforward) to compute the moments of X directly, rather than first obtain the mgf and then differentiate it? The answer is that for many problems the latter approach *is* more easily followed. The following examples will illustrate this.

EXAMPLE 10.7. Suppose that X has a binomial distribution with parameters n and p. Hence (Example 10.2), $M_X(t) = [pe^t + q]^n$. Thus

$$M'(t) = n(pe^t + q)^{n-1}pe^t,$$
$$M''(t) = np[e^t(n-1)(pe^t + q)^{n-2}pe^t + (pe^t + q)^{n-1}e^t].$$

Therefore $E(X) = M'(0) = np$, which agrees with our previous result. Also, $E(X^2) = M''(0) = np[(n-1)p + 1]$. Hence

$$V(X) = M''(0) - [M'(0)]^2 = np(1 - p),$$

which again agrees with our previous finding.

EXAMPLE 10.8. Suppose that X has distribution $N(\alpha, \beta^2)$. Therefore (Example 10.5), $M_X(t) = \exp(\alpha t + \frac{1}{2}\beta^2 t^2)$. Thus

$$M'(t) = e^{\alpha t + \beta^2 t^2/2}(\beta^2 t + \alpha),$$
$$M''(t) = e^{\beta^2 t^2/2 + \alpha t}\beta^2 + (\beta^2 t + \alpha)^2 e^{\beta^2 t^2/2 + \alpha t},$$

and $M'(0) = \alpha$, $M''(0) = \beta^2 + \alpha^2$, yielding $E(X) = \alpha$ and $V(X) = \beta^2$ as before.

Let us use the method of mgf's to evaluate the expectation and variance of a random variable with a geometric probability distribution, Eq. (8.5).

EXAMPLE 10.9. Let X have a geometric probability distribution. That is, $P(X = k) = q^{k-1}p$, $k = 1, 2, \ldots$ $(p + q = 1)$. Thus

$$M_X(t) = \sum_{k=1}^{\infty} e^{tk}q^{k-1}p = \frac{p}{q}\sum_{k=1}^{\infty} (qe^t)^k.$$

If we restrict ourselves to those values of t for which $0 < qe^t < 1$ [that is, $t < \ln(1/q)$] then we may sum the above series as a geometric series and obtain

$$M_X(t) = \frac{p}{q} qe^t[1 + qe^t + (qe^t)^2 + \cdots]$$

$$= \frac{p}{q} \frac{qe^t}{1 - qe^t} = \frac{pe^t}{1 - qe^t}.$$

Therefore

$$M'(t) = \frac{(1 - qe^t)pe^t - pe^t(-qe^t)}{(1 - qe^t)^2} = \frac{pe^t}{(1 - qe^t)^2};$$

$$M''(t) = \frac{(1 - qe^t)^2 pe^t - pe^t 2(1 - qe^t)(-qe^t)}{(1 - qe^t)^4} = \frac{pe^t(1 + qe^t)}{(1 - qe^t)^3}.$$

Hence

$$E(X) = M'(0) = p/(1 - q)^2 = 1/p,$$
$$E(X^2) = M''(0) = p(1 + q)/(1 - q)^3 = (1 + q)/p^2,$$

and

$$V(X) = (1 + q)/p^2 - (1/p)^2 = q/p^2.$$

Thus we have a verification of Theorem 8.5.

The following two theorems will be of particular importance in our application of the mgf.

Theorem 10.2. Suppose that the random variable X has mgf M_X. Let $Y = \alpha X + \beta$. Then M_Y, the mgf of the random variable Y, is given by

$$M_Y(t) = e^{\beta t} M_X(\alpha t). \tag{10.12}$$

In words: To find the mgf of $Y = \alpha X + \beta$, evaluate the mgf of X at αt (instead of t) and multiply by $e^{\beta t}$.

Proof

$$M_Y(t) = E(e^{Yt}) = E[e^{(\alpha X + \beta)t}]$$
$$= e^{\beta t} E(e^{\alpha t X}) = e^{\beta t} M_X(\alpha t).$$

Theorem 10.3. Let X and Y be two random variables with mgf's, $M_X(t)$ and $M_Y(t)$, respectively. If $M_X(t) = M_Y(t)$ for all values of t, then X and Y have the same probability distribution.

Proof: The proof of this theorem is too difficult to be given here. However, it is very important to understand exactly what the theorem states. It says that if two random variables have the same mgf, then they have the same probability distribution. That is, the mgf uniquely determines the probability distribution of the random variable.

EXAMPLE 10.10. Suppose that X has distribution $N(\mu, \sigma^2)$. Let $Y = \alpha X + \beta$. Then Y is again normally distributed. From Theorem 10.2, the mgf of Y is $M_Y(t) = e^{\beta t}M_X(\alpha t)$. However, from Example 10.5 we have that

$$M_X(t) = e^{\mu t + \sigma^2 t^2/2}.$$

Hence

$$M_Y(t) = e^{\beta t}[e^{\alpha \mu t + (\alpha \sigma)^2 t^2/2}]$$
$$= e^{(\beta + \alpha \mu)t}e^{(\alpha \sigma)^2 t^2/2}.$$

But this is the mgf of a normally distributed random variable with expectation $\alpha \mu + \beta$ and variance $\alpha^2 \sigma^2$. Thus, according to Theorem 10.3, the distribution of Y is normal.

The following theorem also plays a vital role in our subsequent work.

Theorem 10.4. Suppose that X and Y are independent random variables. Let $Z = X + Y$. Let $M_X(t)$, $M_Y(t)$, and $M_Z(t)$ be the mgf's of the random variables X, Y, and Z, respectively. Then

$$M_Z(t) = M_X(t)M_Y(t). \tag{10.13}$$

Proof
$$M_Z(t) = E(e^{Zt}) = E[e^{(X+Y)t}] = E(e^{Xt}e^{Yt})$$
$$= E(e^{Xt})E(e^{Yt}) = M_X(t)M_Y(t).$$

Note: This theorem may be generalized as follows: If $X_1, \ldots, X_n$ are independent random variables with mgf's M_{X_i}, $i = 1, 2, \ldots, n$, then M_Z, the mgf of

$$Z = X_1 + \cdots + X_n,$$

is given by

$$M_Z(t) = M_{X_1}(t) \cdots M_{X_n}(t). \tag{10.14}$$

10.5 Reproductive Properties

There are a number of probability distributions which have the following remarkable and very useful property: If two (or more) independent random variables having a certain distribution are added, the resulting random variable has a distribution of the same type as that of the summands. This property is called a *reproductive property*, and we shall establish it for a number of important distributions with the aid of Theorems 10.3 and 10.4.

EXAMPLE 10.11. Suppose that X and Y are independent random variables with distributions $N(\mu_1, \sigma_1^2)$ and $N(\mu_2, \sigma_2^2)$, respectively. Let $Z = X + Y$. Hence

$$M_Z(t) = M_X(t)M_Y(t) = \exp(\mu_1 t + \sigma_1^2 t^2/2)\exp(\mu_2 t + \sigma_2^2 t^2/2)$$
$$= \exp[(\mu_1 + \mu_2)t + (\sigma_1^2 + \sigma_2^2)t^2/2].$$

However, this represents the mgf of a normally distributed random variable with expected value $\mu_1 + \mu_2$ and variance $\sigma_1^2 + \sigma_2^2$. Thus Z has this normal distribution. (See Theorem 10.3.)

Note: The fact that $E(Z) = \mu_1 + \mu_2$ and that $V(Z) = \sigma_1^2 + \sigma_2^2$ could have been obtained immediately from previous results concerning the properties of expectation and variance. But to establish the fact that Z is again *normally* distributed required the use of the mgf. (There is another approach to this result which will be mentioned in Chapter 12.)

EXAMPLE 10.12. The length of a rod is a normally distributed random variable with mean 4 inches and variance 0.01 inch^2. Two such rods are placed end to end and fitted into a slot. The length of this slot is 8 inches with a tolerance of ± 0.1 inch. What is the probability that the two rods will fit?

Letting L_1 and L_2 represent the lengths of rod 1 and rod 2, we have that $L = L_1 + L_2$ is normally distributed with $E(L) = 8$ and $V(L) = 0.02$. Hence

$$P[7.9 \le L \le 8.1] = P\left[\frac{7.9 - 8}{0.14} \le \frac{L - 8}{0.14} \le \frac{8.1 - 8}{0.14}\right]$$

$$= \Phi(+0.714) - \Phi(-0.714) = 0.526,$$

from the tables of the normal distribution.

We can generalize the above result in the following theorem.

Theorem 10.5 (*the reproductive property of the normal distribution*).   Let $X_1, X_2, \ldots, X_n$ be n independent random variables with distribution $N(\mu_i, \sigma_i^2)$, $i = 1, 2, \ldots, n$. Let $Z = X_1 + \cdots + X_n$. Then Z has distribution $N(\sum_{i=1}^{n} \mu_i, \sum_{i=1}^{n} \sigma_i^2)$.

The Poisson distribution also possesses a reproductive property.

Theorem 10.6.   Let $X_1, \ldots, X_n$ be independent random variables. Suppose that X_i has a Poisson distribution with parameter α_i, $i = 1, 2, \ldots, n$. Let $Z = X_1 + \cdots + X_n$. Then Z has a Poisson distribution with parameter

$$\alpha = \alpha_1 + \cdots + \alpha_n.$$

Proof: First let us consider the case of $n = 2$:

$$M_{X_1}(t) = e^{\alpha_1(e^t - 1)}, \qquad M_{X_2}(t) = e^{\alpha_2(e^t - 1)}.$$

Hence $M_Z(t) = e^{(\alpha_1 + \alpha_2)(e^t - 1)}$. But this is the mgf of a random variable with Poisson distribution having parameter $\alpha_1 + \alpha_2$. We can now complete the proof of the theorem with the aid of mathematical induction.

EXAMPLE 10.13. Suppose that the number of calls coming into a telephone exchange between 9 a.m. and 10 a.m., say X_1, is a random variable with Poisson distribution with parameter 3. Similarly, the number of calls arriving between 10 a.m. and 11 a.m., say X_2, also has a Poisson distribution, with parameter 5. If X_1 and X_2 are independent, what is the probability that more than 5 calls come in between 9 a.m. and 11 a.m.?

Let $Z = X_1 + X_2$. From the above theorem, Z has a Poisson distribution with parameter $3 + 5 = 8$. Hence

$$P(Z > 5) = 1 - P(Z \le 5) = 1 - \sum_{k=0}^{5} \frac{e^{-8}(8)^k}{k!}$$

$$= 1 - 0.1912 = 0.8088.$$

Another distribution with a reproductive property is the chi-square distribution.

Theorem 10.7. Suppose that the distribution of X_i is $\chi^2_{n_i}$, $i = 1, 2, \ldots, k$, where the X_i's are independent random variables. Let $Z = X_1 + \cdots + X_k$. Then Z has distribution χ^2_n, where $n = n_1 + \cdots + n_k$.

Proof: From Eq. (10.10) we have $M_{X_i}(t) = (1 - 2t)^{-n_i/2}$, $i = 1, 2, \ldots, k$. Hence

$$M_Z(t) = M_{X_1}(t) \cdots M_{X_k}(t) = (1 - 2t)^{-(n_1 + \cdots + n_k)/2}.$$

But this is the mgf of a random variable having χ^2_n distribution.

We can now indicate one of the reasons for the great importance of the chi-square distribution. In Example 9.9 we found that if X has distribution $N(0, 1)$, X^2 has distribution χ^2_1. Combining this with Theorem 10.7, we have the following result.

Theorem 10.8. Suppose that $X_1, \ldots, X_k$ are independent random variables, each having distribution $N(0, 1)$. Then $S = X_1^2 + X_2^2 + \cdots + X_k^2$ has distribution χ^2_k.

EXAMPLE 10.14. Suppose that $X_1, \ldots, X_n$ are independent random variables, each with distribution $N(0, 1)$. Let $T = \sqrt{X_1^2 + \cdots + X_n^2}$. From our previous discussion we know that T^2 has distribution χ^2_n.

To find the pdf of T, say h, we proceed as usual:

$$H(t) = P(T \le t) = P(T^2 \le t^2)$$

$$= \int_0^{t^2} \frac{1}{2^{n/2}\Gamma(n/2)} z^{n/2-1} e^{-z/2} \, dz.$$

Hence, we have

$$h(t) = H'(t)$$

$$= \frac{2t}{2^{n/2}\Gamma(n/2)} (t^2)^{n/2-1} e^{-t^2/2}$$

$$= \frac{2t^{n-1}e^{-t^2/2}}{2^{n/2}\Gamma(n/2)} \qquad \text{if} \quad t \geq 0.$$

Notes: (a) If $n = 2$, the above distribution is known as a *Rayleigh distribution*. (See Problem 9.7.)

(b) If $n = 3$, the above distribution is known as *Maxwell's distribution* (or, sometimes Maxwell's velocity distribution) and has the following important interpretation. Suppose that we have a gas in a closed container. Let (X, Y, Z) represent the velocity components of a randomly chosen molecule. We shall assume that X, Y, and Z are independent random variables each having distribution $N(0, \sigma^2)$. (Assuming the same distribution for X, Y, and Z means that the pressure in the gas is the same in all directions. Assuming the expectations to be equal to zero means that the gas is not flowing.) Hence the *speed* of the molecule (that is, the magnitude of its velocity) is given by $S = \sqrt{X^2 + Y^2 + Z^2}$. We note that X/σ, Y/σ, and Z/σ are distributed according to $N(0, 1)$. Thus $S/\sigma = \sqrt{(X/\sigma)^2 + (Y/\sigma)^2 + (Z/\sigma)^2}$ has Maxwell's distribution. Therefore g, the pdf of the speed S, is given by

$$g(s) = \sqrt{\frac{2}{\pi}} \frac{s^2}{\sigma^3} e^{-s^2/2\sigma^2}, \qquad s \geq 0.$$

The graph of g is indicated in Fig. 10.2 for $\sigma = 2$. Note that very large and very small values of S are quite unlikely. (It may be shown that the constant σ appearing as a parameter of the above distribution has the following physical interpretation: $\sigma = \sqrt{kT/M}$, where T is the absolute temperature, M is the mass of the molecule, and k is known as Boltzmann's constant.)

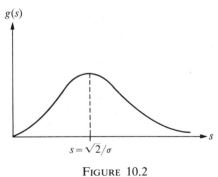

$g(s)$

$s = \sqrt{2}/\sigma$

FIGURE 10.2

We have discussed a number of distributions for which a reproductive property holds. Let us consider the exponential distribution for which, strictly speaking, a reproductive property does not hold, but which nevertheless has a somewhat analogous property.

Let X_i, $i = 1, 2, \ldots, r$ be r independent, identically distributed exponential random variables with parameter α. Hence from Eq. (10.7) we have that

$$M_{X_i}(t) = \alpha/(\alpha - t).$$

Thus if $Z = X_1 + \cdots + X_r$, we have $M_Z(t) = [\alpha/\alpha - t]^r$, which is precisely the moment-generating function of the Gamma distribution with parameters α and r (Eq. 10.9). Unless $r = 1$, this is not an mgf of an exponential distribution, so this distribution does *not* possess a reproductive property. But we do have a very interesting characterization of the Gamma distribution which we summarize in the following theorem.

Theorem 10.9. Let $Z = X_1 + \cdots + X_r$, where the X_i's are r independent, identically distributed random variables, each having an exponential distribution with the (same) parameter α. Then Z has a Gamma distribution with parameters α and r.

Notes: (a) Theorem 10.9 is not true if the parameters of the various exponential distributions are different. This becomes evident when we consider the mgf of the resulting sum of random variables.

(b) The following *corollary* of the above theorem has considerable importance in certain statistical applications: The random variable $W = 2\alpha Z$ has distribution χ^2_{2r}. This is an immediate consequence of the fact that $M_W(t) = M_Z(2\alpha t) = [\alpha/(\alpha - 2\alpha t)]^r = (1 - 2t)^{-2r/2}$. Comparing this with Eq. (10.10) yields the above corollary. Thus we may use the tabulated chi-square distribution in order to evaluate certain probabilities associated with Z. For example, $P(Z \le 3) = P(2\alpha Z \le 6\alpha)$. This latter probability may be obtained directly from the tables of the chi-square distribution, if α and r are given.

10.6 Sequences of Random Variables

Suppose that we have a sequence of random variables $X_1, X_2, \ldots, X_n, \ldots$ Each of these random variables may be described in terms of F_i, its cdf, where $F_i(t) = P(X_i \le t)$, $i = 1, 2, \ldots$ Quite often we are interested in what happens to F_i as $i \to \infty$. That is, is there some *limiting distribution function F* corresponding to some random variable X such that in some sense, the random variables X_i converge to X? The answer is yes, in many cases, and there is a fairly straightforward procedure for determining F.

Such a situation may arise when we consider n independent observations of a random variable X, say $X_1, \ldots, X_n$. We might be interested in the arithmetic mean of these observations, say $\bar{X}_n = (1/n)(X_1 + \cdots + X_n)$. $\bar{X}_n$ is again a random variable. Let $\bar{F}_n$ be the cdf of $\bar{X}_n$. It might be of interest to learn what happens to the probability distribution of $\bar{X}_n$ as n becomes large. Thus our problem involves the limiting behavior of $\bar{F}_n$ as $n \to \infty$. The following theorem, stated without proof, will enable us to solve this and similar problems.

Theorem 10.10. Let $X_1, \ldots, X_n, \ldots$ be a sequence of random variable with cdf's $F_1, \ldots, F_n; \ldots$ and mgf's $M_1, \ldots, M_n, \ldots$ Suppose that $\lim_{n \to \infty} M_n(t) = M(t)$, where $M(0) = 1$. Then $M(t)$ is the mgf of the random variable X whose cdf F is given by $\lim_{n \to \infty} F_n(t)$.

Note: Theorem 10.10 says that to obtain the limiting distribution sought, it suffices to study the moment-generating functions of the random variables under consideration. We obtain the limiting form of the sequence $M_1, \ldots, M_n, \ldots$, say $M(t)$. Because of the uniqueness property of mgf's, there exists only one probability distribution corresponding to the mgf $M(t)$. We may either recognize M as the mgf of a known distribution (such as normal, Poisson, etc.) or we may have to use more advanced methods to determine the probability distribution from M. Just as we are able to obtain the mgf from a knowledge of the pdf so we are able to obtain (under fairly general conditions) the pdf from a knowledge of the mgf. This would involve certain inversion theorems and we shall not pursue this further.

10.7 Final Remark

We have seen that the mgf can be a very powerful tool for studying various aspects of probability distributions. In particular, we found the use of the mgf very helpful in studying sums of independent, identically distributed random variables, and obtaining various reproductive laws. We shall study sums of independent random variables again in Chapter 12, without the use of the mgf, but with methods similar to those we used while studying the product and quotient of independent random variables in Chapter 6.

PROBLEMS

10.1.   Suppose that X has pdf given by

$$f(x) = 2x, \quad 0 \le x \le 1.$$

(a) Determine the mgf of X.
(b) Using the mgf, evaluate $E(X)$ and $V(X)$ and check your answer. (See Note, p. 224.)

10.2.   (a) Find the mgf of the voltage (*including* noise) as discussed in Problem 7.25.
(b) Using the mgf, obtain the expected value and variance of this voltage.

10.3.   Suppose that X has the following pdf:

$$f(x) = \lambda e^{-\lambda(x-a)}, \quad x \ge a.$$

(This is known as a *two-parameter exponential distribution*.)
(a) Find the mgf of X.
(b) Using the mgf, find $E(X)$ and $V(X)$.

10.4.   Let X be the outcome when a fair die is tossed.
(a) Find the mgf of X.
(b) Using the mgf, find $E(X)$ and $V(X)$.

10.5.   Find the mgf of the random variable X of Problem 6.7. Using the mgf, find $E(X)$ and $V(X)$.

10.6.   Suppose that the continuous random variable X has pdf

$$f(x) = \tfrac{1}{2}e^{-|x|}, \quad -\infty < x < \infty.$$

(a) Obtain the mgf of X.
(b) Using the mgf, find $E(X)$ and $V(X)$.

10.7. Use the mgf to show that if X and Y are independent random variables with distribution $N(\mu_x, \sigma_x^2)$ and $N(\mu_y, \sigma_y^2)$, respectively, then $Z = aX + bY$ is again normally distributed, where a and b are constants.

10.8. Suppose that the mgf of a random variable X is of the form

$$M_X(t) = (0.4e^t + 0.6)^8.$$

(a) What is the mgf of the random variable $Y = 3X + 2$?
(b) Evaluate $E(X)$.
(c) Can you check your answer to (b) by some other method? [Try to "recognize" $M_X(t)$.]

10.9. A number of resistances, R_i, $i = 1, 2, \ldots, n$, are put into a series arrangement in a circuit. Suppose that each resistance is normally distributed with $E(R_i) = 10$ ohms and $V(R_i) = 0.16$.

(a) If $n = 5$, what is the probability that the resistance of the circuit exceeds 49 ohms?
(b) How large should n be so that the probability that the total resistance exceeds 100 ohms is approximately 0.05?

10.10. In a circuit n resistances are hooked up into a series arrangement. Suppose that each resistance is uniformly distributed over $[0, 1]$ and suppose, furthermore, that all resistances are independent. Let R be the total resistance.

(a) Find the mgf of R.
(b) Using the mgf, obtain $E(R)$ and $V(R)$. Check your answers by direct computation.

10.11. If X has distribution χ_n^2, using the mgf, show that $E(X) = n$ and $V(X) = 2n$.

10.12. Suppose that V, the velocity (cm/sec) of an object, has distribution $N(0, 4)$. If $K = mV^2/2$ ergs is the kinetic energy of the object (where $m = $ mass), find the pdf of K. If $m = 10$ grams, evaluate $P(K \leq 3)$.

10.13. Suppose that the life length of an item is exponentially distributed with parameter 0.5. Assume that 10 such items are installed successively, so that the ith item is installed "immediately" after the $(i - 1)$-item has failed. Let T_i be the time to failure of the ith item, $i = 1, 2, \ldots, 10$, always measured from the time of installation. Hence $S = T_1 + \cdots + T_{10}$ represents the total time of functioning of the 10 items. Assuming that the T_i's are independent, evaluate $P(S \geq 15.5)$.

10.14. Suppose that $X_1, \ldots, X_{80}$ are independent random variables, each having distribution $N(0, 1)$. Evaluate $P[X_1^2 + \cdots + X_{80}^2 > 77]$. [*Hint:* Use Theorem 9.2.]

10.15. Show that if X_i, $i = 1, 2, \ldots, k$, represents the number of successes in n_i repetitions of an experiment, where $P(\text{success}) = p$, for all i, then $X_1 + \cdots + X_k$ has a binomial distribution. (That is, the binomial distribution possesses the reproductive property.)

10.16. (*The Poisson and the multinomial distribution.*) Suppose that X_i, $i = 1, 2, \ldots, n$ are independently distributed random variables having a Poisson distribution with parameters α_i, $i = 1, \ldots, n$. Let $X = \sum_{i=1}^{n} X_i$. Then the joint conditional probability distribution of $X_1, \ldots, X_n$ given $X = x$ is given by a multinomial distribution. That is, $P(X_1 = x_1, \ldots, X_n = x_n | X = x) = x!/(x_1! \ldots x_n!)(\alpha_1/\sum_{i=1}^{n}\alpha_i)^{x_1} \cdots (\alpha_n/\sum_{i=1}^{n}\alpha_i)^{x_n}$.

10.17. Obtain the mgf of a random variable having a geometric distribution. Does this distribution possess a reproductive property under addition?

10.18.   If the random variable X has an mgf given by $M_X(t) = 3/(3 - t)$, obtain the standard deviation of X.

10.19.   Find the mgf of a random variable which is uniformly distributed over $(-1, 2)$.

10.20.   A certain industrial process yields a large number of steel cylinders whose lengths are distributed normally with mean 3.25 inches and standard deviation 0.05 inch. If two such cylinders are chosen at random and placed end to end, what is the probability that their combined length is less than 6.60 inches?

Note: In evaluating $M'_X(t)$ at $t = 0$, an indeterminate form may arise. That is, $M'_X(0)$ may be of the form $0/0$. In such cases we must try to apply l'Hôpital's rule. For example, if X is uniformly distributed over $[0, 1]$, we easily find that $M_X(t) = (e^t - 1)/t$ and $M'_X(t) = (te^t - e^t + 1)/t^2$. Hence at $t = 0$, $M'_X(t)$ is indeterminate. Applying l'Hôpital's rule, we find that $\lim_{t \to 0} M'_X(t) = \lim_{t \to 0} te^t/2t = \frac{1}{2}$. This checks, since $M'_X(0) = E(X)$, which equals $\frac{1}{2}$ for the random variable described here.

11

Applications to Reliability Theory

11.1 Basic Concepts

We shall investigate in this chapter a very important and growing area of application of some of the concepts introduced in the previous chapters.

Suppose that we are considering a component (or an entire complex of components assembled into a system) which is put under some sort of "stress." This might be a steel beam under a load, a fuse inserted into a circuit, an airplane wing under the influence of forces, or an electronic device put into service. Suppose that for any such component (or system) a state which we shall denote as "failure" can be defined. Thus the steel beam may crack or break, the fuse may burn out, the wing may buckle, or the electronic device may fail to function.

If such a component is put under stress conditions at some specified time, say $t = 0$, and observed until it fails (that is, until it ceases to function properly under the stress applied), the *time to failure* or *life length*, say T, may be considered as a continuous random variable with some pdf f. There is considerable empirical evidence to indicate that the value of T cannot be predicted from a deterministic model. That is, "identical" components subjected to "identical" stress will fail at different and unpredictable times. Some will fail quite early in their life and others at later stages. Of course, "the manner of failing" will vary with the type of item being considered. For example, a fuse will fail rather suddenly, in the sense that one moment it will be working perfectly and the next moment it will not function at all. On the other hand, a steel beam under a heavy load will probably become weaker over a long period of time. In any case, the use of a probabilistic model, with T considered as a random variable, seems to be the only realistic approach. We now introduce the following important concept.

> **Definition.** The *reliability* of a component (or system) at time t, say $R(t)$, is
> defined as $R(t) = P(T > t)$, where T is the life length of the component. R
> is called the *reliability function*.

Note: Although the term "reliability" has many different technical meanings, the above use is becoming more commonly accepted. The definition given here simply says that the reliability of a component equals the probability that the component does not fail during the interval $[0, t]$ (or, equivalently, reliability equals the probability that the component is still functioning at time t). For example, if for a particular item, $R(t_1) = 0.90$, this means that approximately 90 percent of such items, used under certain conditions, will still be functioning at time t_1.

In terms of the pdf of T, say f, we have

$$R(t) = \int_t^\infty f(s)\, ds.$$

In terms of the cdf of T, say F, we have

$$R(t) = 1 - P(T \le t) = 1 - F(t).$$

In addition to the reliability function R, another function plays an important role in describing the failure characteristics of an item.

Definition. The (instantaneous) *failure rate* Z (sometimes called *hazard function*) associated with the random variable T is given by

$$Z(t) = \frac{f(t)}{1 - F(t)} = \frac{f(t)}{R(t)}, \tag{11.1}$$

defined for $F(t) < 1$.

Note: In order to interpret $Z(t)$ consider the conditional probability

$$P(t \le T \le t + \Delta t \mid T > t),$$

that is, the probability that the item will fail during the next Δt time units, given that the item is functioning properly at time t. Applying the definition of conditional probability, we may write the above as

$$P(t \le T \le t + \Delta t \mid T > t) = \frac{P(t < T \le t + \Delta t)}{P(T > t)}$$

$$= \int_t^{t+\Delta t} f(x)\, dx / P(T > t) = \Delta t f(\xi)/R(t),$$

where $t \le \xi \le t + \Delta t$.

The last expression is (for small Δt and supposing that f is continuous at t^+) approximately equal to $\Delta t Z(t)$. Thus, speaking informally, $\Delta t Z(t)$ represents the proportion of items that will fail between t and $t + \Delta t$, among those items which are still functioning at time t.

From the above we note that f, the pdf of T, uniquely determines the failure rate Z. We shall now indicate that the converse also holds: the failure rate Z uniquely determines the pdf f.

Theorem 11.1. If T, the time to failure, is a continuous random variable with pdf f and if $F(0) = 0$ where F is the cdf of T, then f may be expressed in terms of the failure rate Z as follows:

$$f(t) = Z(t)e^{-\int_0^t Z(s)\,ds}. \tag{11.2}$$

Proof: Since $R(t) = 1 - F(t)$, we have $R'(t) = -F'(t) = -f(t)$. Hence

$$Z(t) = \frac{f(t)}{R(t)} = \frac{-R'(t)}{R(t)}.$$

We integrate both sides from 0 to t:

$$\int_0^t Z(s)\, ds = -\int_0^t \frac{R'(s)}{R(s)}\, ds = -\ln R(s)\big|_0^t$$

$$= -\ln R(t) + \ln R(0) = -\ln R(t),$$

provided that $\ln R(0) = 0$ which holds if and only if $R(0) = 1$. [This last condition is satisfied if $F(0) = 0$. This simply says that the probability of *initial* failure equals zero; we shall make this assumption in the remainder of the discussion.] Hence

$$R(t) = e^{-\int_0^t Z(s)\, ds}.$$

Thus

$$f(t) = F'(t) = \frac{d}{dt}[1 - R(t)] = Z(t)e^{-\int_0^t Z(s)\, ds}.$$

Therefore we have shown that the failure rate Z uniquely determines the pdf f.

An interesting relationship exists between the reliability function R and the mean time to failure, $E(T)$.

Theorem 11.2. If $E(T)$ is finite, then

$$E(T) = \int_0^\infty R(t)\, dt. \tag{11.3}$$

Proof: Consider

$$\int_0^\infty R(t)\, dt = \int_0^\infty \left[\int_t^\infty f(s)\, ds\right] dt.$$

Let us integrate this by parts, letting $\int_t^\infty f(s)\, ds = u$ and $dt = dv$. Hence $v = t$ and $du = -f(t)\, dt$. Thus

$$\int_0^\infty R(t)\, dt = t\int_t^\infty f(s)\, ds\Big|_0^\infty + \int_0^\infty tf(t)\, dt.$$

The second integral on the right-hand side represents $E(T)$. Hence the proof is complete if we can show that $t\int_t^\infty f(s)\, ds$ vanishes at $t = 0$ and as $t \to \infty$. The vanishing at $t = 0$ is immediate. Using the finiteness of $E(T)$, the reader can complete the proof.

The concepts of reliability and failure rate are among the important tools needed for a thorough study of "failure models." We shall be chiefly concerned with the following questions:

(a) What underlying "failure laws" are reasonable to assume? (That is, what form should the pdf of T have?)

(b) Suppose that we have two components, say C_1 and C_2, with known failure laws. Assume that these components are combined in series,

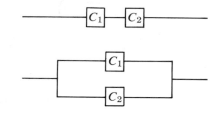

or in parallel,

to form a system. What is the failure law (or reliability) of the system?

The question of what is a "reasonable" failure law brings us back to a problem we have discussed before: What is a reasonable mathematical model for the description of some observational phenomenon? From a strictly mathematical point of view, we may assume practically any pdf for T and then simply study the consequences of this assumption. However, if we are interested in having the model represent (as accurately as possible) the actual failure data available, our choice of model must take this into account.

11.2 The Normal Failure Law

There are many types of components whose failure behavior may be represented by the normal distribution. That is, if T is the life length of an item, its pdf is given by

$$f(t) = \frac{1}{\sqrt{2\pi}\,\sigma} \exp\left(-\frac{1}{2}\left[\frac{t-\mu}{\sigma}\right]^2\right).$$

[Again we note that the time to failure, T, must be greater than (or equal) to zero. Hence in order for the above model to be applicable we must insist that $P(T < 0)$ be essentially zero.] As the shape of the normal pdf indicates, a normal failure law implies that most of the items fail around the mean failure time, $E(T) = \mu$ and the number of failures decreases (symmetrically) as $|T - \mu|$ increases. A normal failure law means that about 95.72 percent of the failures take place for those values of t satisfying $\{t \mid |t - \mu| < 2\sigma\}$. (See Fig. 11.1.)

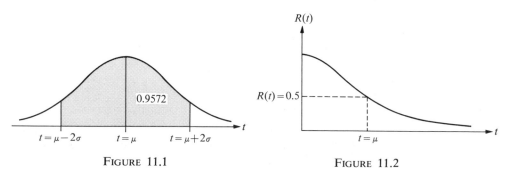

FIGURE 11.1   FIGURE 11.2

The reliability function of the normal failure law may be expressed in terms of the tabulated normal cumulative distribution function Φ, as follows:

$$R(t) = P(T > t) = 1 - P(T \leq t)$$

$$= 1 - \frac{1}{\sqrt{2\pi}\,\sigma} \int_{-\infty}^{t} \exp\left(-\frac{1}{2}\left[\frac{x-\mu}{\sigma}\right]^2\right) dx$$

$$= 1 - \Phi\left(\frac{t-\mu}{\sigma}\right).$$

Figure 11.2 shows a general reliability curve for a normal failure law. Note that in order to achieve a high reliability (say 0.90 or greater) the operating time must be considerably less than μ, the expected life length.

EXAMPLE 11.1. Suppose that the life length of a component is normally distributed with standard deviation equal to 10 (hours). If the component has a reliability of 0.99 for an operation period of 100 hours, what should its expected life length be?

The above equation becomes

$$0.99 = 1 - \Phi\left(\frac{100-\mu}{10}\right).$$

From the tables of the normal distribution this yields $(100 - \mu)/10 = -2.33$. Hence $\mu = 123.3$ hours.

The normal failure law represents an appropriate model for components in which failure is due to some "wearing" effect. It is not, however, among the most important failure laws encountered.

11.3 The Exponential Failure Law

One of the most important failure laws is the one whose time to failure is described by the exponential distribution. We can characterize this in a number of ways, but probably the simplest way is to suppose that the failure rate is *constant*. That is, $Z(t) = \alpha$. An immediate consequence of this assumption is, by Eq. (11.2), that the pdf associated with the time to failure T, is given by

$$f(t) = \alpha e^{-\alpha t}, \qquad t > 0.$$

The converse of this is also immediate: If f has the above form, $R(t) = 1 - F(t) = e^{-\alpha t}$ and hence $Z(t) = f(t)/R(t) = \alpha$. Thus we have the following important result.

Theorem 11.3. Let T, the time to failure, be a continuous random variable assuming all nonnegative values. Then T has an exponential distribution if and only if it has constant failure rate.

Note: The assumption of constant failure rate may be interpreted to mean that after the item has been in use, its probability of failing has not changed. More informally stated, there is no "wearing-out" effect when the exponential model is stipulated. There is another way of saying this which makes this point even more striking.

Consider for $\Delta t > 0$, $P(t \leq T \leq t + \Delta t \mid T > t)$. This represents the probability that the item will fail during the next Δt units, given that it has not failed at time t. Applying the definition of conditional probability, we find that

$$P(t \leq T \leq t + \Delta t \mid T > t) = \frac{e^{-\alpha t} - e^{-\alpha(t+\Delta t)}}{e^{-\alpha t}} = 1 - e^{-\alpha \Delta t}.$$

Hence this conditional probability is independent of t, depending only on Δt. It is in this sense that we may say that an exponential failure law implies that the probability of failure is independent of past history. That is, so long as the item is still functioning it is "as good as new."

If we expand the right-hand side of the above expression in a Maclaurin series we obtain

$$P(t \leq T \leq t + \Delta t \mid T > t) = 1 - \left[1 - \alpha\Delta t + \frac{(\alpha\Delta t)^2}{2!} - \frac{(\alpha\Delta t)^3}{3!} + \cdots\right]$$

$$= \alpha\Delta t + h(\Delta t),$$

where $h(\Delta t)$ becomes negligible for Δt small. Thus for sufficiently small Δt, the above probability is directly proportional to Δt.

For many types of components the assumption leading to the exponential failure law is not only intuitively appealing but is in fact supported by empirical evidence. For instance, it is quite reasonable to suppose that a fuse or a jeweled bearing is "as good as new" while it is still functioning. That is, if the fuse has not melted, it is in practically new condition. Nor will the bearing change much due to wear. In cases such as this, the exponential failure law represents an appropriate model with which to study the failure characteristics of the item.

However, a word of caution should be inserted here. There are many situations involving failure studies for which the underlying assumptions leading to an exponential distribution will not be satisfied. For example, if a piece of steel is ex-

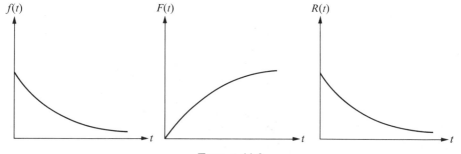

FIGURE 11.3

posed to continuing stress, there will obviously be some deterioration, and hence a model other than the exponential should be considered.

Although we have previously discussed the various properties of the exponential distribution, let us summarize these again in order to have them available for the present purpose. (See Fig. 11.3.) If T, the time to failure, is exponentially distributed (with parameter α), we have

$$E(T) = 1/\alpha; \qquad\qquad V(T) = 1/\alpha^2;$$
$$F(t) = P(T \leq t) = 1 - e^{-\alpha t}; \qquad R(t) = e^{-\alpha t}.$$

EXAMPLE 11.2. If the parameter α is given and $R(t)$ is specified, we can find t, the number of hours, say, of operation. Thus, if $\alpha = 0.01$ and $R(t)$ is to be equal to 0.90, we have

$$0.90 = e^{-0.01 t}.$$

Hence $t = -100 \ln (0.90) = 10.54$ hours. Therefore, if each of 100 such components is operating for 10.54 hours, approximately 90 will not fail during that period.

Notes: (a) It is very important to realize that in the exponential case we may identify operating *time* (from some arbitrary fixed initial value) with operating *age*. For in the exponential case, an item which has not failed is as good as new, and hence its failure behavior during any particular period of service depends only on the length of that period and not on its past history. However, when a nonexponential failure law is assumed (such as the normal law or one of the distributions we shall consider shortly), the past history does have an effect on the item's performance. Hence, while we may define T as the time in service (up to failure) for the exponential case, we must define T as the *total* life length up to failure for the nonexponential cases.

(b) The exponential distribution, which we have introduced in the context of life length for components, has many other important applications. In fact, whenever a continuous random variable T assuming nonnegative values satisfies the assumption $P(T > s + t \mid T > s) = P(T > t)$ for all s and t, then T will have an exponential distribution. Thus, if T represents the time it takes for a radioactive atom to disintegrate, we may suppose that T is exponentially distributed, since the above assumption seems to be satisfied.

EXAMPLE 11.3. It is not unreasonable to suppose that it costs more to produce an item with a large expected life length than one with a small life expectancy. Specifically suppose that the cost C of producing an item is the following function of μ, the mean time to failure,

$$C = 3\mu^2.$$

Assume that a profit of D dollars is realized for every hour the item is in service. Thus the profit per item is given by

$$P = DT - 3\mu^2,$$

where T is the number of hours the item is functioning properly. Hence the expected profit is given by

$$E(P) = D\mu - 3\mu^2.$$

To find for which value of μ this quantity is maximized, simply set $dE(P)/d\mu$ equal to zero and solve for μ. The result is $\mu = D/6$, and therefore the maximum expected profit per item equals $E(P)_{\max} = D^2/12$.

EXAMPLE 11.4. Let us reconsider Example 11.3, making the following additional assumptions. Suppose that T, the time to failure, is exponentially distributed with parameter α. Thus μ, the expected time to failure, is given by $\mu = 1/\alpha$. Suppose furthermore that if the item does not function properly for at least a specified number of hours, say t_0, a penalty is assessed equal to $K(t_0 - T)$ dollars, where $T(T < t_0)$ is the time at which failure takes place. Hence the profit per item is given by

$$P = DT - 3\mu^2 \quad \text{if } T > t_0,$$
$$= DT - 3\mu^2 - K(t_0 - T) \quad \text{if } T < t_0.$$

Therefore the expected profit (per item) may be expressed as

$$E(P) = D\int_{t_0}^{\infty} t\alpha e^{-\alpha t}\, dt - 3\mu^2 e^{-\alpha t_0}$$

$$+ (D + K)\int_0^{t_0} t\alpha e^{-\alpha t}\, dt - (3\mu^2 + Kt_0)(1 - e^{-\alpha t_0}).$$

After straightforward integrations the above may be written as

$$E(P) = D\mu - 3\mu^2 + K[\mu - \mu e^{-t_0/\mu} - t_0].$$

Note that if $K = 0$, this reduces to the result obtained in Example 11.3. We could ask ourselves an analogous question to the one raised in the previous example: For what values of μ does $E(P)$ assume its maximum value? We shall not pursue the details of this problem since they involve the solution of a transcendental equation which must be solved numerically.

11.4 The Exponential Failure Law and the Poisson Distribution

There is a very close connection between the exponential failure law described in the previous section and a Poisson process. Suppose that failure occurs because of the appearance of certain "random" disturbances. These may be caused by external forces such as sudden gusts of wind or a drop (rise) in voltage or by internal causes such as a chemical disintegration or a mechanical malfunctioning. Let X_t be equal to the number of such disturbances occurring during a time interval of length t and suppose that X_t, $t \geq 0$, constitutes a *Poisson process*. That is, for any fixed t, the random variable X_t has a Poisson distribution with parameter αt. Suppose that failure during $[0, t]$ is caused if and only if at least one such disturbance occurs. Let T be the time to failure, which we shall assume to be a continuous random variable. Then

$$F(t) = P(T \leq t) = 1 - P(T > t).$$

Now $T > t$ if and only if *no* disturbance occurs during $[0, t]$. This happens if and only if $X_t = 0$. Hence

$$F(t) = 1 - P(X_t = 0) = 1 - e^{-\alpha t}.$$

This represents the cdf of an exponential failure law. Thus we find that the above "cause" of failure implies an exponential failure law.

The above ideas may be *generalized* in two ways.

(a) Suppose again that disturbances appear according to a Poisson process. Assume furthermore that whenever such a disturbance does appear, there is a constant probability p that it will not cause failure. Therefore, if T is the time to failure, we have, as before,

$$F(t) = P(T \leq t) = 1 - P(T > t).$$

This time, $T > t$ if and only if (during $[0, t]$) no disturbance occurs, *or* one disturbance occurs *and* no failure resulted, *or* two disturbances occurred *and* no failures resulted, or ... Hence,

$$F(t) = 1 - \left[e^{-\alpha t} + (\alpha t)e^{-\alpha t}p + (\alpha t)^2 \frac{e^{-\alpha t}}{2!} p^2 + \cdots \right]$$

$$= 1 - e^{-\alpha t} \sum_{k=0}^{\infty} \frac{(\alpha t p)^k}{k!} = 1 - e^{-\alpha t}e^{\alpha t p} = 1 - e^{-\alpha(1-p)t}.$$

Thus T has an exponential failure law with parameter $\alpha(1 - p)$. (Note that if $p = 0$, we have the case discussed previously.)

(b) Suppose again that disturbances appear according to a Poisson process. This time we shall assume that failure occurs whenever r or more disturbances $(r \geq 1)$ occur during an interval of length t. Therefore, if T is time to failure, we have, as before,

$$F(t) = 1 - P(T > t).$$

In this case, $T > t$ if and only if $(r - 1)$ or fewer disturbances occur. Therefore

$$F(t) = 1 - \sum_{k=0}^{r-1} \frac{(\alpha t)^k e^{-\alpha t}}{k!}.$$

According to Eq. (9.17) the above equals $\int_0^t [\alpha/(r - 1)!](\alpha s)^{r-1} e^{-\alpha s} \, ds$ and hence represents the cdf of a Gamma distribution. Thus we find that the above "cause" of failure leads to the conclusion that the time to failure follows a *Gamma failure law*. (Of course, if $r = 1$, this becomes an exponential distribution.)

11.5   The Weibull Failure Law

Let us modify the notion of constant failure rate which leads to the exponential failure law. Suppose that the failure rate Z, associated with T, the life length of an item, has the following form:

$$Z(t) = (\alpha\beta)t^{\beta-1}, \tag{11.4}$$

where α and β are positive constants. From Eq. (11.2) we obtain the following expression for the pdf of T:

$$f(t) = (\alpha\beta)t^{\beta-1}e^{-\alpha t^{\beta}}, \qquad t > 0, \, \alpha, \, \beta > 0. \tag{11.5}$$

The random variable having pdf given by Eq. (11.5) is said to have a *Weibull distribution*. Figure 11.4 shows the pdf for $\alpha = 1$ and $\beta = 1, 2, 3$. The reliability function R is given by $R(t) = e^{-\alpha t^{\beta}}$ which is a decreasing function of t.

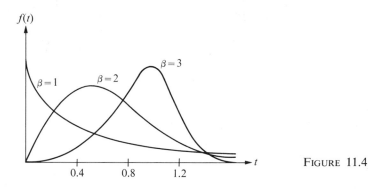

FIGURE 11.4

Note: The exponential distribution is a special case of the Weibull distribution since we obtain the exponential distribution if we let $\beta = 1$ in Eq. (11.4). The assumption (Eq. 11.4) states that $Z(t)$ is not a constant, but is proportional to powers of t. For instance, if $\beta = 2$, Z is a linear function of t; if $\beta = 3$, Z is a quadratic function of t; etc. Thus, Z is an increasing, decreasing, or constant function of t, depending on the value of β, as indicated in Fig. 11.5.

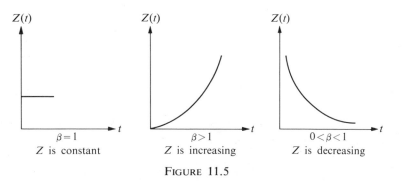

FIGURE 11.5

Theorem 11.4. If the random variable T has a Weibull distribution with pdf given by Eq. (11.5), we have

$$E(T) = \alpha^{-1/\beta}\Gamma\left(\frac{1}{\beta} + 1\right), \tag{11.6}$$

$$V(T) = \alpha^{-2/\beta}\left\{\Gamma\left(\frac{2}{\beta} + 1\right) - \left[\Gamma\left(\frac{1}{\beta} + 1\right)\right]^2\right\}. \tag{11.7}$$

Proof: See Problem 11.8.

Note: The Weibull distribution represents an appropriate model for a failure law whenever the system is composed of a number of components and failure is essentially due to the "most severe" flaw among a large number of flaws in the system. Also, using a Weibull distribution, we may obtain both an increasing and a decreasing failure rate by simply making the appropriate choice for the parameter β.

We have by no means exhausted the number of reasonable failure laws. However, those we have mentioned certainly are extremely important so far as representing meaningful models for the study of failure characteristics of components or systems of components.

11.6 Reliability of Systems

Now that we have considered a number of important failure distributions we can turn to the second question posed in Section 11.1: How can we evaluate the reliability of a system if we know the reliability of its components? This can be a very difficult problem and we shall only discuss a few simple (but relatively important) cases.

Suppose that two components are hooked up in series.

This means that in order for the above system to work, *both* components must be functioning. If, in addition, we assume that the components function *independently*, we may obtain the reliability of the system, say $R(t)$, in terms of the reliabilities of the components, say $R_1(t)$ and $R_2(t)$, as follows:

$$
\begin{aligned}
R(t) &= P(T > t) &&\text{(where } T \text{ is the time to failure of the system)}\\
&= P(T_1 > t \text{ and } T_2 > t) &&\text{(where } T_1 \text{ and } T_2 \text{ are the times to failure}\\
&&&\text{of components } C_1 \text{ and } C_2, \text{ respectively)}\\
&= P(T_1 > t)P(T_2 > t) = R_1(t)R_2(t).
\end{aligned}
$$

Thus we find that $R(t) \leq \min[R_1(t), R_2(t)]$. That is, for a system made up of two independent components in series, the reliability of the system is less than the reliability of any of its parts.

The above discussion may obviously be generalized to n components and we obtain the following theorem.

Theorem 11.5. If n components, functioning independently, are connected in series, and if the ith component has reliability $R_i(t)$, then the reliability of the entire system, $R(t)$, is given by

$$R(t) = R_1(t) \cdot R_2(t) \cdots R_n(t). \tag{11.8}$$

In particular, if T_1 and T_2 have exponential failure laws with parameters α_1 and α_2, Eq. (11.8) becomes

$$R(t) = e^{-\alpha_1 t} e^{-\alpha_2 t} = e^{-(\alpha_1 + \alpha_2)t}.$$

Hence the pdf of the time to failure of the system, say T, is given by

$$f(t) = -R'(t) = (\alpha_1 + \alpha_2)e^{-(\alpha_1 + \alpha_2)t}.$$

Thus we have established the following result.

Theorem 11.6. If two independently functioning components having exponential failures with parameters α_1 and α_2 are connected in *series*, the failure law of the resulting system is again exponential with parameter equal to $\alpha_1 + \alpha_2$.

(This theorem may obviously be generalized to n such components in series.)

EXAMPLE 11.5.   (Taken from I. Bazovsky, *Reliability Theory and Practice*, Prentice-Hall, Inc., Englewood Cliffs, N. J., 1961.) Consider an electronic circuit consisting of 4 silicon transistors, 10 silicon diodes, 20 composition resistors, and 10 ceramic capacitors in continuous *series* operation. Suppose that under certain stress conditions (that is, prescribed voltage, current, and temperature), each of these items has the following *constant* failure rates.

silicon diodes:	0.000002
silicon transistors:	0.00001
composition resistors:	0.000001
ceramic capacitors:	0.000002

Because of the assumed *constant* failure rate, the exponential distribution represents the failure law for each of the above components. Because of the series connection, the time to failure for the entire circuit is again exponentially distributed with parameter (failure rate) equal to

$$10(0.000002) + 4(0.00001) + 20(0.000001) + 10(0.000002) = 0.0001.$$

Hence the reliability for the circuit is given by $R(t) = e^{-0.0001t}$. Thus, for a 10-hour period of operation, the probability that the circuit will not fail is given by $e^{-0.0001(10)} = 0.999$. The *expected* time to failure for the circuit equals $1/0.0001 = 10,000$ hours.

Another important system is a *parallel* system in which the components are connected in such a way that the system fails to function only if all the components fail to function. If only two components are involved, the system may be depicted as in Fig. 11.6. Again assuming that the components function *independently* of each other, the reliability of the system, say $R(t)$, may be expressed in terms of the reliabilities of the components, say $R_1(t)$ and $R_2(t)$, as follows.

$$R(t) = P(T > t) = 1 - P(T \leq t)$$
$$= 1 - P[T_1 \leq t \text{ and } T_2 \leq t]$$
$$= 1 - P(T_1 \leq t)P(T_2 \leq t)$$
$$= 1 - \{[1 - P(T_1 > t)][1 - P(T_2 > t)]\}$$
$$= 1 - [1 - R_1(t)][1 - R_2(t)]$$
$$= R_1(t) + R_2(t) - R_1(t)R_2(t).$$

The last form indicates that $R(t) \geq$ maximum $[R_1(t), R_2(t)]$. That is, a system composed of two independently functioning components operating in parallel will be more reliable than either of the components.

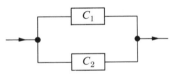

FIGURE 11.6

All the ideas discussed above for two components may be generalized in the following theorem.

Theorem 11.7. If n components functioning independently are operating in parallel, and if the ith component has reliability $R_i(t)$, then the reliability of the system, say $R(t)$, is given by

$$R(t) = 1 - [1 - R_1(t)][1 - R_2(t)] \cdots [1 - R_n(t)]. \tag{11.9}$$

It often happens that all the components have *equal* reliabilities, say $R_i(t) = r(t)$ for all i. In this case the above expression becomes

$$R(t) = 1 - [1 - r(t)]^n. \tag{11.10}$$

Let us consider, in particular, two components in parallel, each of whose failure time is exponentially distributed. Then

$$R(t) = R_1(t) + R_2(t) - R_1(t)R_2(t) \doteq e^{-\alpha_1 t} + e^{-\alpha_2 t} - e^{-(\alpha_1 + \alpha_2)t}.$$

Thus the pdf of the failure time of the parallel system, say T, is given by

$$f(t) = -R'(t) = \alpha_1 e^{-\alpha_1 t} + \alpha_2 e^{-\alpha_2 t} - (\alpha_1 + \alpha_2)e^{-(\alpha_1 + \alpha_2)t}.$$

Hence T is *not* exponentially distributed. The expected value of T equals

$$E(T) = \frac{1}{\alpha_1} + \frac{1}{\alpha_2} - \frac{1}{\alpha_1 + \alpha_2}.$$

Whereas a series operation is often mandatory (that is, a number of components *must* function in order for the system to function), we often use a parallel operation in order to increase the reliability of the system. The following example illustrates this point.

EXAMPLE 11.6. Suppose that three units are operated in parallel. Assume that each has the same constant failure rate $\alpha = 0.01$. (That is, the time to failure for each unit is exponentially distributed with parameter $\alpha = 0.01$.) Hence the reliability for each unit is $R(t) = e^{-0.01t}$, and thus the reliability for a period of operation of 10 hours equals $e^{-0.1} = 0.905$ or about 90 percent. How much of an improvement can be obtained (in terms of increasing the reliability) by operating three such units in parallel?

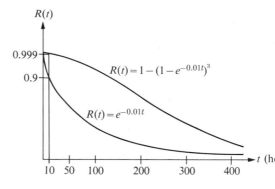

FIGURE 11.7

The reliability of three units operating in parallel for 10 hours would be

$$R(10) = 1 - [1 - 0.905]^3 = 1 - 0.00086$$
$$= 0.99914, \quad \text{or about 99.9 percent.}$$

In Fig. 11.7 we see the reliability curves for the single unit versus the three units in parallel. For the single unit, $R(t) = e^{-\alpha t}$, while for the three units in parallel, $R(t) = 1 - (1 - e^{-\alpha t})^3$, with $\alpha = 0.01$.

We have, so far, considered only the simplest ways of combining individual units into a system, namely the series and parallel operations of components. There are many other ways of combining components, and we shall simply list

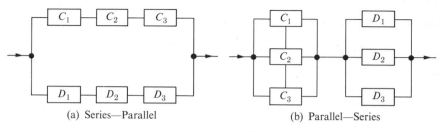

FIGURE 11.8

a few of these. (See Fig. 11.8.) Some of the questions that arise in connection with these combinations will be considered in the problems at the end of the chapter.

(a) Series-parallel. (Here we consider parallel groups of components as in a circuit, for example, each having m components in series.)
(b) Parallel-series.
(c) Standby system. Here we consider two components in which the second component "stands by" and functions if and only if the first component fails. In this case, the second component takes over (instantaneously) and functions in place of the first component.

Let us briefly discuss the concept of *safety factor*. Suppose that the stress S applied to a structure is considered as a (continuous) random variable. Similarly, the resistance of the structure, say R, may also be considered as a continuous random variable. We define the safety factor of the structure as the ratio of R to S,

$$T = R/S.$$

If R and S are independent random variables with pdf's g and h, respectively, then the pdf of T is given by

$$f(t) = \int_0^\infty g(ts)h(s)s \, ds.$$

(See Theorem 6.5.) Now the structure will fail if $S > R$, that is, if $T < 1$. Hence the probability of failure $P_F = \int_0^1 f(t) \, dt$.

PROBLEMS

11.1. Suppose that T, the time to failure, of an item is normally distributed with $E(T) = 90$ hours and standard deviation 5 hours. In order to achieve a reliability of 0.90, 0.95, 0.99, how many hours of operation may be considered?

11.2. Suppose that the life length of an electronic device is exponentially distributed. It is known that the reliability of the device (for a 100-hour period of operation) is 0.90. How many hours of operation may be considered to achieve a reliability of 0.95?

11.3. Suppose that the life length of a device has constant failure rate C_0 for $0 < t < t_0$ and a different constant failure rate C_1 for $t \geq t_0$. Obtain the pdf of T, the time to failure, and sketch it.

11.4. Suppose that the failure rate Z is given by,

$$Z(t) = 0, \quad 0 < t < A,$$
$$= C, \quad t \geq A.$$

(This implies that no failures occur before $T = A$.)

(a) Find the pdf associated with T, the time to failure.
(b) Evaluate $E(T)$.

11.5. Suppose that the failure law of a component has the following pdf:

$$f(t) = (r + 1)A^{r+1}/(A + t)^{r+2}, \quad t > 0.$$

(a) For what values of A and r is the above a pdf?
(b) Obtain an expression for the reliability function and hazard function.
(c) Show that the hazard function is decreasing in t.

11.6. Suppose that the failure law of a component is a linear combination of k exponential failure laws. That is, the pdf of the failure time is given by

$$f(t) = \sum_{j=1}^{k} c_j \beta_j e^{-\beta_j t}, \quad t > 0, \quad \beta_j > 0.$$

(a) For what values of c_j is the above a pdf?
(b) Obtain an expression for the reliability function and hazard function.
(c) Obtain an expression for the mean time to failure.
(d) Answer (b) and (c) if $\beta_j = \beta$ for all j.

11.7. Each of the six tubes of a radio set has a life length (in years) which may be considered as a random variable. Suppose that these tubes function independently of one another. What is the probability that no tubes will have to be replaced during the first two months of service if:

(a) The pdf of the time to failure is $f(t) = 50te^{-25t^2}, t > 0$?
(b) The pdf of the time to failure is $f(t) = 25te^{-25t}, t > 0$?

11.8. Prove Theorem 11.4.

11.9. The life length of a satellite is an exponentially distributed random variable with expected life time equal to 1.5 years. If three such satellites are launched simultaneously, what is the probability that at least two will still be in orbit after 2 years?

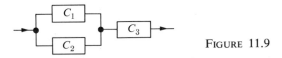

FIGURE 11.9

11.10. Three independently functioning components are connected into a single system as indicated in Fig. 11.9. Suppose that the reliability for each of the components for an operational period of t hours is given by

$$R(t) = e^{-0.03t}.$$

If T is the time to failure of the entire system (in hours), what is the pdf of T? What is the reliability of the system? How does it compare with $e^{-0.03t}$?

11.11.   Suppose that n independently functioning components are connected in a series arrangement.   Assume that the time to failure for each component is normally distributed with expectation 50 hours and standard deviation 5 hours.

(a) If $n = 4$, what is the probability that the system will still be functioning after 52 hours of operation?

(b) If n components are placed *in parallel*, how large should n be in order that the probability of failure during the first 55 hours is approximately equal to 0.01?

11.12.   (Taken from Derman and Klein, *Probability and Statistical Inference*. Oxford University Press, New York, 1959.) The life length (L) in months of a certain vacuum tube used in radar sets has been found to be exponentially distributed with parameter $\beta = 2$. In carrying out its preventive maintenance program, a company wishes to decide how many months (m) after installation a tube should be replaced to minimize the expected cost per tube. The cost per tube in dollars is denoted by C. The shortest practicable elapsed time between installation and replacement is 0.01 month. Subject to this restriction, what value of m minimized $E(C)$, the expected cost, in each of the following situations where the cost C is the designated function of L and m?

(a) $C(L, m) = 3|L - m|$.   (b) $C(L, m) = 3$ if $L < m$,

$\qquad\qquad\qquad\qquad\qquad\qquad\qquad\qquad\ = 5(L - m)$ if $L \geq m$.

(c) $C(L, m) = 2$ if $L < m$,

$\qquad\qquad\ = 5(L - m)$ if $L \geq m$.

(In each case, draw a graph of $E(C)$ as a function of m.)

Note: C clearly is a random variable since it is a function of L which is a random variable. $E(C)$ is a function of m, and the problem simply asks to find that value of m which minimized $E(C)$, subject to the restriction that $m \geq 0.01$.

11.13.   Suppose that the failure rate associated with the life length T of an item is given by the following function:

$$Z(t) = C_0, \qquad 0 \leq t < t_0,$$
$$\quad = C_0 + C_1(t - t_0), \qquad t \geq t_0.$$

Note: This represents another generalization of the exponential distribution.   The above reduces to constant failure rate (and hence the exponential distribution) if $C_1 = 0$.

(a) Obtain the pdf of T, the time to failure.

(b) Obtain an expression for the reliability $R(t)$ and sketch its graph.

11.14.   Suppose that each of three electronic devices has a failure law given by an exponential distribution with parameters β_1, β_2, and β_3. Suppose that these three devices function independently and are connected in parallel to form a single system.

(a) Obtain an expression for $R(t)$, the reliability of the system.

(b) Obtain an expression for the pdf of T, the time to failure of the system. Sketch the pdf.

(c) Find the mean time to failure of the system.

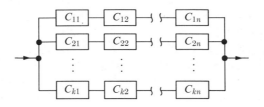

FIGURE 11.10

11.15. (a) Suppose that n components are hooked up in a series arrangement. Then k such series connections are hooked up in parallel to form an entire system. (See Figure 11.10.) If each component has the same reliability, say R, for a certain period of operation, find an expression for the reliability of the entire system (for that same period of operation).

(b) Suppose that each of the above components obeys an exponential failure law with failure rate 0.05. Suppose furthermore that the time of operation is 10 hours and that $n = 5$. Determine the value of k in order that the reliability of the entire system equals 0.99.

11.16. Suppose that k components are connected in parallel. Then n such parallel connections are hooked up in series into a single system. (See Figure 11.11.) Answer (a) and (b) of Problem 11.15 for this situation.

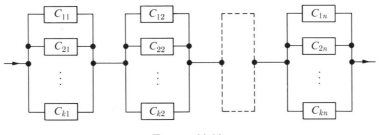

FIGURE 11.11

11.17. Suppose that n components, each having the same constant failure rate λ, are connected in parallel. Find an expression for the mean time to failure of the resulting system.

11.18. (a) An aircraft propulsion system consists of three engines. Assume that the constant failure rate for each engine is $\lambda = 0.0005$ and that the engines fail independently of one another. The engines are connected in parallel. What is the reliability of this propulsion system for a mission requiring 10 hours if at least two engines must survive?

(b) Answer the above question for a mission requiring 100 hours; 1000 hours. (This problem is suggested by a discussion in I. Bazovsky, *Reliability Theory and Practice.* Prentice-Hall, Inc., Englewood Cliffs, N.J., 1961.)

11.19. Consider components A, A', B, B', and C connected as indicated in Figs. 11.12 (a) and (b). (Component C may be thought of as representing a "safeguard" should both A and B fail to function.) Letting R_A, $R_{A'}$, R_B, $R_{B'}$, and R_C represent the reliabilities of the individual components (and assuming that the components function independently of one another), obtain an expression for the reliability of the entire

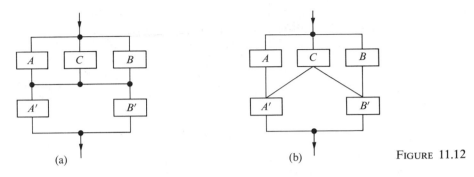

(a) (b) FIGURE 11.12

system in each of the two cases. [*Hint*: In the second case (Fig. 11.12(b)), use conditional probability considerations.]

11.20. If all the components considered in Problem 11.19 have the same constant failure rate λ, obtain an expression for the reliability $R(t)$ for the system indicated in Fig. 11.12(b). Also find the mean time to failure of this system.

11.21. Component A has reliability 0.9 when used for a particular purpose. Component B which may be used in place of component A has a reliability of only 0.75. What is the minimum number of components of type B that would have to be hooked up in parallel in order to achieve the same reliability as component A has by itself?

11.22. Suppose that two independently functioning components, each with the same constant failure rate are connected in parallel. If T is the time to failure of the resulting system, obtain the mgf of T. Also determine $E(T)$ and $V(T)$, using the mgf.

11.23. Whenever we have considered a system made up of several components, we have always supposed that the components function independently of one another. This assumption has simplified our calculations considerably. However, it may not always be a realistic assumption. In many cases it is known that the behavior of one component may affect the behavior of others. This is, in general, a very difficult problem with which to cope, and we shall only consider a special case here. Specifically, suppose that two components, say C_1 and C_2, always fail together. That is, C_1 fails if and only if C_2 fails. Show that in this case, $P(C_1 \text{ fails and } C_2 \text{ fails}) = P(C_1 \text{ fails}) = P(C_2 \text{ fails})$.

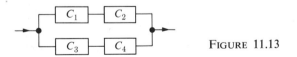

FIGURE 11.13

11.24. Consider four components C_1, C_2, C_3, and C_4 hooked up as indicated in Fig. 11.13. Suppose that the components function independently of one another with the exception of C_1 and C_2 which always fail together as described in Problem 11.23. If T_i, the time to failure of component C_i, is exponentially distributed with parameter β_i, obtain the reliability $R(t)$ of the entire system. Also obtain the pdf of T, the time to failure of the system.

11.25. Consider the same system as described in Problem 11.24 except that this time components C_1 and C_3 fail together. Answer the questions of Problem 11.24.

12

Sums of Random Variables

12.1 Introduction

In this chapter we want to be precise about something that we have been hinting at throughout the text. Namely, that as the number of repetitions of an experiment increases, f_A, the relative frequency of some event A, converges (in a probabilistic sense to be described) to the theoretical probability $P(A)$. It is this fact which allows us to "identify" the relative frequency of an event, based on a large number of repetitions, with the probability of the event.

For example, if a new item is produced and if we have no previous knowledge about how probable it is that the item is defective, we might proceed by inspecting a large number of these items, say N, count the number of defective items among these, say n, and then use n/N as an approximation for the probability of an item being defective. The number n/N is a random variable, and its value depends essentially on two things. First, the value of n/N depends on the underlying (but presumably unknown) probability p that an item is defective. Second, n/N depends on the *particular* N items that we happened to inspect. What we shall show is that if the method of choosing the N items is "random," then the quotient n/N will be close to p (in a sense to be described). (Clearly, the random choice of the N items is important. If we were to choose only those items which displayed some external physical flaw, for example, we might prejudice our computation severely.)

12.2 The Law of Large Numbers

With the aid of Chebyshev's inequality (Eq. 7.20) we are able to derive the result referred to above. Again, let us consider an example. Suppose that a guided missile has a probability of 0.95 of functioning properly during a certain period of operation. Hence, if we released N missiles having the above reliability, and if X is the number of missiles *not* functioning properly, we have $E(X) = 0.05N$, since we may assume X to be binomially distributed. That is, we would expect about 1 missile out of 20 to fail. As N, the number of missiles launched, is increased, X, the total number of missile failures divided by N, should converge in some sense to the number 0.05. This important result can be stated more precisely as the Law of Large Numbers.

The Law of Large Numbers (Bernoulli's form). Let $\mathcal{E}$ be an experiment and let A be an event associated with $\mathcal{E}$. Consider n independent repetitions of $\mathcal{E}$, let n_A be the number of times A occurs among the n repetitions, and let $f_A = n_A/n$. Let $P(A) = p$ (which is assumed to be the same for all repetitions).

Then for every positive number ϵ, we have

$$\text{Prob}\,[|f_A - p| \geq \epsilon] \leq \frac{p(1-p)}{n\epsilon^2}$$

or, equivalently,

$$\text{Prob}\,[|f_A - p| < \epsilon] \geq 1 - \frac{p(1-p)}{n\epsilon^2}. \tag{12.1}$$

Proof: Let n_A be the number of times the event A occurs. This is a binomially distributed random variable. Then $E(n_A) = np$ and $V(n_A) = np(1-p)$. Now $f_A = n_A/n$, and hence $E(f_A) = p$ and $V(f_A) = p(1-p)/n$.

Applying Chebyshev's inequality to the random variable f_A, we obtain

$$P\left[|f_A - p| < k\sqrt{\frac{p(1-p)}{n}}\right] \geq 1 - \frac{1}{k^2}.$$

Let $\epsilon = k\sqrt{p(1-p)/n}$. Then $k^2 = (n\epsilon^2)/[p(1-p)]$, and thus

$$P[|f_A - p| < \epsilon] \geq 1 - \frac{p(1-p)}{n\epsilon^2}.$$

Notes: (a) The above result may be stated in a number of equivalent alternative ways. It is clear that the above immediately implies that

$$\lim_{n \to \infty} P[|f_A - p| < \epsilon] = 1 \qquad \text{for all} \quad \epsilon > 0.$$

It is in this sense that we say that the relative frequency f_A "converges" to $P(A)$.

(b) It is important to note the difference between the convergence referred to above (called *convergence in probability*) and the type of convergence often referred to in the calculus. When we say that 2^{-n} converges to zero as $n \to \infty$ we mean that for n sufficiently large, 2^{-n} becomes and remains arbitrarily close to zero. When we say that $f_A = n_A/n$ converges to $P(A)$ we mean that the *probability* of the event

$$\{|n_A/n - P(A)| < \epsilon\}$$

can be made arbitrarily close to one by taking n sufficiently large.

(c) Still another form of the Law of Large Numbers is obtained when we ask ourselves the following question: How many repetitions of $\mathcal{E}$ should we perform in order to have a probability of at least 0.95, say, that the relative frequency differs from $p = P(A)$ by less than 0.01, say? That is, for $\epsilon = 0.01$ we wish to choose n so that $1 - p(1-p)/[n(0.01)^2] = 0.95$. Solving for n we obtain $n = p(1-p)/(0.01)^2(0.05)$. Replacing the specific values of 0.05 and 0.01 by δ and ϵ, respectively, we have

$$P[|f_A - p| < \epsilon] \geq 1 - \delta \qquad \text{whenever} \quad n \geq \frac{p(1-p)}{\epsilon^2\delta}.$$

Again it should be stressed that taking $n \geq p(1 - p)/\epsilon^2 \delta$ does not *guarantee* anything about $|f_A - p|$. It only makes it probable that $|f_A - p|$ will be very small.

EXAMPLE 12.1. How many times would we have to toss a fair die in order to be at least 95 percent sure that the relative frequency of having a six come up is within 0.01 of the theoretical probability $\frac{1}{6}$?

Here $p = \frac{1}{6}$, $1 - p = \frac{5}{6}$, $\epsilon = 0.01$, and $\delta = 0.05$. Hence from this relationship we find that $n \geq (\frac{1}{6})(\frac{5}{6})/(0.01)^2(0.05) = 27{,}778$.

Notes: (a) Recall that f_A is a random variable and not just an observed value. If we actually toss a die 27,778 times and then compute the relative frequency of having a six come up, this number either is or is not within 0.01 of $\frac{1}{6}$. The point of the above example is that if we were to throw a die 27,778 times in each of 100 rooms, in about 95 of the rooms, the observed relative frequency *would be* within 0.01 of $\frac{1}{6}$.

(b) In many problems we do not know the value of $p = P(A)$ and hence would be unable to use the above bound on n. In that case we can use the fact that $p(1 - p)$ assumes its maximum value when $p = \frac{1}{2}$ and this maximum value equals $\frac{1}{4}$. Hence we would certainly be on the safe side if we stated that for $n \geq 1/4\epsilon^2\delta$ we have

$$P[|f_A - p| < \epsilon] \geq 1 - \delta.$$

EXAMPLE 12.2. Items are produced in such a manner that the probability of an item being defective is p (assumed unknown). A large number of items, say n, are classified as defective or nondefective. How large should n be so that we may be 99 percent sure that the relative frequency of defectives differs from p by less than 0.05?

Since we do not know the value of p, we must apply the last-stated form of the Law of Large Numbers. Thus with $\epsilon = 0.05$, $\delta = 0.01$ we find that if $n \geq 1/4(0.05)^2(0.01) = 10{,}000$, the required condition is satisfied.

As in our example for Chebyshev's inequality, we shall find that additional knowledge about the probability distribution will give us an "improved" statement. (For instance, we might be able to get by with a smaller number of repetitions and still make the same statement concerning the proximity of f_A to p.)

Note: Another form of the Law of Large Numbers may be obtained as follows. Suppose that $X_1, \ldots, X_n$ are identically distributed, independent random variables with finite mean and variance. Let $E(X_i) = \mu$ and $V(X_i) = \sigma^2$. Define $\overline{X} = (1/n)(X_1 + \cdots + X_n)$. Now $\overline{X}$ is a function of $X_1, \ldots, X_n$, namely their arithmetic mean, and hence is again a random variable. (We shall study this random variable in more detail in Chapter 13. For the moment let us simply say that we may think of $X_1, \ldots, X_n$ as independent measurements on a numerical characteristic X, yielding the arithmetic mean $\overline{X}$.) From the properties of expectation and variance we have immediately, $E(\overline{X}) = \mu$ and $V(\overline{X}) = \sigma^2/n$. Let us apply Chebyshev's inequality to the random variable $\overline{X}$:

$$P\left[|\overline{X} - \mu| < \frac{k\sigma}{\sqrt{n}}\right] \geq 1 - \frac{1}{k^2}.$$

Let $k\sigma/\sqrt{n} = \epsilon$. Then $k = \sqrt{n}\,\epsilon/\sigma$ and we may write

$$P[|\bar{X} - \mu| < \epsilon] \geq 1 - \frac{\sigma^2}{\epsilon^2 n}. \tag{12.2}$$

As $n \to \infty$, the right-hand side of the above inequality approaches one. It is in this sense that the arithmetic mean "converges" to $E(X)$.

EXAMPLE 12.3. A large number of electronic tubes are tested. Let T_i be the time to failure of the ith tube. Suppose, furthermore, that all the tubes come from the same stockpile and that all may be assumed to be exponentially distributed with the same parameter α.

Hence $E(T_i) = \alpha^{-1}$. Let $\bar{T} = (T_1 + \cdots + T_n)/n$. The above form of the Law of Large Numbers says that if n is quite large, it would be "very probable" that the value obtained for the arithmetic mean of a large number of failure times would be close to α^{-1}.

12.3 Normal Approximation to the Binomial Distribution

The Law of Large Numbers as stated above deals essentially with the binomially distributed random variable X. For X was defined as the number of successes in n independent repetitions of an experiment; and we need simply associate "success" with the occurrence of the event A in order to recognize this relationship. Thus the above result may be stated informally by saying that as the number of repetitions of an experiment is increased, the relative frequency of success, X/n, converges to the probability of success p, in the sense indicated previously.

However, knowing that X/n will be "close" to p for large n does not tell us how this "closeness" is achieved. In order to investigate this question we must study the probability distribution of X when n is large.

For example, suppose that a manufacturing process produces washers, about 5 percent of which are defective (say, too large). If 100 washers are inspected, what is the probability that fewer than 4 are defective?

Letting X be the number of defective washers found, the Law of Large Numbers simply tells us that $X/100$ should be "close" to 0.05. However, it does not tell us how to evaluate the desired probability. The *exact* value of this probability is given by

$$P(X < 4) = \sum_{k=0}^{3} \binom{100}{k} (0.05)^k (0.95)^{100-k}.$$

This probability would be rather difficult to compute directly. We have already studied one method of approximation for the binomial probabilities, namely the Poisson approximation. We shall now consider another important approximation to such probabilities which is applicable whenever n is sufficiently large.

Consider then $P(X = k) = \binom{n}{k}p^k(1 - p)^{n-k}$. This probability depends on n in a rather complicated way and it is by no means clear what happens to the above expression if n is large. In order to investigate this probability, we need to use

TABLE 12.1

n	$n!$	$\sqrt{2\pi}\,e^{-n}n^{n+(1/2)}$	Difference	Difference $n!$
1	1	0.922	0.078	0.08
2	2	1.919	0.081	0.04
5	120	118.019	1.981	0.02
10	$(3.6288)10^6$	$(3.5986)10^6$	$(0.0302)10^6$	0.008
100	$(9.3326)10^{157}$	$(9.3249)10^{157}$	$(0.0077)10^{157}$	0.0008

Stirling's formula, a well-known approximation to $n!$ This formula states that for large n,

$$n! \sim \sqrt{2\pi}\,e^{-n}n^{n+1/2}, \tag{12.3}$$

in the sense that $\lim_{n\to\infty} (n!)/(\sqrt{2\pi}\,e^{-n}n^{n+1/2}) = 1$. (A proof of this approximation may be found in most texts on advanced calculus.) Table 12.1 may give the reader an idea of the accuracy of this approximation. This table is taken from W. Feller, *Probability Theory and Its Applications*, John Wiley and Sons, Inc., 1st Edition, New York, 1950.

Note: Although the difference between $n!$ and its approximation becomes larger as $n \to \infty$, the important thing to observe in Table 12.1 is that the percentage error (last column) becomes smaller.

Using Stirling's formula for the various factorials appearing in the expression for $P(X = k)$, it may be shown (after considerable manipulation), that for large n,

$$P(X = k) = \binom{n}{k} p^k (1 - p)^{n-k}$$

$$\simeq \frac{1}{\sqrt{2\pi np(1 - p)}} \exp\left(-\frac{1}{2}\left[\frac{k - np}{\sqrt{np(1 - p)}} \right]^2 \right) \tag{12.4}$$

Finally we can show that for large n,

$$P(X \leq k) = P\left[\frac{X - np}{\sqrt{np(1 - p)}} \leq \frac{k - np}{\sqrt{np(1 - p)}} \right]$$

$$\simeq \frac{1}{\sqrt{2\pi}} \int_{-\infty}^{(k-np)/\sqrt{np(1-p)}} e^{-t^2/2}\, dt. \tag{12.5}$$

Thus we have the following important result (known as the DeMoivre-Laplace approximation to the binomial distribution):

Normal approximation to the binomial distribution. If X has a binomial distribution with parameters n and p and if

$$Y = \frac{X - np}{[np(1 - p)]^{1/2}},$$

then, for large n, Y has approximately a $N(0, 1)$ distribution in the sense that $\lim_{n \to \infty} P(Y \leq y) = \Phi(y)$. This approximation is valid for values of $n > 10$ provided p is close to $\frac{1}{2}$. If p is close to 0 or 1, n should be somewhat larger to insure a good approximation.

Notes: (a) The above result is not only of considerable theoretical interest but also of great practical importance. It means that we may use the extensively tabulated normal distribution to evaluate probabilities arising from the binomial distribution.

(b) In Table 12.2 the accuracy of the approximation (12.4) is shown for various values of n, k and p.

TABLE 12.2

k	$n = 8, p = 0.2$ Approximation	Exact	$n = 8, p = 0.5$ Approximation	Exact	$n = 25, p = 0.2$ Approximation	Exact
0	0.130	0.168	0.005	0.004	0.009	0.004
1	0.306	0.336	0.030	0.031	0.027	0.024
2	0.331	0.294	0.104	0.109	0.065	0.071
3	0.164	0.147	0.220	0.219	0.121	0.136
4	0.037	0.046	0.282	0.273	0.176	0.187
5	0.004	0.009	0.220	0.219	0.199	0.196
6	0+	0.001	0.104	0.109	0.176	0.163
7	0+	0+	0.030	0.031	0.121	0.111
8	0+	0+	0.005	0.004	0.065	0.062
9	0+	0+	0+	0+	0.027	0.029
10	0+	0+	0+	0+	0.009	0.012
11	0+	0+	0+	0+	0.002	0.004

Returning to the example above, we note that

$$E(X) = np = 100(0.05) = 5,$$
$$V(X) = np(1 - p) = 4.75.$$

Hence we may write

$$P(X \leq 3) = P\left(\frac{0 - 5}{\sqrt{4.75}} \leq \frac{X - 5}{\sqrt{4.75}} \leq \frac{3 - 5}{\sqrt{4.75}}\right)$$
$$= \Phi(-0.92) - \Phi(-2.3) = 0.168,$$

from the tables of the normal distribution.

Note: In using the normal approximation to the binomial distribution, we are approximating the distribution of a discrete random variable with the distribution of a continuous random variable. Hence some care must be taken with the endpoints of the intervals involved. For example, for a continuous random variable, $P(X = 3) = 0$, while for a discrete random variable this probability may be positive.

The following *corrections for continuity* have been found to improve the above approximation:

(a) $P(X = k) \simeq P(k - \frac{1}{2} \leq X \leq k + \frac{1}{2})$,
(b) $P(a \leq X \leq b) \simeq P(a - \frac{1}{2} \leq X \leq \frac{1}{2} + b)$.

Using this latter correction for the above evaluation of $P(X \leq 3)$, we have

$$P(X \leq 3) = P(0 \leq X \leq 3) = P(-\tfrac{1}{2} \leq X \leq 3\tfrac{1}{2})$$
$$\simeq \Phi(-0.69) - \Phi(-2.53) = 0.239.$$

EXAMPLE 12.4. Suppose that a system is made up of 100 components each of which has a reliability equal to 0.95. (That is, the probability that the component functions properly during a specified time equals 0.95.) If these components function independently of one another, and if the entire system functions properly when at least 80 components function, what is the reliability of the system?

Letting X be the number of components functioning, we must evaluate

$$P(80 \leq X \leq 100).$$

We have

$$E(X) = 100(0.95) = 95; \quad V(X) = 100(0.95)(0.05) = 4.75.$$

Hence, using the correction for continuity, we obtain

$$P(80 \leq X \leq 100) \simeq P(79.5 \leq X \leq 100.5)$$
$$= P\left(\frac{79.5 - 95}{2.18} \leq \frac{X - 95}{2.18} \leq \frac{100.5 - 95}{2.18}\right)$$
$$\simeq \Phi(2.52) - \Phi(-7.1) = 0.994.$$

12.4 The Central Limit Theorem

The above approximation represents only a special case of a general result. In order to realize this, let us recall that the binomially distributed random variable X may be represented as the *sum* of the following independent random variables:

$$X_i = 1 \quad \text{if success occurs on the } i\text{th repetition};$$
$$= 0 \quad \text{if failure occurs on the } i\text{th repetition}.$$

Hence $X = X_1 + X_2 + \cdots + X_n$. (See Example 7.14.) For *this* random variable we have shown that $E(X) = np$, $V(X) = np(1 - p)$ and, furthermore, that for large n, $(X - np)/\sqrt{np(1 - p)}$ has the approximate distribution $N(0, 1)$.

If a random variable X may be represented as a *sum of any n independent random variables* (satisfying certain conditions which hold in most applications), then this sum, for sufficiently large n, is approximately normally distributed. This remarkable result is known as the Central Limit Theorem. One form of this theorem may be stated as follows.

> **Central Limit Theorem.** Let $X_1, X_2, \ldots, X_n, \ldots$ be a sequence of independent random variables with $E(X_i) = \mu_i$ and $V(X_i) = \sigma_i^2$, $i = 1, 2, \ldots$ Let $X = X_1 + X_2 + \cdots + X_n$. Then under certain general conditions (which we shall not explicitly state here),
>
> $$Z_n = \frac{X - \sum_{i=1}^{n} \mu_i}{\sqrt{\sum_{i=1}^{n} \sigma_i^2}}$$
>
> has approximately the distribution $N(0, 1)$. That is, if G_n is the cdf of the random variable Z_n, we have $\lim_{n \to \infty} G_n(z) = \Phi(z)$.

Notes: (a) This theorem represents an obvious generalization of the DeMoivre-Laplace approximation. For the independent random variables X_i assuming only the values 1 and 0 have been replaced by random variables possessing any kind of distribution (so long as they have finite expectation and finite variance). The fact that the X_i's may have (essentially) any kind of distribution, and yet the sum $X = X_1 + \cdots + X_n$ may be approximated by a normally distributed random variable, represents the basic reason for the importance of the normal distribution in probability theory. It turns out that in many problems, the random variable under consideration may be represented as the sum of n independent random variables, and hence its distribution may be approximated by the normal distribution.

For example, the electricity consumption in a city at any given time is the sum of the demands of a large number of individual consumers. The quantity of water in a reservoir may be thought of as representing the sum of a very large number of individual contributions. And the error of measurement in a physical experiment is composed of many unobservable small errors which may be considered additive. The molecular bombardment which a particle suspended in a liquid is undergoing causes it to be displaced in a random direction and random magnitude, and its position (after a specified length of time) may be considered as a sum of individual displacements.

(b) The general conditions referred to in the above statement of the Central Limit Theorem may be summarized informally as follows: The individual terms in the sum contribute a negligible amount to the variation of the sum, and it is very unlikely that any single term makes a very large contribution to the sum. (Errors of measurement seem to have this characteristic. The final error we make may be represented as a sum of many small contributions none of which contribute very much to the entire error.)

(c) We have already established (Theorem 10.5) that the sum of any finite number of independent normally distributed random variables is again normally distributed. The Central Limit Theorem states that the summands need not be normally distributed in order for the sum to be approximated by a normal distribution.

(d) We cannot prove the above theorem, without exceeding the intended level of presentation. However there is a special, important case of this theorem which we shall state and for which we can give at least an indication of proof.

Theorem 12.1. Let $X_1, \ldots, X_n$ be n independent random variables all of which have the *same* distribution. Let $\mu = E(X_i)$ and $\sigma^2 = V(X_i)$ be the common expectation and variance. Let $S = \sum_{i=1}^{n} X_i$. Then $E(S) = n\mu$ and $V(S) = n\sigma^2$, and we have, for large n, that $T_n = (S - n\mu)/\sqrt{n}\,\sigma$ has approximately the distribution $N(0, 1)$ in the sense that $\lim_{n \to \infty} P(T_n \leq t) = \Phi(t)$.

Proof (outline): (The reader would do well to review the basic concepts of the mgf as discussed in Chapter 10.) Let M be the (common) mgf of the X_i's. Since the X_i's are independent, M_S, the mgf of S, is given by $M_S(t) = [M(t)]^n$, and since T_n is a linear function of S, the mgf of T_n is given by (using Theorem 10.2)

$$M_{T_n}(t) = e^{-(\sqrt{n}\mu/\sigma)t} \left[M\left(\frac{t}{\sqrt{n}\,\sigma} \right) \right]^n.$$

Thus

$$\ln M_{T_n}(t) = \frac{-\sqrt{n}\,\mu}{\sigma} t + n \ln M\left(\frac{t}{\sqrt{n}\,\sigma} \right).$$

(Let us note at this point that the idea of the proof consists of investigating $M_{T_n}(t)$ for large values of n.)

Let us expand $M(t)$ in a Maclaurin series:

$$M(t) = 1 + M'(0)t + \frac{M''(0)t^2}{2} + R,$$

where R is the remainder term. Recalling that $M'(0) = \mu$ and $M''(0) = \mu^2 + \sigma^2$, we obtain

$$M(t) = 1 + \mu t + \frac{(\mu^2 + \sigma^2)t^2}{2} + R.$$

Hence

$$\ln M_{T_n}(t) = -\frac{\sqrt{n}\,\mu t}{\sigma} + n \ln\left[1 + \frac{\mu t}{\sqrt{n}\,\sigma} + \frac{(\mu^2 + \sigma^2)t^2}{2n\sigma^2} + R \right].$$

We shall now use the Maclaurin expansion for $\ln (1 + x)$:

$$\ln (1 + x) = x - \frac{x^2}{2} + \frac{x^3}{3} + \cdots$$

(This expansion is valid for $|x| < 1$. In our case

$$x = \frac{\mu t}{\sqrt{n}\,\sigma} + \frac{(\mu^2 + \sigma^2)t^2}{2n\sigma^2} + R;$$

and for n sufficiently large, the absolute value of this expression will be less than one.) Thus we obtain

$$\ln M_{T_n}(t) = -\frac{\sqrt{n}\,\mu}{\sigma} + n\left[\left(\frac{\mu t}{\sqrt{n}\,\sigma} + (\mu^2 + \sigma^2)\frac{t^2}{2n\sigma^2} + R\right)\right.$$
$$\left. -\frac{1}{2}\left(\frac{\mu t}{\sqrt{n}\,\sigma} + (\mu^2 + \sigma^2)\frac{t^2}{2n\sigma^2} + R\right)^2 + \cdots\right].$$

Since we are simply outlining the main steps of the proof without giving all the details, let us omit some algebraic manipulations and simply state what we are doing. We want to investigate the above expression $\left(\ln M_{T_n}(t)\right)$ as $n \to \infty$. Any term having a positive power of n in the denominator (such as $n^{-1/2}$, for example) will approach zero as $n \to \infty$. It may also be shown that all terms involving R approach zero as $n \to \infty$. After fairly straightforward but tedious algebra we find that

$$\lim_{n\to\infty} \ln M_{T_n}(t) = t^2/2.$$

Hence we have

$$\lim_{n\to\infty} M_{T_n}(t) = e^{t^2/2}.$$

This is the mgf of a random variable with distribution $N(0, 1)$. Because of the uniqueness property of the mgf (see Theorem 10.3) we may conclude that the random variable T_n converges in distribution (as $n \to \infty$) to the distribution $N(0,1)$.

Notes: (a) Although the above does not represent a rigorous proof, it should never-theless give the reader some feeling for the derivation of this important theorem. The more general form of the Central Limit Theorem (as originally stated) can be proved using an approach similar to the one employed here.

(b) The special form of the Central Limit Theorem as stated above says that the arithmetic mean $(1/n)\sum_{i=1}^{n} X_i$ of n observations from the same random variable has, for large n, approximately a normal distribution.

(c) Although a mathematical proof should establish the validity of a theorem, it may not contribute greatly to the intuitive feeling of the result. We therefore present the following example for those who are more numerically oriented.

EXAMPLE 12.5. Consider an urn containing three kinds of things identified as 0, 1, and 2. Suppose that there are 20 zeros, 30 one's, and 50 two's. An item is chosen at random and its value, say X, is recorded. Assume that X has the follow-ing distribution. (See Fig. 12.1.)

X	0	1	2
$P(X = x)$	0.2	0.3	0.5

Suppose that the item chosen first is replaced and then a second item is chosen

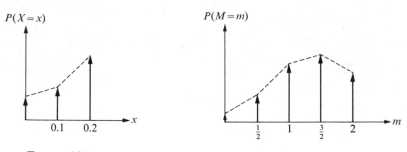

FIGURE 12.1   FIGURE 12.2

and its value, say Y, is recorded. Consider the random variable $M = (X + Y)/2$ and its distribution (Fig. 12.2).

M	0	$\frac{1}{2}$	1	$\frac{3}{2}$	2
$P(M = m)$	0.04	0.12	0.29	0.30	0.25

We obtained the above values for $P(M)$ as follows:

$$P(M = 0) = P(X = 0, Y = 0) = (0.2)^2 = 0.04;$$
$$P(M = \tfrac{1}{2}) = P(X = 0, Y = 1) + P(X = 1, Y = 0)$$
$$= (0.2)(0.3) + (0.3)(0.2) = 0.12, \quad \text{etc.}$$

Finally, suppose that after the second item has also been replaced, a third item is chosen and its value, say Z, is recorded. Consider the random variable $N = (X + Y + Z)/3$ and its distribution (Fig. 12.3):

N	0	$\frac{1}{3}$	$\frac{2}{3}$	1	$\frac{4}{3}$	$\frac{5}{3}$	2
$P(N = n)$	0.008	0.036	0.114	0.207	0.285	0.225	0.125

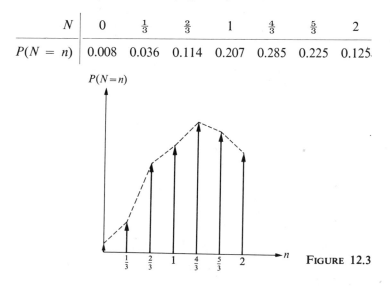

FIGURE 12.3

The probability distributions of the random variables M and N are already showing signs of "normality." That is, the bell-shaped appearance of the distribution
curve is beginning to be apparent. Starting with the distribution of X, which is
quite asymmetric, we find that the mean of only three observations has a distribution which is already showing "indications of normality."

The above example *proves* nothing, of course. However, it does represent a
numerical illustration of the result discussed previously in a more mathematical
manner. The reader should continue this example by adding one additional
observation and then find the probability distribution of the mean of the four
observations obtained. (See Problem 12.10.)

EXAMPLE 12.6. Suppose that we have a number of independent noise voltages,
say V_i, $i = 1, 2, \ldots, n$, which are received in what is called an "adder." (See
Fig. 12.4.) Let V be the sum of the voltages received. That is, $V = \sum_{i=1}^{n} V_i$.
Suppose that each of the random variables V_i is uniformly distributed over the
interval $[0, 10]$. Hence $E(V_i) = 5$ volts and var $(V_i) = 100/12$.

According to the Central Limit Theorem, if n is
sufficiently large, the random variable

$$S = (V - 5n)\sqrt{12}/10\sqrt{n}$$

has approximately the distribution $N(0, 1)$. Thus, if
$n = 20$, we can compute the probability that the total
incoming voltage exceeds 105 volts, as follows:

$$P(V > 105) = P\left(\frac{V - 100}{(10/\sqrt{12})\sqrt{20}} > \frac{105 - 100}{(10/\sqrt{12})\sqrt{20}}\right)$$

$$\simeq 1 - \Phi(0.388) = 0.352.$$

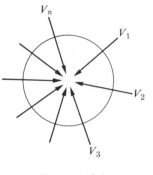

FIGURE 12.4

12.5 Other Distributions Approximated by the Normal Distribution: Poisson, Pascal, and Gamma

There are a number of important distributions other than the binomial discussed in Section 12.3 which may be approximated by the normal distribution.
And in each case, as we shall note, the random variable whose distribution we
shall approximate may be represented as a sum of independent random variables,
thus giving us an application of the Central Limit Theorem as discussed in
Section 12.4.

(a) *The Poisson distribution.* Recall that a Poisson random variable arises when
(subject to certain assumptions) we are interested in the total number of occurrences of some event in a time interval of length t, with intensity (i.e., rate of occurrence per unit time) of α (see Section 8.2). As was indicated in that section, we
may consider the total number of occurrences as the *sum* of the occurrences in

smaller, nonoverlapping intervals, thus making the results of the previous section applicable.

EXAMPLE 12.7.   Suppose that calls come into a particular telephone exchange at the rate of 2 per minute.   What is the probability that 22 or fewer calls are received during a 15-minute period?   (We assume, of course, that the intensity remains the same during the period considered.)   If X = number of calls received, then $E(X) = 2(15) = 30$.   To evaluate $P(X \leq 22)$, and applying the correction for continuity (see Eq. (b) just preceding Example 12.4), we have

$$P(X \leq 22) \cong P\left(Y \leq \frac{22 + \frac{1}{2} - 30}{\sqrt{30}}\right)$$

where Y has $N(0, 1)$ distribution.   Hence the above probability equals $\Phi(-1.37) = 0.0853$.

(b) *The Pascal distribution.*   If Y = number of Bernoulli trials required to obtain r successes, then Y has a Pascal distribution and may be represented as the sum of r independent random variables (see Section 8.5).   Thus for r sufficiently large, the results of the previous section apply.

EXAMPLE 12.8.   Find an approximate value for the probability that it will take 150 or fewer trials to obtain 48 successes when $P(\text{success}) = 0.25$.   Letting X = number of trials required, we have (see Eq. 8.8) $E(X) = r/p = 48/0.25 = 192$, and Var $X = rq/p^2 = (48)(0.75)/(0.25)^2 = 576$.   Hence,

$$P(X \leq 150) \cong \Phi\left(\frac{150 + \frac{1}{2} - 192}{\sqrt{576}}\right) = \Phi(-1.73) = 0.0418.$$

(c) *The Gamma distribution.*   As indicated in Theorem 10.9, a random variable having a Gamma distribution (with parameters α and r) may be represented as a sum of r independent exponentially distributed random variables.   Hence for large r, the Central Limit Theorem again applies.

12.6 The Distribution of the Sum of a Finite Number of Random Variables

Example 12.6 serves to motivate the following discussion.   We know that the sum of any finite number of independent, normally distributed random variables is again normally distributed.   From the Central Limit Theorem we may conclude that for large n, the sum of n independent random variables is approximately normally distributed.   The following question remains to be answered: Suppose that we consider $X_1 + \cdots + X_n$, where the X_i's are independent random variables (not necessarily normal) and n is not sufficiently large to justify the use of the Central Limit Theorem.   What is the distribution of this sum?   For example, what is the distribution of the incoming voltage V (Example 12.6), if $n = 2$ or $n = 3$?

We shall first consider the important case of the sum of two random variables. The following result may be established.

Theorem 12.2. Suppose that X and Y are independent, continuous random variables with pdf's g and h, respectively. Let $Z = X + Y$ and denote the pdf of Z by s. Then,

$$s(z) = \int_{-\infty}^{+\infty} g(w)h(z - w)\, dw. \tag{12.6}$$

Proof: Since X and Y are independent, their joint pdf f may be factored:

$$f(x, y) = g(x)h(y).$$

Consider the transformation:

$$z = x + y, \qquad w = x.$$

Hence $x = w$ and $y = z - w$. The Jacobian of this transformation is

$$J = \begin{vmatrix} 0 & 1 \\ 1 & -1 \end{vmatrix} = -1.$$

Thus the absolute value of J is 1 and hence the joint pdf of $Z = X + Y$ and $W = X$ is

$$k(z, w) = g(w)h(z - w).$$

The pdf of Z is now obtained by integrating $k(z, w)$ from $-\infty$ to ∞ with respect to w, thus yielding the above result.

Notes: (a) The above integral involving the functions g and h occurs in many other mathematical contexts. It is often referred to as the *convolution* integral of g and h and is sometimes written as $g * h$.

(b) The evaluation of the above integral must be carried out with considerable care. In fact, the same difficulty that arose in the evaluation of the pdf of a product or a quotient arises again. The functions g and h will often be nonzero only for certain values of their argument. Hence the integrand in the above integral will be nonzero only for those values of the variable of integration w for which *both* factors of the integrand are nonzero.

(c) The above formula Eq. (12.6) may be used repeatedly (with increasing difficulty, however) to obtain the pdf of the sum of any finite number of independent, continuous random variables. For example, if $S = X + Y + W$, we may write this as $S = Z + W$, where $Z = X + Y$. We may then use the above approach to obtain the pdf of Z, and then, knowing the pdf of Z, use this method again to obtain the pdf of S.

(d) We can derive Eq. (12.6) in another manner, without using the notion of a Jacobian. Let S denote the cdf of the random variable $Z = X + Y$. Then

$$S(z) = P(Z \le z) = P(X + Y \le z) = \iint_R g(x)h(y)\, dx\, dy,$$

where

$$R = \{(x, y) \mid x + y \leq z\}.$$

(See Fig. 12.5.) Hence

$$S(z) = \int_{-\infty}^{+\infty} \int_{-\infty}^{z-x} g(x)h(y)\, dx\, dy$$

$$= \int_{-\infty}^{+\infty} g(x)\left[\int_{-\infty}^{z-x} h(y)\, dy\right] dx.$$

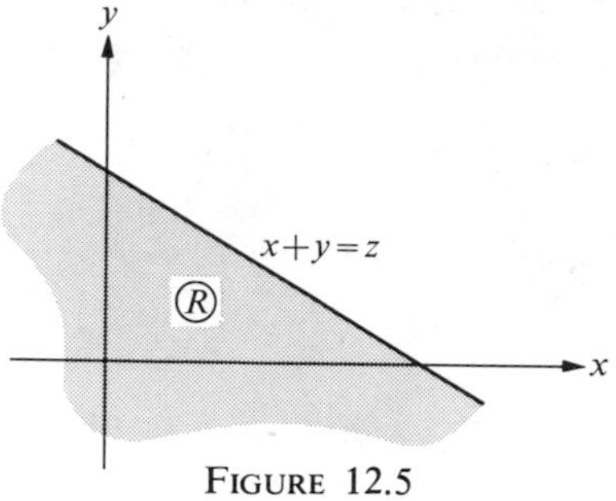

FIGURE 12.5

Differentiating $S(z)$ with respect to z (under the integral sign, which can be justified) we obtain

$$s(z) = S'(z) = \int_{-\infty}^{+\infty} g(x)h(z - x)\, dx,$$

which agrees with Eq. (12.6).

(e) Since the distribution of $X + Y$ should presumably be the same as the distribution of $Y + X$, we should be able to verify that $\int_{-\infty}^{+\infty} g(x)h(z - x)\, dx = \int_{-\infty}^{+\infty} h(y)g(z - y)\, dy$. Simply letting $z - x = y$ in the first integral will yield the second form. We sometimes indicate this property by writing $g * h = h * g$.   See Note (a) above.

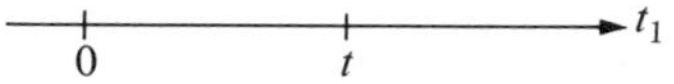

FIGURE 12.6

EXAMPLE 12.9.   Consider two electronic devices, say D_1 and D_2. Suppose that D_1 has a life length which may be represented by a random variable T_1 having exponential distribution with parameter α_1, while D_2 has a life length which may be represented by a random variable T_2 having an exponential distribution with parameter α_2. Suppose that D_1 and D_2 are hooked up in such a way that D_2 begins to function at the moment that D_1 ceases to function. Then $T = T_1 + T_2$ represents the total time that the system made up of these two devices is functioning. Assuming T_1 and T_2 to be independent, we may apply the above result to obtain

$$g(t_1) = \alpha_1 e^{-\alpha_1 t_1}, \qquad t_1 \geq 0,$$
$$h(t_2) = \alpha_2 e^{-\alpha_2 t_2}, \qquad t_2 \geq 0.$$

(For all other values of t_1 and t_2, the functions g and h are assumed to be zero!) Hence using Eq. (12.6), we find that the pdf of $T_1 + T_2 = T$ is given by

$$s(t) = \int_{-\infty}^{+\infty} g(t_1)h(t - t_1)\, dt_1, \qquad t \geq 0.$$

The integrand is positive if and only if *both* factors of the integrand are positive; that is, whenever $t_1 \geq 0$ *and* $t - t_1 \geq 0$. This is equivalent to $t_1 \geq 0$ *and* $t_1 \leq t$, which in turn is equivalent to $0 \leq t_1 \leq t$. (See Fig. 12.6.) Hence the above integral becomes

$$s(t) = \int_0^t \alpha_1 e^{-\alpha_1 t_1} \alpha_2 e^{-\alpha_2(t-t_1)} dt_1$$

$$= \alpha_1 \alpha_2 e^{-\alpha_2 t} \int_0^t e^{-t_1(\alpha_1-\alpha_2)} dt_1$$

$$= \frac{\alpha_1 \alpha_2}{\alpha_2 - \alpha_1} (e^{-t\alpha_1} - e^{-t\alpha_2}), \qquad \text{for } t \geq 0.$$

Notes: (a) We note that the sum of two independent, exponentially distributed random variables is *not* exponentially distributed.

(b) For $\alpha_1 > \alpha_2$, the graph of the pdf of T is shown in Fig. 12.7.

(c) The above expression for the pdf is not defined for $\alpha_1 = \alpha_2$, that is, for the case where T_1 and T_2 have the same exponential distribution. In order to take care of this special case, consider the first integral expression for $s(t)$ and let $\alpha = \alpha_1 = \alpha_2$. We obtain

$$s(t) = \int_0^t \alpha e^{-\alpha t_1} \alpha e^{-\alpha(t-t_1)} dt_1$$

$$= \alpha^2 e^{-\alpha t} \int_0^t dt_1 = \alpha^2 t e^{-\alpha t}, \qquad t \geq 0.$$

This represents a Gamma distribution (see Eq. 9.16). The graph of this pdf is given in Fig. 12.8. The maximum occurs for $t = 1/\alpha = E(T_1) = E(T_2)$.

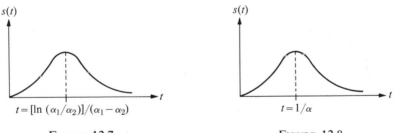

$t = [\ln (\alpha_1/\alpha_2)]/(\alpha_1-\alpha_2)$ | $t = 1/\alpha$

FIGURE 12.7 | FIGURE 12.8

EXAMPLE 12.10.   Let us reconsider Example 12.6, dealing with the addition of two independent random voltages V_1 and V_2, each of which is uniformly distributed over $[0, 10]$. Thus

$$f(v_1) = \tfrac{1}{10}, \qquad 0 \leq v_1 \leq 10,$$
$$g(v_2) = \tfrac{1}{10}, \qquad 0 \leq v_2 \leq 10.$$

(Again recall that the functions f and g are zero elsewhere.) If $V = V_1 + V_2$, we have

$$s(v) = \int_{-\infty}^{+\infty} f(v_1)g(v - v_1) dv_1.$$

Reasoning as in Example 12.8, we note that the integrand is nonzero only for those values of v_1 satisfying $0 \leq v_1 \leq 10$ and $0 \leq v - v_1 \leq 10$. These conditions are equivalent to $0 \leq v_1 \leq 10$ *and* $v - 10 \leq v_1 \leq v$.

(a)   (b)

FIGURE 12.9

Two cases arise, as is indicated in Fig. 12.9.

(a) $v - 10 \leq 10$ and $0 \leq v \leq 10$ which together imply that $0 \leq v \leq 10$.
(b) $0 \leq v - 10 \leq 10$ and $v \geq 10$ which together imply that $10 \leq v \leq 20$.

In case (a), v_1 may assume values between 0 and v, while in case (b), v_1 may assume values between $v - 10$ and 10.
    Thus we obtain

$$\text{for } 0 \leq v \leq 10: \qquad s(v) = \int_0^v \frac{1}{10} \frac{1}{10} \, dv_1 = \frac{v}{100},$$

$$\text{for } 10 \leq v \leq 20: \qquad s(v) = \int_{v-10}^{10} \frac{1}{10} \frac{1}{10} \, dv_1 = \frac{20 - v}{100}.$$

Thus the pdf of V has the graph shown in Fig. 12.10.

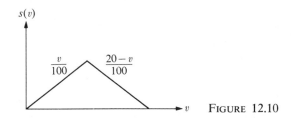

FIGURE 12.10

EXAMPLE 12.11.   As a final illustration of sums of random variables, let us reconsider a result we have already proved using the more indirect method of moment generating functions (see Example 10.11), namely that the sum of two independent normal random variables is again normally distributed. In order to avoid some involved algebraic manipulations, let us consider only a special case.
    Suppose that $Z = X + Y$, where X and Y are independent random variables, each with distribution $N(0, 1)$. Then

$$f(x) = \frac{1}{\sqrt{2\pi}} e^{-x^2/2}, \qquad -\infty < x < \infty,$$

$$g(y) = \frac{1}{\sqrt{2\pi}} e^{-y^2/2}, \qquad -\infty < y < \infty.$$

Hence the pdf of Z is given by

$$s(z) = \int_{-\infty}^{+\infty} f(x)g(z - x)\, dx = \frac{1}{2\pi} \int_{-\infty}^{+\infty} e^{-x^2/2} e^{-(z-x)^2/2}\, dx$$

$$= \frac{1}{2\pi} \int_{-\infty}^{+\infty} e^{-(1/2)[x^2 + z^2 - 2zx + x^2]}\, dx$$

$$= \frac{1}{2\pi} e^{-z^2/2} \int_{-\infty}^{+\infty} e^{-(x^2 - zx)}\, dx.$$

Completing the square in the exponent of the integrand, we obtain

$$(x^2 - zx) = \left[\left(x - \frac{z}{2}\right)^2 - \frac{z^2}{4}\right].$$

Hence

$$s(z) = \frac{1}{2\pi} e^{-z^2/2} e^{z^2/4} \int_{-\infty}^{+\infty} e^{-(1/2)[\sqrt{2}(x - z/2)]^2}\, dx.$$

Let $\sqrt{2}(x - z/2) = u$; then $dx = du/\sqrt{2}$ and we obtain

$$s(z) = \frac{1}{\sqrt{2\pi}\sqrt{2}} e^{-z^2/4} \frac{1}{\sqrt{2\pi}} \int_{-\infty}^{+\infty} e^{-u^2/2}\, du.$$

The above integral (including the factor $1/\sqrt{2\pi}$) equals one. Thus

$$s(z) = \frac{1}{\sqrt{2\pi}\sqrt{2}} e^{-(1/2)(z/\sqrt{2})^2}.$$

But this represents the pdf of a random variable with distribution $N(0, 2)$, which was to be shown.

In discussing the distribution of the sum of two independent random variables, we have limited ourselves to continuous random variables. In the discrete case the problem is somewhat simpler, at least in certain cases, as the following theorem indicates.

Theorem 12.3. Suppose that X and Y are independent random variables each of which may assume only nonnegative integral values. Let $p(k) = P(X = k)$, $k = 0, 1, 2, \ldots$, and let $q(r) = P(Y = r)$, $r = 0, 1, 2, \ldots$ Let $Z = X + Y$ and let $w(i) = P(Z = i)$. Then,

$$w(i) = \sum_{k=0}^{i} p(k)q(i - k), \qquad i = 0, 1, 2, \ldots$$

Proof

$$w(i) = P(Z = i)$$
$$= P[X = 0, Y = i \text{ or } X = 1, Y = i - 1 \text{ or } \cdots \text{ or } X = i, Y = 0]$$
$$= \sum_{k=0}^{i} P[X = k, Y = i - k] = \sum_{k=0}^{i} p(k)q(i - k)$$

since X and Y are independent.

Note: Observe the similarity between this sum and the convolution integral derived in Theorem 12.2.

EXAMPLE 12.12.   Let X and Y represent the number of emissions of α-particles from two different sources of radioactive materials during a specified time period of length t.   Assume that X and Y have Poisson distributions with parameters $\beta_1 t$ and $\beta_2 t$, respectively.   Let $Z = X + Y$ represent the total number of emitted particles from the two sources.   Using Theorem 12.3, we obtain

$$P(Z = k) = \sum_{k=0}^{i} p(k)q(i-k)$$
$$= e^{-(\beta_1+\beta_2)t} \sum_{k=0}^{i} \frac{(\beta_1 t)^k (\beta_2 t)^{i-k}}{k!(i - k)!}$$
$$= e^{-(\beta_1+\beta_2)t} \frac{(\beta_1 t + \beta_2 t)^i}{i!}.$$

(The last equality is obtained by applying the binomial theorem to the above sum.) The last expression represents the probability that a random variable having a Poisson distribution with parameter $\beta_1 t + \beta_2 t$ assumes the value i.   Thus we have verified what we already knew: The sum of two independent Poisson random variables has a Poisson distribution.

PROBLEMS

12.1.   (a) Items are produced in such a manner that 2 percent turn out to be defective. A large number of such items, say n, are inspected and the relative frequency of defectives, say f_D, is recorded.   How large should n be in order that the probability is at least 0.98 that f_D differs from 0.02 by less than 0.05?

(b) Answer (a) above if 0.02, the probability of obtaining a defective item, is replaced by p which is assumed to be unknown.

12.2.   Suppose that a sample of size n is obtained from a very large collection of bolts, 3 percent of which are defective.   What is the probability that at most 5 percent of the chosen bolts are defective if:

(a) $n = 6$?   (b) $n = 60$?   (c) $n = 600$?

12.3.   (a) A complex system is made up of 100 components functioning independently. The probability that any one component will fail during the period of operation equals 0.10. In order for the entire system to function, at least 85 of the components must be working. Evaluate the probability of this.

(b) Suppose that the above system is made up of n components each having a reliability of 0.90. The system will function if at least 80 percent of the components function properly. Determine n so that the system has reliability of 0.95.

12.4.   Suppose that 30 electronic devices, say $D_1, \ldots, D_{30}$, are used in the following manner: As soon as D_1 fails D_2 becomes operative. When D_2 fails D_3 becomes operative, etc. Assume that the time to failure of D_i is an exponentially distributed random variable with parameter $\beta = 0.1$ hour^{-1}. Let T be the total time of operation of the 30 devices. What is the probability that T exceeds 350 hours?

12.5.   A computer, in adding numbers, rounds each number off to the nearest integer. Suppose that all rounding errors are independent and uniformly distributed over $(-0.5, 0.5)$.

(a) If 1500 numbers are added, what is the probability that the magnitude of the total error exceeds 15?

(b) How many numbers may be added together in order that the magnitude of the total error is less than 10, with probability 0.90?

- 12.6.   Suppose that $X_i, i = 1, 2, \ldots, 50$, are independent random variables each having a Poisson distribution with parameter $\lambda = 0.03$. Let $S = X_1 + \cdots + X_{50}$.

(a) Using the Central Limit Theorem, evaluate $P(S \geq 3)$.

(b) Compare the answer in (a) with the exact value of this probability.

12.7.   In a simple circuit two resistors R_1 and R_2 are connected in series. Hence the total resistance is given by $R = R_1 + R_2$. Suppose that R_1 and R_2 are independent random variables each having the pdf

$$f(r_i) = \frac{10 - r_i}{50}, \qquad 0 < r_i < 10, \qquad i = 1, 2.$$

Find the pdf of R, the total resistance, and sketch its graph.

12.8.   Suppose that the resistors in Problem 12.7 are connected in parallel. Find the pdf of R, the total resistance of the circuit (set up only, in integral form). [*Hint:* The relationship between R, and R_1, and R_2 is given by $1/R = 1/R_1 + 1/R_2$.]

12.9.   In measuring T, the life length of an item, an error may be made which may be assumed to be uniformly distributed over $(-0.01, 0.01)$. Thus the recorded time (in hours) may be represented as $T + X$, where T has an exponential distribution with parameter 0.2 and X has the above-described uniform distribution. If T and X are independent, find the pdf of $T + X$.

12.10.   Suppose that X and Y are independent, identically distributed random variables. Assume that the pdf of X (and hence of Y) is given by

$$f(x) = a/x^2, \qquad x > a, \quad a > 0,$$
$$= 0, \qquad \text{elsewhere.}$$

Find the pdf of $X + Y$. [*Hint:* Use partial fractions for the integration.]

12.11.   Perform the calculations suggested at the end of Example 12.5.

12.12.   (a) An electronic device has failure time T whose distribution is given by $N(100, 4)$. Suppose that in recording T an error is made whose value may be represented as a random variable X, uniformly distributed over $(-1, 1)$. If X and T are independent, obtain the pdf of $S = X + T$ in terms of Φ, the cdf of the distribution $N(0, 1)$.

   (b) Evaluate $P(100 \leq S \leq 101)$.   [*Hint:* Use Simpson's rule to approximate the integral.]

12.13.   Suppose a new apparatus is tested repeatedly under certain stress conditions until it fails. The probability of failure on any trial is p_1. Let X be equal to the number of trials required up to and including the first failure. A second apparatus is also tested repeatedly until it fails. Suppose that a constant probability of failure of p_2 is associated with it. Let Y equal the number of trials required up to and including its first failure. Suppose that X and Y are independent and let $Z = X + Y$. Hence Z is equal to the number of trials required until both apparatuses have failed.

   (a) Find the probability distribution of Z.
   (b) Evaluate $P(Z = 4)$ if $p_1 = 0.1$, $p_2 = 0.2$.
   (c) Discuss (a) if $p_1 = p_2$.

13

Samples and Sampling Distributions

13.1 Introduction

Let us again consider a problem discussed previously. Suppose that we have a source of radioactive material which is emitting α-particles. Suppose that the assumptions stated in Chapter 8 hold, so that the random variable X defined as the number of particles emitted during a specified time period t has a Poisson distribution with parameter λt.

In order to "use" this probabilistic model describing the emission of α-particles, we need to know the value of λ. The assumptions we made only lead to the conclusion that X has a Poisson distribution with some parameter λt. But if we wish to compute $P(X > 10)$, for example, the answer will be in terms of λ unless we know its numerical value. Similarly, the important parameters associated with the distribution such as $E(X)$ and $V(X)$ are all functions of λ.

To seek a numerical value for λ we must, for the moment at least, leave the world of our theoretical mathematical model and enter into the empirical world of observations. That is, we must actually observe the emission of particles, obtain numerical values of X, and then use these values in some systematic way in order to obtain some relevant information about λ.

It is important for the reader to have a clear idea about the interplay between empirical verification and mathematical deduction that arises in many areas of applied mathematics. This is particularly important when we construct probabilistic models for the study of observational phenomena.

Let us consider a trivial example from elementary trigonometry. A typical problem may involve the computation of the height of a tree. A mathematical model for this problem might be obtained by postulating that the relationship between the unknown height h, the shadow cast s, and the angle α is of the form $h = s \tan \alpha$. (We are assuming that the tree stands straight and perpendicular to the ground, Fig. 13.1.) Hence, if s and α are known we can, with the aid of a suitable table, evaluate h. The important point we make here is that s and α must be *known* before we can evaluate h. That is, someone must have measured s and α.

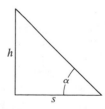

FIGURE 13.1

The mathematical deduction which leads to the relationship $h = s \tan \alpha$ is quite independent of the means by which we measure s and α. If these measurements are accurate, then $s \tan \alpha$ will represent an accurate value of h (assuming that the model is valid). Saying it differently, we cannot simply deduce the value of h from our knowledge of trigonometry and with the aid of trigonometric tables. We must leave our sanctuary (wherever it may be) and make some measurements! And just how these measurements are made, although an important problem to be resolved, in no way influences the *validity* of our mathematical deduction.

In using probabilistic models we will again be required to enter the empirical world and make some measurements. For example, in the case being considered we are using the Poisson distribution as our probabilistic model and hence need to know the value of the parameter λ. In order to obtain some information about λ we must make some measurements and then use these measurements in a systematic way for the purpose of estimating λ. Just how this is done will be described in Chapter 14.

Two final points should be emphasized here. First, the measurements required to obtain information about λ will usually be easier to obtain than would direct measurements for $e^{-\lambda t}(\lambda t)^k/k!$ (just as it is easier to obtain measurements for the length of the shadow s and the angle α than for the height h). Second, the way in which we obtain measurements for λ and the manner in which we use these measurements in no way invalidates (or confirms) the applicability of the Poisson model.

The above example is typical of a large class of problems. In many situations it is relatively natural (and appropriate) to hypothesize that a random variable X has a particular probability distribution. We have already seen a number of examples indicating that fairly simple assumptions about the probabilistic behavior of X will lead to a definite type of distribution such as the binomial, exponential, normal, Poisson, and others. Each of these distributions depends on certain parameters. In some cases the value of one or more of these parameters may be known. (Such knowledge may come about because of previous study of the random variable.) More often, however, we do not know the value of all the parameters involved. In such cases we must proceed as suggested above and obtain some empirical values of X and then use these values in an appropriate way. How this is done will occupy our attention in Chapter 14.

13.2 Random Samples

We have previously discussed the notion of random sampling with or without replacement from a finite set of objects or *population* of objects. We have under consideration a specified population of objects (people, manufactured items, etc.) about which we want to make some inference without looking at every single object. Thus we "sample." That is, we try to consider some "typical" objects, from which we hope to extract some information that in some sense is characteristic of the entire population. Let us be more precise.

Suppose that we label each member of the finite population with a number, consecutively say, so that without loss of generality a population consisting of N objects may be represented as $1, 2, \ldots, N$. Now choose n items, in a way to be described below. Define the following random variables.

X_i = population value obtained when the ith item is chosen, $i = 1, 2, \ldots, n$.

The probability distribution of the random variables $X_1, \ldots, X_n$ obviously depends on how we go about sampling. If we sample with replacement, each time choosing an object at random, the random variables $X_1, \ldots, X_n$ are independent and identically distributed. That is, for each $X_i, i = 1, 2, \ldots, n$ we have

$$P(X_i = j) = 1/N, \quad j = 1, 2, \ldots, N.$$

If we sample without replacement, the random variables $X_1, \ldots, X_n$ are no longer independent. Instead, their joint probability distribution is given by

$$P[X_1 = j_1, \ldots, X_n = j_n] = \frac{1}{N(N-1)\cdots(N-n+1)},$$

where $j_1, \ldots, j_n$ are any n values from $(1, \ldots, N)$. (We may show that the marginal distribution of any X_i, irrespective of the values taken by $X_1, \ldots, X_{i-1}, X_{i+1}, \ldots, X_n$, is the same as above when we sampled with replacement.)

In our discussion so far, we have supposed that there exists an underlying population, say $1, 2, \ldots, N$ which is finite and about which we want to obtain some information based on a sample of size $n < N$. In many situations there is no finite population from which we are sampling; in fact we may have difficulty in defining an underlying population of any kind. Consider the following examples.

(a) A coin is tossed. Define the random variable X_1 = number of heads obtained. In a sense we could think of X_1 as a sample of size one from the "population" of all possible tossings of that coin. If we tossed the coin a second time and defined the random variable X_2 as the number of heads obtained on the second toss, X_1, X_2 could presumably be considered as a sample of size two from the *same* population.

(b) The total yearly rainfall at a certain locality for the year 1970 could be defined as a random variable X_1. During successive years random variables $X_2, \ldots, X_n$ could be defined analogously. Again we may consider $(X_1, \ldots, X_n)$ as a sample of size n, obtained from the population of all possible yearly rainfalls at the specified locality. And it might be realistically supposed that the X_i's are independent, identically distributed random variables.

(c) The life length of a light bulb manufactured by a certain process at a certain factory is studied by choosing n bulbs and measuring their life lengths, say $T_1, \ldots, T_n$. We may consider $(T_1, \ldots, T_n)$ as a random sample from the population of all possible life lengths of bulbs manufactured in the specified way.

Let us formalize these notions as follows.

Definition. Let X be a random variable with a certain probability distribution. Let $X_1, \ldots, X_n$ be n independent random variables each having the same distribution as X. We then call $(X_1, \ldots, X_n)$ a *random sample from the random variable X.*

Notes: (a) We state the above more informally: A random sample of size n from a random variable X corresponds to n repeated measurements on X, made under essentially unchanged conditions. As we have mentioned a number of times before in other contexts, the mathematically idealized notion of a random sample can at best be only approximated by actual experimental conditions. In order for X_1 and X_2 to have the same distribution, all "relevant" conditions under which the experiment is performed must be the same when X_1 is observed and when X_2 is observed. Of course experimental conditions can never be identically duplicated. The point is that those conditions which are different should have little or no effect on the outcome of the experiment. Nevertheless, some care should be taken to ensure that we are really obtaining a random sample.

For example, suppose that the random variable under consideration is X, the number of calls coming into a telephone exchange between 4 p.m. and 5 p.m. on Wednesday. In order to obtain a random sample from X, we would presumably choose n Wednesday's at random and record the value of $X_1, \ldots, X_n$. We would have to be certain that all the Wednesday's are "typical" Wednesday's. For instance, we might not want to include a particular Wednesday if it happened to coincide with Christmas Day.

Again, if we try to obtain a random sample from the random variable X defined as the life length of an electronic device which has been manufactured subject to certain specifications, we would want to make sure that a sample value has not been obtained from an item which was produced at a time when the production process was faulty.

(b) If X is a continuous random variable with pdf f and if $X_1, \ldots, X_n$ is a random sample from X, then g, the joint pdf of $X_1, \ldots, X_n$, may be written as $g(x_1, \ldots, x_n) = f(x_1) \cdots f(x_n)$. If X is a discrete random variable and $p(x_i) = P(X = x_i)$, then

$$P[X_1 = x_1, \ldots, X_n = x_n] = p(x_1) \cdots p(x_n).$$

(c) As we have done previously we shall use capital letters for the random variable and lower-case letters for the value of the random variable. Thus the values assumed by a particular sample $(X_1, \ldots, X_n)$ will be denoted by $(x_1, \ldots, x_n)$. We shall often talk of the *sample point* $(x_1, \ldots, x_n)$. By this we simply mean that we can consider $(x_1, \ldots, x_n)$ as the coordinates of a point in n-dimensional Euclidean space.

13.3 Statistics

Once we have obtained the values of a random sample we usually want to use these sample values in order to make some inference about the population represented by the sample which in the present context means the probability distribution of the random variable being sampled. Since the various parameters which characterize a probability distribution are numbers, it is only natural that we would want to compute certain pertinent numerical characteristics obtainable from the sample values, which might help us in some way to make appropriate statements about the parameter values which are often not known. Let us define the following important concept.

Definition. Let $X_1, \ldots, X_n$ be a random sample from a random variable X and let $x_1, \ldots, x_n$ be the values assumed by the sample. Let H be a function defined for the n-tuple $(x_1, \ldots, x_n)$. We define $Y = H(X_1, \ldots, X_n)$ to be a *statistic*, assuming the value $y = H(x_1, \ldots, x_n)$.

In words: A statistic is a real-valued function of the sample. Sometimes we use the term statistic to refer to the value of the function. Thus we may speak of the statistic $y = H(x_1, \ldots, x_n)$ when we really should say that y is the value of the statistic $Y = H(X_1, \ldots, X_n)$.

Notes: (a) The above is a very special, but generally accepted, use of the term statistic. Note that we are using it in the singular.

(b) According to the above definition, a statistic is a random variable! It is very important to keep this in mind. Hence it will be meaningful to consider the probability distribution of a statistic, its expectation, and its variance. When a random variable is in fact a statistic, that is a function of a sample, we often speak of its *sampling distribution* rather than its probability distribution.

As we have suggested at the beginning of this chapter, we shall use the information obtained from a sample for the purpose of estimating certain unknown parameters associated with a probability distribution. We shall find that certain statistics play an important role in the solution of this problem. Before we consider this in more detail (in Chapter 14), let us study a number of important statistics and their properties.

13.4 Some Important Statistics

There are certain statistics which we encounter frequently. We shall list a few of them and discuss some of their important properties.

Definition. Let $(X_1, \ldots, X_n)$ be a random sample from a random variable X. The following statistics are of interest.

(a) $\bar{X} = (1/n)\sum_{i=1}^{n} X_i$ is called the *sample mean*.

(b) $S^2 = [1/(n-1)]\sum_{i=1}^{n} (X_i - \bar{X})^2$ is called the *sample variance*. (We shall indicate shortly why we divide by $(n-1)$ rather than by the more obvious choice n.)

(c) $K = \min (X_1, \ldots, X_n)$ is called the *minimum of the sample*. (K simply represents the smallest observed value.)

(d) $M = \max (X_1, \ldots, X_n)$ is called the *maximum of the sample*. (M represents the largest observed value.)

(e) $R = M - K$ is called the *sample range*.

(f) $X_n^{(j)} = j$th largest observation in the sample, $j = 1, 2, \ldots, n$. (We have $X_n^{(1)} = M$ while $X_n^{(n)} = K$.)

Notes: (a) The random variables $X_n^{(j)}$, $j = 1, 2, \ldots, n$, are called the *order statistics* associated with the random sample $X_1, \ldots, X_n$. If X is a continuous random variable we may suppose that $X_n^{(1)} > X_n^{(2)} > \cdots > X_n^{(n)}$.

(b) The extreme values of the sample (in the above notation, $X_n^{(1)}$ and $X_n^{(n)}$) are often of considerable interest. For example, in the construction of dams for flood control, the greatest height a particular river has reached over the past 50 years may be very important.

Of course, there are many other important statistics that we encounter but certainly the ones mentioned above play an important role in many statistical applications. We shall now state (and prove) a few theorems concerning the above statistics.

Theorem 13.1. Let X be a random variable with expectation $E(X) = \mu$ and variance $V(X) = \sigma^2$. Let $\overline{X}$ be the sample mean of a random sample of size n. Then

(a) $E(\overline{X}) = \mu$.
(b) $V(\overline{X}) = \sigma^2/n$.
(c) For large n, $(\overline{X} - \mu)/(\sigma/\sqrt{n})$ has approximately the distribution $N(0, 1)$.

Proof: (a) and (b) follow immediately from the previously established properties of expectation and variance:

$$E(\overline{X}) = E\left(\frac{1}{n}\sum_{i=1}^{n} X_i\right) = \frac{1}{n}\sum_{i=1}^{n} E(X_i) = \frac{1}{n}n\mu = \mu.$$

Since the X_i's are independent,

$$V(\overline{X}) = V\left(\frac{1}{n}\sum_{i=1}^{n} X_i\right) = \frac{1}{n^2}\sum_{i=1}^{n} V(X_i) = \frac{1}{n^2}n\sigma^2 = \frac{\sigma^2}{n}.$$

(c) follows from a direct application of the Central Limit Theorem. We may write $\overline{X} = (1/n)X_i + \cdots + (1/n)X_n$ as the sum of independently distributed random variables.

Notes: (a) As the sample size n increases, the sample mean $\overline{X}$ tends to vary less and less. This is intuitively clear and corresponds to our experience with numerical data. Consider, for example, the following set of 18 numbers:

$$-1, \ 3, \ 2, \ -4, \ -5, \ 6, \ 7, \ 2, \ 0, \ 1, \ -2, \ -3, \ 8, \ 9, \ 6, \ -3, \ 0, \ 5.$$

If we take the average of these numbers, two at a time in the order listed, we obtain the following set of averages:

$$1, \ -1, \ 0.5, \ 4.5, \ 0.5, \ -2.5, \ 8.5, \ 1.5, \ 2.5.$$

If we average the original set of numbers, three at a time, we obtain

$$1.3, \ -1, \ 3, \ -1.3, \ 7.7, \ 0.7.$$

Finally, if we average the numbers, six at a time, we obtain

$$0.2, \ 0.8, \ 4.1.$$

The variance in each of these sets of averages is less than in the previous set, because in each case the average is based on a larger number of numbers. The above theorem indicates precisely how the variation of $\overline{X}$ (measured in terms of its variance) decreases with increasing n. (In this connection see the Law of Large Numbers, Section 12.2 and in particular Eq. 12.2.)

(b) If n is not sufficiently large to warrant the application of the Central Limit Theorem, we may try to find the exact probability distribution of $\overline{X}$ by direct (but usually quite involved) means. In Section 12.5 we suggested a method by which we can find the probability distribution of the sum of random variables. By a repeated application of this method we may be able to obtain the probability distribution of $\overline{X}$, particularly if n is relatively small.

(c) Theorem 13.1 indicates that for sufficiently large n, the sample mean $\overline{X}$ is approximately normally distributed (with expectation μ and variance σ^2/n).

We find that not only $\overline{X}$ but most "well-behaved" functions of $\overline{X}$ have this property. At this level of presentation, we cannot give a careful development of this result. However, the result is of sufficient importance in many applications to warrant at least a heuristic, intuitive argument.

Suppose that $Y = r(\overline{X})$ and that r may be expanded in a Taylor series about μ. Thus $r(\overline{X}) = r(\mu) + (\overline{X} - \mu)r'(\mu) + R$, where R is a remainder term and may be expressed as $R = [(\overline{X} - \mu)^2/2]r''(z)$, where z is some value between $\overline{X}$ and μ. If n is sufficiently large, $\overline{X}$ will be close to μ, and hence $(\overline{X} - \mu)^2$ will be small compared to $(\overline{X} - \mu)$. For large n, we may therefore consider the remainder to be negligible and *approximate* $r(\overline{X})$ as follows:

$$r(\overline{X}) \simeq r(\mu) + r'(\mu)(\overline{X} - \mu).$$

We see that for n sufficiently large, $r(\overline{X})$ may be approximated by a *linear* function of $\overline{X}$. Since $\overline{X}$ will be approximately normal (for n large), we find that $r(\overline{X})$ will also be approximately normal, since a linear function of a normally distributed random variable is normally distributed.

From the above representation of $r(\overline{X})$ we find that

$$E\big(r(\overline{X})\big) \simeq r(\mu), \qquad V\big(r(\overline{X})\big) \simeq \frac{[r'(\mu)]^2 \sigma^2}{n}.$$

Thus for sufficiently large n, we see that (under fairly general conditions on the function r) the distribution of $r(\overline{X})$ is approximately $N(r(\mu), [r'(\mu)]^2 \sigma^2/n)$.

Theorem 13.2. Let X be a continuous random variable with pdf f and cdf F. Let $X_1, \ldots, X_n$ be a random sample of X and let K and M be the minimum and maximum of the sample, respectively. Then:

(a) the pdf of M is given by $g(m) = n[F(m)]^{n-1}f(m)$,
(b) the pdf of K is given by $h(k) = n[1 - F(k)]^{n-1}f(k)$.

Proof: Let $G(m) = P(M \leq m)$ be the cdf of M. Now $\{M \leq m\}$ is equivalent to the event $\{X_i \leq m, \text{ all } i\}$. Hence since the X_i's are independent, we find

$$G(m) = P[X_1 \leq m \text{ and } X_2 \leq m \cdots \text{ and } X_n \leq m] = [F(m)]^n.$$

Therefore

$$g(m) = G'(m) = n[F(m)]^{n-1}f(m).$$

The derivation for the pdf of K will be left to the reader. (See Problem 13.1.)

EXAMPLE 13.1. An electronic device has a life length T which is exponentially distributed with parameter $\alpha = 0.001$; that is, its pdf is $f(t) = 0.001e^{-0.001t}$. Suppose that 100 such devices are tested, yielding observed values $T_1, \ldots, T_{100}$.
(a) What is the probability that $950 < \bar{T} < 1100$? Since the sample size is quite large, we could apply the Central Limit Theorem and proceed as follows:

$$E(\bar{T}) = \frac{1}{0.001} = 1000, \qquad V(\bar{T}) = \frac{1}{100}(0.001)^{-2} = 10,000.$$

Hence $(\bar{T} - 1000)/100$ has approximately the distribution $N(0, 1)$. Thus

$$P(950 < \bar{T} < 1100) = P\left(-0.5 < \frac{\bar{T} - 1000}{100} < 1\right)$$
$$= \Phi(1) - \Phi(-0.5)$$
$$= 0.532,$$

from the tables of the normal distribution.

Note: In the present case we can actually obtain the exact distribution of $\bar{T}$ without resorting to the Central Limit Theorem. We proved in Theorem 10.9 that the sum of independent exponentially distributed random variables has a Gamma distribution; that is,

$$g(s) = \frac{(0.001)^{100}s^{99}e^{-0.001s}}{99!},$$

where g is the pdf of $T_1 + \cdots + T_{100}$. Hence the pdf of $\bar{T}$ is given by

$$f(\bar{t}) = \frac{(0.1)^{100}\bar{t}^{99}e^{-0.1\bar{t}}}{99!}.$$

Thus $\bar{T}$ has a Gamma distribution with parameters 0.1 and 100.

(b) What is the probability that the largest observed value exceeds 7200 hours? We require that $P(M > 7200) = 1 - P(M \leq 7200)$. Now the maximum value will be less than 7200 if and only if every sample value is less than 7200. Hence

$$P(M > 7200) = 1 - [F(7200)]^{100}.$$

To evaluate $F(7200)$, recall that for the exponentially distributed random variable with parameter 0.001, $F(t) = 1 - e^{-0.001t}$. Hence

$$F(7200) = 1 - e^{-0.001(7\,200)} = 1 - e^{-7.2} = 0.99925.$$

Thus the required probability is $1 - (0.99925)^{100}$, which equals 0.071.

(c) What is the probability that the shortest time to failure is less than 10 hours? We require that $P(K < 10) = 1 - P(K \geq 10)$.

Now the minimum of the sample is greater than or equal to 10 if and only if every sample value is greater than or equal to 10. Hence

$$P(K < 10) = 1 - [1 - F(10)]^{100}.$$

Using the expression for F as given in (b) above, we have

$$1 - F(10) = e^{-0.001(10)} = e^{-0.01} = 0.99005.$$

Hence

$$P(K < 10) = 1 - (0.99005)^{100} = 0.63.$$

The last part of the above example may be generalized as the following theorem indicates.

Theorem 13.3. Let X be exponentially distributed with parameter α and let $(X_1, \ldots, X_n)$ be a random sample from X. Let $K = \min(X_1, \ldots, X_n)$. Then K is also exponentially distributed with parameter $n\alpha$.

Proof: Let H be the cdf of K. Then

$$H(k) = P(K \leq k) = 1 - P(K > k) = 1 - [1 - F(k)]^n,$$

where F is the cdf of X. Now $F(x) = 1 - e^{-\alpha x}$. Thus $H(k) = 1 - e^{-n\alpha k}$. Taking the derivative of $H(k)$ with respect to k yields, $h(k) = n\alpha e^{-n\alpha k}$.

Note: This theorem may be extended as follows. If $X_1, \ldots, X_n$ are independent random variables and if X_i has exponential distribution with parameter α_i, $i = 1, \ldots, n$, then $K = \min(X_1, \ldots, X_n)$ has exponential distribution with parameter $\alpha_1 + \cdots + \alpha_n$. For a proof of this, see Problem 13.2.

Theorem 13.4 gives us some information about the statistic S^2.

Theorem 13.4. Suppose that $X_1, \ldots, X_n$ is a random sample from a random variable X with expectation μ and variance σ^2. Let

$$S^2 = \frac{1}{n-1} \sum_{i=1}^{n} (X_i - \bar{X})^2,$$

where $\bar{X}$ is the sample mean. Then we have the following:

(a) $E(S^2) = \sigma^2$,
(b) If X is normally distributed, $[(n-1)/\sigma^2]S^2$ has chi-square distribution with $(n-1)$ degrees of freedom.

Proof: (a) Write

$$\sum_{i=1}^{n} (X_i - \bar{X})^2 = \sum_{i=1}^{n} (X_i - \mu + \mu - \bar{X})^2$$

$$= \sum_{i=1}^{n} [(X_i - \mu)^2 + 2(\mu - \bar{X})(X_i - \mu) + (\mu - \bar{X})^2]$$

$$= \sum_{i=1}^{n} (X_i - \mu)^2 + 2(\mu - \bar{X}) \sum_{i=1}^{n} (X_i - \mu) + n(\mu - \bar{X})^2$$

$$= \sum_{i=1}^{n} (X_i - \mu)^2 - 2n(\mu - \bar{X})^2 + n(\mu - \bar{X})^2$$

$$= \sum_{i=1}^{n} (X_i - \mu)^2 - n(\bar{X} - \mu)^2.$$

Hence

$$E\left(\frac{1}{n-1} \sum_{i=1}^{n} (X_i - \bar{X})^2\right) = \frac{1}{n-1}\left[n\sigma^2 - n\frac{\sigma^2}{n}\right] = \sigma^2.$$

Note: If we had divided by n rather than by $(n-1)$ in defining S^2, the above property would not hold.

(b) We shall not prove (b), but only make its validity plausible by considering the following special case. Consider $\sum_{i=1}^{n} (X_i - \bar{X})^2$ for $n = 2$. Then

$$(X_1 - \bar{X})^2 + (X_2 - \bar{X})^2 = [X_1 - \tfrac{1}{2}(X_1 + X_2)]^2 + [X_2 - \tfrac{1}{2}(X_1 + X_2)]^2$$
$$= \tfrac{1}{4}[2X_1 - X_1 - X_2]^2 + \tfrac{1}{4}[2X_2 - X_1 - X_2]^2$$
$$= \tfrac{1}{4}[(X_1 - X_2)^2 + (X_2 - X_1)^2] = \frac{[X_1 - X_2]^2}{2}.$$

Since X_1 and X_2 are each independently distributed with distribution $N(\mu, \sigma^2)$, we find that $(X_1 - X_2)$ has distribution $N(0, 2\sigma^2)$. Hence

$$\left[\frac{X_1 - X_2}{\sqrt{2}\,\sigma}\right]^2 = \frac{1}{\sigma^2} \sum_{i=1}^{2} (X_i - \bar{X}) = \frac{1}{\sigma^2} S^2$$

has chi-square distribution with one degree of freedom.   (See Theorem 10.8.) The proof for general n follows along similar lines: We must show that $\sum_{i=1}^{n} (X_i - \overline{X})^2/\sigma^2$ may be decomposed into the sum of squares of $(n - 1)$ independent random variables, each with distribution $N(0, 1)$.

Note: Although S^2 is defined as the sum of squares of n terms, these n terms are not independent.   In fact, $(X_1 - \overline{X}) + (X_2 - \overline{X}) + \cdots + (X_n - \overline{X}) = \sum_{i=1}^{n} X_i - n\overline{X} = 0$. Hence there is a linear relationship among these n terms which means that as soon as any $(n - 1)$ of these are known, the nth one is determined.

Finally, let us state (without proof) a result concerning the probability distribution of the sample range R.

Theorem 13.5. Let X be a continuous random variable with pdf f. Let $R = M - K$ be the sample range based on a random sample of size n.   Then the pdf of R is given by

$$g(r) = n(n - 1) \int_{s=-\infty}^{+\infty} \left[\int_{x=s}^{s+r} f(x) \, dx \right]^{n-2} f(s) f(s + r) \, ds, \qquad \text{for} \quad r \geq 0.$$

EXAMPLE 13.2.   A random voltage V is uniformly distributed over $[0, 1]$.   A sample of size n is obtained, say $V_1, \ldots, V_n$, and the sample range R is computed.   The pdf of R is then found to be

$$g(r) = n(n - 1) \int_{s=-\infty}^{+\infty} r^{n-2} f(s) f(s + r) \, ds.$$

We have $f(s) = f(s + r) = 1$ whenever $0 \leq s \leq 1$ and $0 \leq s + r \leq 1$, which together imply that $0 \leq s \leq 1 - r$. Hence

$$k(r) = n(n - 1) \int_{0}^{1-r} r^{n-2} \, ds$$

$$= n(n - 1) r^{n-2}(1 - r), \qquad 0 \leq r \leq 1.$$

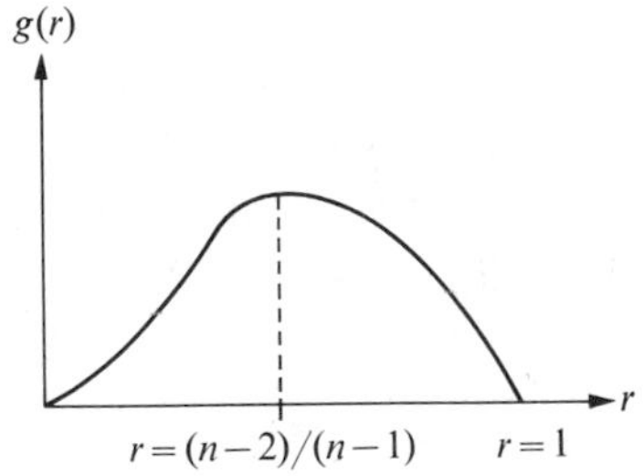

$g(r)$

$r = (n-2)/(n-1) \qquad r = 1$

FIGURE 13.2

For $n > 2$, the graph of the pdf of R has the shape shown in Fig. 13.2.   Note that as $n \to \infty$, the value of r at which the maximum occurs shifts to the right. Thus, as the sample size is increased, it becomes more and more likely that the range R is close to 1, which is intuitively what we would expect.

13.5 The Integral Transform

A sample from a random variable X may be used to obtain some information about unknown parameters associated with the probability distribution of X. We can use a sample for a different purpose, however. We might take some observations on a random variable whose distribution is completely specified and then use these sample values to approximate certain probabilities which would be very difficult to obtain by straightforward mathematical manipulation. For example, suppose that X has distribution $N(0, 1)$ and we want to study the random variable $Y = e^{-X} \sin X$. In particular, assume that we want to evaluate $P(0 \le Y \le \frac{1}{2})$. In order to obtain the exact answer, we need to find G, the cdf of Y and then evaluate $G(\frac{1}{2}) - G(0)$. We would encounter considerable difficulty in doing this. However, we can take another approach which is based on the idea of *simulating* the experiment giving rise to the random variable Y. We then use the relative frequency as an approximation for the desired probability. If this relative frequency is based on a sufficiently large number of observations, the Law of Large Numbers justifies our procedure.

Specifically, suppose that we have a random sample from the above random variable X whose distribution is completely specified, say $X_1, \ldots, X_n$. For each X_i, define the random variable $Y_i = e^{-X_i} \sin X_i$. We then evaluate the relative frequency n_A/n, where n_A is equal to the number of Y_i values, say y_i, satisfying $0 \le y_i \le \frac{1}{2}$. Hence n_A/n is the relative frequency of the event $0 \le Y \le \frac{1}{2}$, and if n is large, this relative frequency will be "close" to $P[0 \le Y \le \frac{1}{2}]$ in the sense of the Law of Large Numbers.

To carry out the above procedure, we must find a means of "generating" a random sample $X_1, \ldots, X_n$ from the random variable whose distribution is $N(0, 1)$. Before indicating how this is done, let us briefly discuss one distribution for which this task has essentially been performed for us because of the availability of tables. Suppose that X is uniformly distributed over the interval $[0, 1]$. In order to obtain a random sample from such a random variable, we need only turn to a *Table of Random Numbers* (see Appendix). These tables have been compiled in a way to make them suitable for this purpose. To use them, we simply select a location at random in the table and then obtain numbers by either proceeding along rows or columns. If we want to use these tabulated numbers to represent values between 0 and 1, we need only insert a decimal point at the beginning of the number. Thus the number 4573 as listed would be used to represent the number 0.4573, etc.

The availability of these tables of random numbers makes the task of obtaining a random sample from an arbitrary distribution relatively simple because of the result contained in the following theorem.

Theorem 13.6. Let X be a random variable with pdf f and cdf F. [Assume that $f(x) = 0$, $x \notin (a, b)$.] Let Y be the random variable defined by $Y = F(X)$. Then Y is uniformly distributed over $[0, 1]$. (Y is called the *integral transform* of X.)

Proof: Since X is a continuous random variable, the cdf F is a continuous, strictly monotone function with an inverse, say F^{-1}. That is, $Y = F(X)$ may be solved for X in terms of Y: $X = F^{-1}(Y)$. (See Fig. 13.3.) [If $F(x) = 0$ for $x \le a$, define $F^{-1}(0) = a$. Similarly, if $F(x) = 1$ for $x \ge b$, define $F^{-1}(1) = b$.]

Let G be the cdf of the random variable Y defined above. Then

$$G(y) = P(Y \leq y) = P(F(X) \leq y) = P(X \leq F^{-1}(y)) = F(F^{-1}(y)) = y.$$

Hence the pdf of Y, $g(y) = G'(y) = 1$. This establishes our result.

Notes: (a) Let us comment briefly on how a *value* of the random variable Y is actually observed. We observe a value of the random variable X say x, and then compute the value of $Y = F(X)$ as $y = F(x)$, where F is the (known) cdf of X.

(b) Theorem 13.6 stated and proved above for continuous random variables also holds for discrete random variables. A slight change must be made in the proof since the cdf of a discrete random variable is a step function and has no unique inverse.

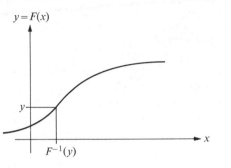

FIGURE 13.3

We can now use the above result in order to generate a random sample from a random variable with specified distribution. Again we will consider only the continuous case. Let X be a random variable with cdf F from which a sample is required. Let y_1 be a value (between 0 and 1) obtained from a table of random numbers. Since $Y = F(X)$ is uniformly distributed over $[0, 1]$, we may consider y_1 as an observation from that random variable. Solving the equation $y_1 = F(x_1)$ for x_1 (which is possible if X is continuous), we obtain a value from a random variable whose cdf is F. Continuing this procedure with the numbers y_2, $\ldots$, y_n obtained from a table of random numbers, we obtain x_i, $i = 1, \ldots, n$, as the solution of the equation $y_i = F(x_i)$, and hence we have our desired sample values.

EXAMPLE 13.3. Suppose that we want to obtain a sample of size five from a random variable with distribution $N(2, 0.09)$. Suppose that we obtain the following values from a table of random numbers: 0.487, 0.722, 0.661, 0.194, 0.336. Define x_1 as follows:

$$0.487 = \frac{1}{\sqrt{2\pi}\,(0.3)} \int_{-\infty}^{x_1} \exp\left[-\frac{1}{2}\left(\frac{t-2}{0.3}\right)^2\right] dt$$

$$= \frac{1}{\sqrt{2\pi}} \int_{-\infty}^{(x_1-2)/0.3} \exp\left(\frac{-s^2}{2}\right) ds = \Phi\left(\frac{x_1-2}{0.3}\right).$$

From the tables of the normal distribution we find that $(x_1 - 2)/0.3 = -0.03$. Hence $x_1 = (-0.03)(0.3) + 2 = 1.991$. This number represents our first sample value from the specified distribution. Proceeding in the same manner with the other values we obtain the following additional sample values: 2.177, 2.124, 1.742, 1.874.

The above procedure may be generalized as follows. To obtain a sample value x_1 from the distribution $N(\mu, \sigma^2)$, we obtain a sample value y_1 (between 0 and 1) from a table of random numbers. The desired value x_1 is defined by the equation $\Phi((x_1 - \mu)/\sigma) = y_1$.

Notes: (a) There are alternative methods to the one suggested above, for generating random samples from a specified distribution. See Problem 13.8.

(b) One of the aspects which makes this approach of "simulation" useful is the possibility of obtaining a very *large* random sample. This is feasible, particularly if an electronic computing machine is available.

EXAMPLE 13.4. The thrust T of a solid propellant rocket system is a very complicated function of a number of variables. If X = throat area, Y = burning rate factor, Z = solid propellant burning area, W = density of the solid propellant, n = burning coefficient, and C_i = constant, $i = 1, 2, 3$, we may express T as follows:

$$T = C_1 \left(\frac{X}{YWZ}\right)^{1/(n-1)} X + C_2 \left(\frac{X}{YWZ}\right)^{1/(n-1)} + C_3.$$

Previous investigations have made the following assumptions plausible: X, Y, W, and Z are independently distributed normal random variables with known means and variances. Any attempt to obtain the probability distribution of the random variable T or even exact expressions for $E(T)$ and $V(T)$ will be thwarted because of the complex relationship between X, Y, W, Z, and T.

If we could generate a (large) random sample of X, Y, Z, and W, say obtaining 4-tuples (X_i, Y_i, Z_i, W_i), we could then generate a large sample of T, say $(T_1, \ldots, T_n)$, and attempt to study the characteristic of the random variable T empirically (in terms of the sample). Suppose, for instance, that we wish to compute $P(a \leq T \leq b)$. Applying the Law of Large Numbers, we need simply obtain the relative frequency of the event $\{a \leq T \leq b\}$ from our (large) sample, and can then be reasonably certain that this relative frequency differs little from the theoretical probability being sought.

So far we have been concerned only with the problem of approximating an exact probability with a relative frequency based on a large number of observations. However, the method we have suggested may be used to obtain approximate solutions to problems which are completely nonprobabilistic in nature. We shall indicate only one of the many types of problems that may be considered in this manner. The general approach is referred to as the "Monte-Carlo" method. A very good description of this method is given in *Modern Mathematics for the*

Engineer, edited by E. F. Beckenbach, McGraw-Hill Book Co., Inc., New York, 1956, Chapter 12. Example 13.5 is taken from there.

EXAMPLE 13.5. Suppose that we wish to evaluate the integral $\int_0^1 x \, dx$ without resorting to the trivial calculation to obtain its value $\frac{1}{2}$. We proceed as follows. Obtain, from a table of random numbers, a random sample from the uniformly distributed random variable over $[0, 1]$. Assume that the sample values are 0.69, 0.37, 0.39, 0.97, 0.66, 0.51, 0.60, 0.41, 0.76, and 0.09. Since the desired integral represents $E(X)$, where X is the uniformly distributed random variable being sampled, it seems reasonable that we might approximate $E(X)$ using the arithmetic mean of the sample values. We find that $\bar{X} = 0.545$. (If we had taken a larger sample, we would have good reason to expect greater accuracy.)

This trivial illustration does indicate the basic idea behind many Monte-Carlo methods. These methods have been used successfully to evaluate multiple integrals over complicated regions and to solve certain differential equations.

Note: The means of obtaining samples from an arbitrary distribution, as described in Section 13.5, can become quite cumbersome. Because of the great importance of the normal distribution, tables are available (See Table 7 in the appendix) which eliminate a major portion of the calculations described above. Table 7 lists, directly, samples from the $N(0, 1)$ distribution. These sample values are called *normal deviates*. If n sample values $x_1, \ldots, x_n$ from a $N(0, 1)$ distribution are desired, they are read directly from Table 7 (choosing the starting place in some suitably random fashion, as was described for the use of the Table of Random Numbers).

In an obvious way, this table may also be used to obtain samples from an arbitrary normal distribution $N(\mu, \sigma^2)$. Simply multiply x_i, the chosen value from the table, by σ and then add μ. That is, form $y_i = \sigma x_i + \mu$. Then y_i will be a sample value from the desired distribution.

PROBLEMS

13.1. Derive the expression for the pdf of the minimum of a sample. (See Theorem 13.2.)

13.2. Show that if $X_1, \ldots, X_n$ are independent random variables, each having an exponential distribution with parameter α_i, $i = 1, 2, \ldots, n$, and if $K = \min (X_1, \ldots, X_n)$, then K has an exponential distribution with parameter $\alpha_1 + \cdots + \alpha_n$. (See Theorem 13.3.)

13.3. Suppose that X has a geometric distribution with parameter p. Let $X_1, \ldots, X_n$ be a random sample from X and let $M = \max (X_1, \ldots, X_n)$ and $K = \min (X_1, \ldots, X_n)$. Find the probability distribution of M and of K. [*Hint:* $P(M = m) = F(m) - F(m - 1)$, where F is the cdf of M.]

13.4. A sample of size 5 is obtained from a random variable with distribution $N(12, 4)$.

(a) What is the probability that the sample mean exceeds 13?

(b) What is the probability that the minimum of the sample is less than 10?

(c) What is the probability that the maximum of the sample exceeds 15?

13.5. The life length (in hours) of an item is exponentially distributed with parameter $\beta = 0.001$. Six items are tested and their times to failure recorded.

(a) What is the probability that no item fails before 800 hours have elapsed?

(b) What is the probability that no item lasts more than 3000 hours?

13.6. Suppose that X has distribution $N(0, 0.09)$. A sample of size 25 is obtained from X, say $X_1, \ldots, X_{25}$. What is the probability that $\sum_{i=1}^{25} X_i^2$ exceeds 1.5?

13.7. Using a table of random numbers, obtain a random sample of size 8 from a random variable having the following distributions:

(a) exponential, with parameter 2,

(b) chi-square with 7 degrees of freedom,

(c) $N(4, 4)$.

13.8. In Section 13.5 we have indicated a method by which random samples from a specified distribution may be generated. There are numerous other methods by which this may be done, some of which are to be preferred to the one given, particularly if computing devices are available. The following is one such method. Suppose that we want to obtain a random sample from a random variable having a chi-square distribution with $2k$ degrees of freedom. Proceed as follows: obtain a random sample of size k (with the aid of a table of random numbers) from a random variable which is uniformly distributed over $(0, 1)$, say $U_1, \ldots, U_k$. Then evaluate $X_1 = -2 \ln (U_1 U_2 \cdots U_k) = -2\sum_{i=1}^{k} \ln (U_i)$. The random variable X_1 will then have the desired distribution, as we shall indicate below. We then continue this scheme, obtaining another sample of size k from a uniformly distributed random variable, and thus finding the second sample value X_2. Note that this procedure requires k observations from a uniformly distributed random variable for every observation from X_{2k}^2. To verify the statement made above proceed as follows.

(a) Obtain the moment-generating function of the random variable $-2 \ln (U_i)$, where U_i is uniformly distributed over $(0, 1)$.

(b) Obtain the moment-generating function of the random variable $-2\sum_{i=1}^{k} \ln (U_i)$, where the U_i's are independent random variables each with the above distribution. Compare this mgf with that of the chi-square distribution and hence obtain the desired conclusion.

13.9. Using the scheme outlined in Problem 13.8, obtain a random sample of size 3 from the distribution X_8^2.

13.10. A continuous random variable X is uniformly distributed over $(-\frac{1}{2}, \frac{1}{2})$. A sample of size n is obtained from X and the sample mean $\overline{X}$ is computed. What is the standard deviation of $\overline{X}$?

13.11. Independent samples of size 10 and 15 are taken from a normally distributed random variable with expectation 20 and variance 3. What is the probability that the means of the two samples differ (in absolute value) by more than 0.3?

13.12. (For this exercise and the following three, read the Note at the end of Chapter 13.) Use the table of normal deviates (Table 7, appendix) and obtain a sample of size

30 from a random variable X having a $N(1, 4)$ distribution. Use this sample to answer the following:

(a) Compare $P(X \geq 2)$ with the relative frequency of that event.

(b) Compare the sample mean $\overline{X}$ and the sample variance S^2 with 1 and 4, respectively.

(c) Construct a graph of $F(t) = P(X \leq t)$. Using the same coordinate system, obtain the graph of the *empirical distribution function* F_n defined as follows:

$$
\begin{aligned}
F_n(t) &= 0 & \text{if} & \quad t < X^{(1)} \\
&= k/n & \text{if} & \quad X^{(k)} \leq t < X^{(k+1)} \\
&= 1 & \text{if} & \quad t \geq X^{(n)},
\end{aligned}
$$

where $X^{(i)}$ is the ith largest observation in the sample (i.e., $X^{(i)}$ is the ith order statistic). [The function F_n is frequently used to approximate the cdf F. It may be shown that under fairly general conditions $\lim_{n \to \infty} F_n(t) = F(t)$.]

13.13   Let X have distribution $N(0, 1)$. From Table 7 in the appendix, obtain a sample of size 20 from this distribution. Let $Y = |X|$.

(a) Use this sample to compare $P[1 < Y \leq 2]$ with the relative frequency of that event.

(b) Compare $E(Y)$ with the sample mean $\overline{Y}$.

(c) Compare the cdf of Y, say $F(t) = P(Y \leq t)$ with F_n, the empirical cdf of Y.

13.14   Suppose that X has distribution $N(2, 9)$. Let $X_1, \ldots, X_{20}$ be a random sample of X obtained with the aid of Table 7. Compute

$$
S^2 = \frac{1}{n-1} \sum_{i=1}^{n} (X_i - \overline{X})^2
$$

and compare it with $E(S^2) = 9$.

13.15   Let X have distribution $N(0, 1)$. Let $X_1, \ldots, X_{30}$ be a random sample of X obtained by using Table 7. Compute $P(X^2 \geq 0.10)$ and compare this value with the relative frequency of that event.

14

Estimation of Parameters

14.1 Introduction

In the previous chapter we suggested that very often a sample from a random variable X may be used for the purpose of estimating one or several (unknown) parameters associated with the probability distribution of X. In this chapter we shall consider this problem in considerable detail.

In order to have a specific example in mind, consider the following situation. A manufacturer has supplied us with 100,000 small rivets. A soundly riveted joint requires that a rivet fits properly into its hole and consequently some trouble will be experienced when the rivet is burred. Before accepting this shipment, we want to have some idea about the magnitude of p, the proportion of defective (that is, burred) rivets. We proceed as follows. We inspect n rivets chosen at random from the lot. Because of the large lot size, we may suppose that we choose with replacement although in actuality this would not be done. We define the following random variables: $X_i = 1$, if the ith item is defective, and 0 otherwise, $i = 1, 2, \ldots, n$. Hence we may consider $X_1, \ldots, X_n$ to be a sample from the random variable X whose distribution is given by $P(X = 1) = p, P(X = 0) = 1 - p$.

The probability distribution of X depends on the unknown parameter p in a very simple way. Can we use the sample $X_1, \ldots, X_n$ in some way in order to estimate the value of p? Is there some statistic H such that $H(X_1, \ldots, X_n)$ may be used as a (point) estimate of p?

It should be obvious that a sample of size n, where $n < 100,000$, can never enable us to reconstruct the actual composition of the shipment no matter how cleverly we use the information obtained from the sample. In other words, unless we inspect every item (that is, take $n = 100,000$) we can never know the true value of p. (This last sentence obviously refers to sampling without replacement.)

Thus, when we propose $\hat{p}$ as an estimate for p we do not really expect $\hat{p}$ to be equal to p. (Recall that $\hat{p}$ is a random variable and hence may assume many values.) This dilemma raises two important questions:

(1) What characteristics do we want a "good" estimate to possess?

(2) How do we decide whether one estimate is "better" than another?

Since this may be the first time the reader has encountered questions of this type, it might be worthwhile to comment briefly about the general nature of this problem. To many mathematical questions, there exists a definite answer. This may be very difficult to find, involving many technical problems, and we may have to be satisfied with an approximation. Yet it is usually clear when we have an answer and when we do not. (For example, suppose that we are asked to find a real root of the equation $3x^5 - 4x^2 + 13x - 7 = 0$. Once we have found a solution, it is simple enough to verify whether it is the correct one: we need only substitute it into the given equation. If we have two approximate answers, say r_1 and r_2, it is also a simple matter to decide which approximation is better.)

However, the present problem, namely the estimation of p, does not admit such a simple analysis. In the first place, since we can never (in all realistic situations, at least) know the true value of p, it does not make sense to say that our estimate $\hat{p}$ is "correct." Second, if we have two estimates of p, say $\hat{p}_1$ and $\hat{p}_2$, we must find some means of deciding which is "better." This means that we must stipulate some criteria which we can apply to decide whether one estimate is to be preferred to another.

14.2 Criteria for Estimates

We shall now define a number of important concepts which will help us to resolve the problems suggested above.

Definition. Let X be a random variable with some probability distribution depending on an unknown parameter θ. Let $X_1, \ldots, X_n$ be a sample of X and let $x_1, \ldots, x_n$ be the corresponding sample values. If $g(X_1, \ldots, X_n)$ is a function of the sample to be used for estimating θ we refer to g as an *estimator* of θ. The value which g assumes, that is $g(x_1, \ldots, x_n)$, will be referred to as an *estimate* of θ and is usually written as $\hat{\theta} = g(x_1, \ldots, x_n)$. (See the following note.)

Note: In this chapter we shall violate a rule which we have observed rather carefully up till now: making a careful distinction between a random variable and its value. That is, we shall often speak of $\hat{\theta}$, the estimate of θ, when we should really be speaking of the estimator $g(X_1, \ldots, X_n)$. Thus we shall write $E(\hat{\theta})$ when, of course, we mean $E[g(X_1, \ldots, X_n)]$. However, the context in which we allow ourselves this freedom should eliminate any possible ambiguity.

Definition. Let $\hat{\theta}$ be an estimate for the unknown parameter θ associated with the distribution of a random variable X. Then $\hat{\theta}$ is an *unbiased estimator* (or unbiased estimate; see the preceding note) for θ if $E(\hat{\theta}) = \theta$ for all θ.

Note: Any good estimate should be "close" to the value it is estimating. "Unbiasedness" means essentially that the average value of the estimate will be close to the true parameter value. For example, if the same estimate is used repeatedly and we average these values, we would expect the average to be close to the true value of the parameter. Although it is desirable that an estimate be unbiased, there may be occasions when we might prefer biased estimate (see below). It is possible (and in fact quite easily done) to find more than one unbiased estimate for an unknown parameter. In order to make a plausible choice in such situations we introduce the following concept.

> **Definition.** Let $\hat{\theta}$ be an unbiased estimate of θ. We say that $\hat{\theta}$ is an *unbiased, minimum variance* estimate of θ if for all estimates θ^* such that $E(\theta^*) = \theta$, we have $V(\hat{\theta}) \leq V(\theta^*)$ for all θ. That is, among all unbiased estimates of θ, $\hat{\theta}$ has the smallest variance.

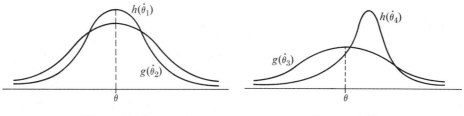

FIGURE 14.1   FIGURE 14.2

Notes: (a) The variance of a random variable measures the variability of the random variable about its expected value. Hence to require an unbiased estimate to have small variance is intuitively appealing. For if the variance is small, then the value of the random variable tends to be close to its mean, which in the case of an unbiased estimate means close to the true value of the parameter. Thus, if $\hat{\theta}_1$ and $\hat{\theta}_2$ are two estimates for θ, whose pdf is sketched in Fig. 14.1, we would presumably prefer $\hat{\theta}_1$ to $\hat{\theta}_2$. Both estimates are unbiased, and $V(\hat{\theta}_1) < V(\hat{\theta}_2)$.

In the case of estimates $\hat{\theta}_3$ and $\hat{\theta}_4$, the decision is not so clear (Fig. 14.2). For $\hat{\theta}_3$ is unbiased while $\hat{\theta}_4$ is not. However, $V(\hat{\theta}_3) > V(\hat{\theta}_4)$. This means that while on the average $\hat{\theta}_3$ will be close to θ, its large variance indicates that considerable deviations from θ would not be surprising. $\hat{\theta}_4$ on the other hand would tend to be somewhat larger than θ, on the average, and yet might be closer to θ than $\hat{\theta}_3$ (see Fig. 14.2).

(b) There exist some general techniques for finding minimum-variance unbiased estimates. We shall not be able to discuss these here, however. We shall make use of this concept mainly in order to choose between two or more available unbiased estimates. That is, if $\hat{\theta}_1$ and $\hat{\theta}_2$ are both unbiased estimates of θ, and if $V(\hat{\theta}_1) < V(\hat{\theta}_2)$, we would prefer $\hat{\theta}_1$.

Another criterion for judging estimates is somewhat more difficult to formulate and is based on the following definition.

> **Definition.** Let $\hat{\theta}$ be an estimate (based on a sample $X_1, \ldots, X_n$) of the parameter θ. We say that $\hat{\theta}$ is a *consistent* estimate of θ if

$$\lim_{n \to \infty} \text{Prob} \left[|\hat{\theta} - \theta| > \epsilon \right] = 0 \qquad \text{for all} \quad \epsilon > 0$$

or, equivalently, if

$$\lim_{n \to \infty} \text{Prob} \left[|\hat{\theta} - \theta| \le \epsilon \right] = 1 \qquad \text{for all} \quad \epsilon > 0.$$

Notes: (a) This definition states that an estimate is consistent if, as the sample size n is increased, the estimate $\hat{\theta}$ converges in the above probabilistic sense to θ. Again, this is an intuitively appealing characteristic for an estimate to possess. For it says that as the sample size increases (which should mean, in most reasonable circumstances, that more information becomes available) the estimate becomes "better" in the sense indicated.

(b) It is relatively easy to check whether an estimate is unbiased or not. It is also quite straightforward to compare the variances of two unbiased estimates. However, to verify consistency by applying the above definition is not as simple. The following theorem is sometimes quite useful.

Theorem 14.1. Let $\hat{\theta}$ be an estimate of θ based on a sample of size n. If $\lim_{n \to \infty} E(\hat{\theta}) = \theta$, and if $\lim_{n \to \infty} V(\hat{\theta}) = 0$, then $\hat{\theta}$ is a consistent estimate of θ.

Proof: We shall use Chebyshev's inequality, Eq. (7.20). Thus we write

$$P[|\hat{\theta} - \theta| \ge \epsilon] \le \frac{1}{\epsilon^2} E[\hat{\theta} - \theta]^2 = \frac{1}{\epsilon^2} E[\hat{\theta} - E(\hat{\theta}) + E(\hat{\theta}) - \theta]^2$$

$$= \frac{1}{\epsilon^2} E\{[\hat{\theta} - E(\hat{\theta})]^2 + 2[\hat{\theta} - E(\hat{\theta})][E(\hat{\theta}) - \theta]$$

$$+ [E(\hat{\theta}) - \theta]^2\}$$

$$= \frac{1}{\epsilon^2} \{\text{Var } \hat{\theta} + 0 + [E(\hat{\theta}) - \theta]^2\}.$$

Hence letting $n \to \infty$ and using the hypotheses of the theorem, we find that $\lim_{n \to \infty} P[|\hat{\theta} - \theta| \ge \epsilon] \le 0$ and thus equal to 0.

Note: If the estimate $\hat{\theta}$ is unbiased, the first condition is automatically satisfied.

A final criterion which is often applied to estimates may be formulated as follows. Suppose that $X_1, \ldots, X_n$ is a sample of X and θ is an unknown parameter. Let $\hat{\theta}$ be a function of $(X_1, \ldots, X_n)$.

Definition. We say that $\hat{\theta}$ is a *best linear unbiased* estimate of θ if:
(a) $E(\hat{\theta}) = \theta$.
(b) $\hat{\theta} = \sum_{i=1}^{n} a_i X_i$. That is, $\hat{\theta}$ is a *linear* function of the sample.
(c) $V(\hat{\theta}) \le V(\theta^*)$, where θ^* is any other estimate of θ satisfying (a) and (b) above.

Subsequently we shall consider a fairly general method which will yield good estimates for a large number of problems, in the sense that they will satisfy one or

more of the above criteria. Before doing this, let us simply consider a few estimates which are intuitively very reasonable, and then check just how good or how
bad these are in terms of the above criteria.

14.3 Some Examples

The above criteria of unbiasedness, minimum variance, consistency, and linearity give us at least some guidelines by which to judge an estimate. Let us now
consider a few examples.

EXAMPLE 14.1. Let us reconsider the above problem. We have sampled n rivets
and found that the sample $(X_1, \ldots, X_n)$ yielded exactly k defectives; that is
$Y = \sum_{i=1}^{n} X_i = k$. Because of our assumptions, Y is a binomially distributed
random variable.

The intuitively most suggestive estimate of the parameter p is $\hat{p} = Y/n$, the
proportion defective found in the sample. Let us apply some of the above criteria
to see how good an estimate $\hat{p}$ is.

$$E(\hat{p}) = E\left(\frac{Y}{n}\right) = \frac{1}{n}(np) = p.$$

Thus $\hat{p}$ is an unbiased estimate of p.

$$V(\hat{p}) = V\left(\frac{Y}{n}\right) = \frac{1}{n^2}(np)(1 - p) = \frac{p(1 - p)}{n}.$$

Thus $V(\hat{p}) \to 0$ as $n \to \infty$, and hence $\hat{p}$ is a consistent estimate. As we have
pointed out earlier, there may be many unbiased estimates for a parameter, some
of which may be very poor indeed. Consider, for instance, in the context of the
present example, the estimate p^* defined as follows: $p^* = 1$ if the first chosen
item is defective, and 0 otherwise. That this is not a very good estimate is clear
when we note that its value is a function only of X_1, rather than of $X_1, \ldots, X_n$.
However p^* is unbiased, for $E(p^*) = 1P(X = 1) + 0P(X = 0) = p$. The
variance of p^* is $p(1 - p)$, which compares very poorly with the variance of p
considered above, namely $p(1 - p)/n$, particularly if n is large.

The result obtained in the above example is a special case of the following general statement.

Theorem 14.2. Let X be a random variable with finite expectation μ and variance
σ^2. Let $\overline{X}$ be the sample mean, based on a random sample of size n. Then
$\overline{X}$ is an unbiased and consistent estimate of μ.

Proof: This follows immediately from Theorem 13.1, where we proved that
$E(\overline{X}) = \mu$ and $V(\overline{X}) = \sigma^2/n$ which approaches 0 as $n \to \infty$.

Note: That the result demonstrated in Example 14.1 is a special case of Theorem 14.2 follows when we note that the proportion defective in the sample, Y/n, may be written as $(1/n)(X_1 + \cdots + X_n)$, where the X_i's assume the values one and zero, depending on whether or not the item inspected is defective.

The sample mean referred to in Theorem 14.2 is a linear function of the sample. That is, it is of the form $a_1 X_1 + a_2 X_2 + \cdots + a_n X_n$, with $a_1 = \cdots = a_n = 1/n$. It is readily seen that $\hat{\mu} = \sum_{i=1}^{n} a_i X_i$ is an unbiased estimate of μ for any choice of coefficients satisfying the condition $\sum_{i=1}^{n} a_i = 1$. The following interesting question arises. For what choice of the a_i's (subject to $\sum_{i=1}^{n} a_i = 1$) is the variance of $\sum_{i=1}^{n} a_i X_i$ smallest? It turns out that the variance is minimized if $a_i = 1/n$ for all i. That is, $\bar{X}$ is the minimum-variance, unbiased, linear estimate.

To see this, consider

$$\hat{\mu} = \sum_{i=1}^{n} a_i X_i, \qquad \sum_{i=1}^{n} a_i = 1.$$

Hence

$$\text{Var } \hat{\mu} = \sigma^2 \sum_{i=1}^{n} a_i^2$$

since the X_i's are independent random variables with common variance σ^2. We write

$$\sum_{i=1}^{n} a_i^2 = (a_1 - 1/n)^2 + \cdots + (a_n - 1/n)^2 + (2/n)(a_1 + \cdots + a_n) - n(1/n^2)$$

$$= (a_1 - 1/n)^2 + \cdots + (a_n - 1/n)^2 + 1/n \qquad \left(\text{since } \sum_{i=1}^{n} a_i = 1\right).$$

Hence this expression is obviously minimized if $a_i = 1/n$ for all i.

EXAMPLE 14.2. Suppose that T, the time to failure of a component, is exponentially distributed. That is, the pdf of T is given by $f(t) = \beta e^{-\beta t}, t \geq 0$. Assume that we test n such components, recording the time to failure of each, say $T_1, \ldots, T_n$. We want an unbiased estimate of the expected time to failure, $E(T) = 1/\beta$, based on the sample $(T_1, \ldots, T_n)$. One such estimate is $\bar{T} = (1/n)\sum_{i=1}^{n} T_i$. From Theorem 14.2 we know that $E(\bar{T}) = 1/\beta$. Since $V(T) = 1/\beta^2$, Theorem 13.1 tells us that $V(\bar{T}) = 1/\beta^2 n$. However $\bar{T}$ is not the only unbiased estimate of $1/\beta$. Consider, in fact, the minimum of the sample, say $Z = \min (T_1, \ldots, T_n)$. According to Theorem 13.3, Z is again exponentially distributed with parameter $n\beta$. Hence $E(Z) = 1/n\beta$. Thus the estimate nZ is also an unbiased estimate for $1/\beta$.

To evaluate its variance we compute

$$V(nZ) = n^2 V(Z) = n^2 \frac{1}{(n\beta)^2} = \frac{1}{\beta^2}.$$

Thus, although the two estimates nZ and $\bar{T}$ are both unbiased, the latter has a smaller variance and hence should be preferred.

However, in this particular situation there is another consideration which might influence our choice between the two suggested estimates. The n components might be tested simultaneously. (For example, we might insert n light bulbs into n sockets and record their burning time.) When we use nZ as our estimate, the test may be terminated as soon as the first component has failed. In using $\bar{T}$ as our estimate we must wait until all the components have failed. It is quite conceivable that a long time span exists between the first and the last failure. Saying it differently, if L is the time it requires to test the n items and we compute the estimate for $1/\beta$, then using nZ, we have $L = \min(T_1, \ldots, T_n)$, while when using $\bar{T}$, we have $L = \max(T_1, \ldots, T_n)$. Thus, if the time required to perform the testing is of any serious consequence (in terms of cost, say), we might prefer the estimate with the larger variance.

EXAMPLE 14.3. Suppose that we want an unbiased estimate of the variance σ^2 of a random variable, based on a sample $X_1, \ldots, X_n$.

Although, intuitively, we might consider $(1/n)\sum_{i=1}^{n}(X_i - \bar{X})^2$, it turns out that this statistic has an expected value equal to $[(n-1)/n]\sigma^2$. (See Theorem 13.4.) Hence an unbiased estimate for σ^2 is obtained by taking

$$\hat{\sigma}^2 = \frac{1}{n-1}\sum_{i=1}^{n}(X_i - \bar{X})^2.$$

Notes: (a) Although dividing by $(n-1)$ instead of n does make a difference when n is relatively small, for larger n it makes little difference which of these estimates we use.

(b) Example 14.3 illustrates a fairly common situation: It may happen that an estimate for a parameter β, say $\hat{\beta}$, is biased in the sense that $E(\hat{\beta}) = k\beta$; in such a case we simply consider the new estimate $\hat{\beta}/k$, which will then be unbiased.

(c) It can be shown, although we shall not do so here, that the above estimate for σ^2 is consistent.

EXAMPLE 14.4. In Table 14.1, we reproduce the data obtained from the famous experiment carried out by Rutherford [Rutherford and Geiger, *Phil. Mag.* **S6, 20,** 698 (1910)] on the emission of α-particles from a radioactive source. In the

TABLE 14.1

k	0	1	2	3	4	5	6	7	8	9	10	11	Total
n_k	57	203	383	525	532	408	273	139	49	27	10	6	2612

table, k is the number of particles observed in a unit of time (unit $= \frac{1}{8}$ minute), while n_k is the number of intervals in which k particles were observed. If we let X be the number of particles emitted during the time interval of length $\frac{1}{8}$ (minute), and if we assume that X follows the Poisson distribution, we have

$$P(X = k) = \frac{e^{-(1/8)\lambda}(\frac{1}{8}\lambda)^k}{k!}.$$

Since $E(\overline{X}) = \frac{1}{8}\lambda$, we can use the sample mean to obtain an unbiased estimate of $\frac{1}{8}\lambda$. For λ, we then obtain the estimate $\hat{\lambda} = 8\overline{X}$.
   To compute $\overline{X}$ we simply evaluate

$$\frac{\sum_{k=0}^{11} k n_k}{\sum_{k=0}^{11} n_k} = 3.87 \text{ particles.}$$

Hence an unbiased estimate of λ, based on the sample mean equals 30.96 (which may be interpreted as the expected number of particles emitted per minute).

EXAMPLE 14.5. In the manufacture of explosives, there may occur at random a certain number of ignitions. Let X be the number of ignitions per day. Assume that X has a Poisson distribution with parameter λ. Table 14.2 gives some data which are to be used for the estimation of λ.

TABLE 14.2

Number of ignitions, k	0	1	2	3	4	5	6	Total
Number of days with k ignitions, n_k	75	90	54	22	6	2	1	250

Again using the sample mean for the estimate of λ, we obtain

$$\hat{\lambda} = \frac{\sum_k k n_k}{\sum_k n_k} = 1.22 \text{ number of ignitions per day.}$$

EXAMPLE 14.6. It has been indicated that the ash content in coal is normally distributed with parameters μ and σ^2. The data in Table 14.3 represent the ash content in 250 samples of coal analyzed. [Data obtained from E. S. Grummel and A. C. Dunningham (1930); *British Standards Institution* **403**, 17.] n_x is the number of samples having x percent of ash content.

In order to estimate the parameters μ and σ^2, we shall use the unbiased estimates discussed before:

$$\hat{\mu} = \frac{\sum_x x n_x}{250} = 16.998, \qquad \hat{\sigma}^2 = \frac{1}{249} \sum_x n_x (x - \hat{\mu})^2 = 7.1.$$

TABLE 14.3

Ash content in coal									
x	9.25	9.75	10.25	10.75	11.25	11.75	12.25	12.75	13.25	13.75
n_x	1	0	2	1	1	2	5	4	7	6
x	14.25	14.75	15.25	15.75	16.25	16.75	17.25	17.75	18.25	18.75
n_x	13	14	15	13	24	15	19	23	22	12
x	19.25	19.75	20.25	20.75	21.25	21.75	22.25	22.75	23.25	23.75
n_x	12	7	6	8	6	4	2	2	0	3
x	24.25	24.75	25.25							
n_x	0	0	1							
Total samples	250									

Note: Suppose that we have an unbiased estimate, say $\hat{\theta}$, for a parameter θ. It may happen that we are really interested in estimating some function of θ, say $g(\theta)$. [For example, if X is exponentially distributed with parameter θ, we will probably be interested in $1/\theta$, namely $E(X)$.] It might be supposed that all we need to do is to consider $1/\hat{\theta}$ or $(\hat{\theta})^2$, for example, as the appropriate unbiased estimates of $1/\theta$ or $(\theta)^2$. This is emphatically not so. In fact, one of the disadvantages of the criterion of unbiasedness is that if we have found an unbiased estimate for θ, we must, in general, start from the beginning to find an estimate for $g(\theta)$. Only if $g(\theta) = a\theta + b$, that is, if g is a linear function of θ, is it true that $E[g(\hat{\theta})] = g[E(\hat{\theta})]$. In general, $E[g(\hat{\theta})] \neq g[E(\hat{\theta})]$. For example, suppose that X is a random variable with $E(X) = \mu$ and $V(X) = \sigma^2$. We have seen that the sample mean $\overline{X}$ is an unbiased estimate of μ. Is $\overline{X}^2$ an unbiased estimate of $(\mu)^2$? The answer is "no," as the following computation indicates. Since $V(\overline{X}) = E(\overline{X})^2 - (E(\overline{X}))^2$, we have

$$E(\overline{X})^2 = V(\overline{X}) + (E(\overline{X}))^2 = \sigma^2/n + (\mu)^2 \neq (\mu)^2.$$

Although the above examples show, quite conclusively, that in general

$$E[g(\hat{\theta})] \neq g[E(\hat{\theta})],$$

it turns out that in many cases equality holds, at least approximately, particularly if the sample size is large. For instance, in the above example we found [with $\hat{\theta} = \overline{X}$ and $g(z) = z^2$] that $E(\overline{X}) = \mu$ and $E(\overline{X})^2 = \mu^2 + \sigma^2/n$, which is approximately equal to μ^2 if n is large.

14.4 Maximum Likelihood Estimates

We have considered only a number of criteria by which we may judge an estimate. That is, given a proposed estimate for an unknown parameter, we may

check whether it is unbiased and consistent, and we can compute (at least in principle) its variance and compare it with the variance of another estimate. However,
we do not have as yet a general procedure with which we can find "reasonable"
estimates. Several such procedures exist, and we shall discuss one of these, namely
the method of maximum likelihood. In many cases this method leads to reasonable estimates.

In order to avoid repeating our discussion for the discrete and the continuous
case, let us agree on the following terminology for the purpose of the present
discussion.

We shall write $f(x; \theta)$ either for the pdf of X (evaluated at x) or for $P(X = x)$ if
X is discrete. We include θ (in the notation) in order to remind us that the probability distribution of X depends on the parameter θ with which we shall be
concerned.

Let $X_1, \ldots, X_n$ be a random sample from the random variable X and let
$x_1, \ldots, x_n$ be the sample values. We define the *likelihood function L* as the following function of the sample and θ.

$$L(X_1, \ldots, X_n; \theta) = f(X_1; \theta)f(X_2; \theta) \cdots f(X_n; \theta). \qquad (14.1)$$

If X is discrete, $L(x_1, \ldots, x_n; \theta)$ represents $P[X_1 = x_1, \ldots, X_n = x_n]$ while if
X is continuous, $L(x_1, \ldots, x_n; \theta)$ represents the joint pdf of $(X_1, \ldots, X_n)$.

If the sample $(X_1, \ldots, X_n)$ has been obtained, the sample values $(x_1, \ldots, x_n)$
are known. Since θ is unknown, we might ask ourselves the following question.
For what value of θ will $L(x_1, \ldots, x_n; \theta)$ be largest? Putting it differently, let us
suppose that we consider two possible values for θ, say θ_1 and θ_2. Let us assume
furthermore that $L(x_1, \ldots, x_n; \theta_1) < L(x_1, \ldots, x_n; \theta_2)$. We would then prefer θ_2 to θ_1 for the *given* sample values $(x_1, \ldots, x_n)$. For, if θ_2 really is the true
value of θ, then the probability of obtaining sample values such as those we did
is greater than if θ_1 were the true value of θ. Informally, we prefer that value of
the parameter which makes as probable as possible that event which in fact did
occur. That is, we wish to choose the most probable value for θ *after* the data are
obtained, assuming that each value for θ was equally likely *before* the data were
obtained. We make the following formal definition.

Definition. *The maximum likelihood* estimate of θ, say $\hat{\theta}$, based on a random
sample $X_1, \ldots, X_n$ is that value of θ which maximizes $L(X_1, \ldots, X_n; \theta)$,
considered as a function of θ for a given sample $X_1, \ldots, X_n$, where L is
defined by Eq. (14.1). (This is usually referred to as the ML estimate.)

Notes: (a) $\hat{\theta}$ will of course be a statistic and hence a random variable, since its value
will depend on the sample $(X_1, \ldots, X_n)$. (We will not consider a constant as a solution.)

(b) In most of our examples θ will represent a single real number. However, it may
happen that the probability distribution of X depends on two or more parameter values
(as does the normal distribution, for example). In such a case, θ may represent a vector,
say $\theta = (\alpha, \beta)$ or $\theta = (\alpha, \beta, \gamma)$, etc.

(c) In order to find the ML estimate we must determine the maximum value of a function. Hence in many problems we may apply some of the standard techniques of the calculus to find this maximum. Since ln x is an *increasing* function of x,

$$\ln L(X_1, \ldots, X_n; \theta)$$

will achieve its maximum value for the same value of θ as will $L(X_1, \ldots, X_n; \theta)$. Hence under fairly general conditions, assuming that θ is a real number and that $L(X_1, \ldots, X_n; \theta)$ is a differentiable function of θ, we can obtain the ML estimate $\hat{\theta}$ by solving what is known as the *likelihood equation*:

$$\frac{\partial}{\partial \theta} \ln L(X_1, \ldots, X_n; \theta) = 0 \tag{14.2}$$

If $\theta = (\alpha, \beta)$, the above equation must be replaced by the simultaneous likelihood equations

$$\frac{\partial}{\partial \alpha} \ln L(X_1, \ldots, X_n; \alpha, \beta) = 0,$$

$$\frac{\partial}{\partial \beta} \ln L(X_1, \ldots, X_n; \alpha, \beta) = 0. \tag{14.3}$$

It should again be emphasized that the above approach does not always work. However, in a large number of important examples (some of which we shall consider shortly), this method does yield the required ML estimate with relative ease.

Properties of maximum likelihood estimates:

(a) The ML estimate may be biased. Quite often such a bias may be removed by multiplying by an appropriate constant.

(b) Under fairly general conditions, ML estimates are consistent. That is, if the sample size upon which these estimates are based is large, the ML estimate will be "close" to the parameter value to be estimated. (The ML estimates possess another very important "large sample size" property which we shall discuss subsequently.)

(c) The ML estimates possess the following very important *invariance property*. Suppose that $\hat{\theta}$ is the ML estimate of θ. Then it may be shown that the ML estimate of $g(\theta)$ is $g(\hat{\theta})$. That is, if statistician A takes his measurement in feet[2] and statistician B measures in feet and if A's ML estimate is $\hat{\theta}$, then B's would be $\sqrt{\hat{\theta}}$. Recall that this property is *not* possessed by unbiased estimates.

We shall now consider a number of important illustrations of ML estimates.

EXAMPLE 14.7. Suppose that the time to failure, say T, of a component has exponential distribution with parameter β. Hence the pdf of T is given by

$$f(t) = \beta e^{-\beta t}, \qquad t \geq 0.$$

Suppose that n such components are tested yielding failure times $T_1, \ldots, T_n$.

Hence the likelihood function of this sample is given by

$$L(T_1, \ldots, T_n; \beta) = \beta^n \exp\left(-\beta \sum_{i=1}^{n} T_i\right).$$

Thus $\ln L = n \ln \beta - \beta \sum_{i=1}^{n} T_i$. Hence

$$\frac{\partial \ln L}{\partial \beta} = \frac{n}{\beta} - \sum_{i=1}^{n} T_i \quad \text{and} \quad \frac{\partial \ln L}{\partial \beta} = 0$$

yields $\hat{\beta} = 1/\overline{T}$, where $\overline{T}$ is the sample mean of the failure times. Since the expected value of T, the mean time to failure, is given by $1/\beta$, we find, using the invariance property of ML estimates, that the ML estimate of $E(T)$ is given by $\overline{T}$, the sample mean. We know that $E(\overline{T}) = 1/\beta$ and hence $\overline{T}$, the ML estimate of $E(T)$, is unbiased.

Note: In general, it is not easy to find the probability distribution of ML estimates, particularly if the sample size is small. (If n is large, we shall find that a general answer is available.) However, in the present example we are able to obtain the distribution of the ML estimate. From the corollary of Theorem 10.9 we find that $2n\beta\overline{T}$ has distribution χ^2_{2n}. Hence $P(\overline{T} \leq t) = P(2n\beta\overline{T} \leq 2n\beta t)$. This probability may be obtained directly from the table of the chi-square distribution if n, β, and t are known.

EXAMPLE 14.8. It is known that a certain (fixed) proportion, say p, of shells is defective. From a large supply of shells, n are chosen at random and are tested. Define the following random variables.

$X_i = 1$ if the ith shell is defective and 0 otherwise, $i = 1, 2, \ldots, n$.

Hence $(X_1, \ldots, X_n)$ is a random sample from the random variable X having probability distribution $P(X = 0) = f(0; p) = 1 - p$, $P(X = 1) = f(1; p) = p$. That is, $f(x, p) = p^x(1 - p)^{1-x}$, $x = 0, 1$. Hence

$$L(X_1, \ldots, X_n; p) = p^k(1 - p)^{n-k},$$

where $k = \sum_{i=1}^{n} X_i = $ total number of defectives. Thus $\ln L(X_1, \ldots, X_n; p) = k \ln p + (n - k) \ln (1 - p)$. Therefore

$$\frac{\partial \ln L}{\partial p} = \frac{k}{p} + \frac{n - k}{1 - p} (-1) = \frac{k}{p} - \frac{n - k}{1 - p}.$$

If $k = 0$ or n we find directly, by considering the expression for L, that the maximum value of L is attained when $p = 0$ or 1, respectively. For $k \neq 0$ or n, we set $\partial \ln L/\partial p = 0$ and find as its solution $\hat{p} = k/n = \overline{X}$, the sample mean. Thus we find again that the ML estimate yields an unbiased estimate of the sought-after parameter.

EXAMPLE 14.9. Suppose that the random variable X is normally distributed with expectation μ and variance 1. That is, the pdf of X is given by

$$f(x) = \frac{1}{\sqrt{2\pi}} e^{-(1/2)(x-\mu)^2}.$$

If $(X_1, \ldots, X_n)$ is a random sample from X, the likelihood function of this sample is

$$L(X_1, \ldots, X_n; \mu) = \frac{1}{(2\pi)^{n/2}} \exp\left[-\frac{1}{2} \sum_{i=1}^{n} (X_i - \mu)^2\right].$$

Hence

$$\ln L = -\frac{n}{2} \ln (2\pi) - \frac{1}{2} \sum_{i=1}^{n} (X_i - \mu)^2 \quad \text{and} \quad \frac{\partial \ln L}{\partial \mu} = \sum_{i=1}^{n} (X_i - \mu).$$

Thus $\partial \ln L/\partial \mu = 0$ yields $\hat{\mu} = \bar{X}$, the sample mean.

EXAMPLE 14.10. So far we have considered situations in which we were able to find the maximum value of L by simply differentiating L (or $\ln L$) with respect to the parameter and setting this derivative equal to zero. That this does not always work is illustrated by the following example.

Suppose that the random variable X is uniformly distributed over the interval $[0, \alpha]$ where α is an unknown parameter. The pdf of X is given by

$$f(x) = 1/\alpha, \quad 0 \le x \le \alpha,$$
$$= 0, \quad \text{elsewhere.}$$

If $(X_1, \ldots, X_n)$ is a sample from X, its likelihood function is given by

$$L(X_1, \ldots, X_n; \alpha) = (1/\alpha)^n, \quad 0 \le X_i \le \alpha \quad \text{all } i,$$

$$= 0, \quad \text{elsewhere.}$$

Considering L as a function of α for given $(X_1, \ldots, X_n)$, we note that we must have $\alpha \ge X_i$ for all i in order that L be nonzero. This is equivalent to requiring $\alpha \ge \max (X_1, \ldots, X_n)$. Thus, if we plot L as a function of α we obtain the graph shown in Fig. 14.3. From this graph it is immediately clear which value of α maximizes L, namely

$$\hat{\alpha} = \max (X_1, \ldots, X_n).$$

Let us study some of the properties of this estimate. From Theorem 13.2, we obtain the pdf of $\hat{\alpha}$: $g(\hat{\alpha}) = n[F(\hat{\alpha})]^{n-1} f(\hat{\alpha})$. But $F(x) = x/\alpha$, $0 \le x \le \alpha$, and $f(x)$ is given above. Hence we obtain

$$g(\hat{\alpha}) = n \left[\frac{\hat{\alpha}}{\alpha}\right]^{n-1} \left(\frac{1}{\alpha}\right) = \frac{n(\hat{\alpha})^{n-1}}{\alpha^n}, \quad 0 \le \hat{\alpha} \le \alpha.$$

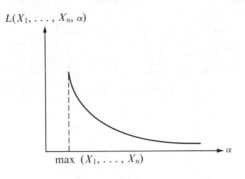

$L(X_1, \ldots, X_n, \alpha)$

$\max (X_1, \ldots, X_n)$

FIGURE 14.3

To find $E(\hat{\alpha})$ we compute

$$E(\hat{\alpha}) = \int_0^\alpha \hat{\alpha} g(\hat{\alpha}) \, d\hat{\alpha} = \int_0^\alpha \hat{\alpha} \frac{n\hat{\alpha}^{n-1}}{\alpha^n} \, d\hat{\alpha}$$

$$= \frac{n}{\alpha^n} \frac{\hat{\alpha}^{n+1}}{n+1} \Big|_0^\alpha = \frac{n}{n+1} \alpha.$$

Thus $\hat{\alpha}$ is not an unbiased estimate of α; $\hat{\alpha}$ tends to "underestimate" α. If we want an unbiased estimate, we can use

$$\hat{\hat{\alpha}} = \frac{n+1}{n} \max (X_1, \ldots, X_n).$$

Note that although $E(\hat{\alpha}) \neq \alpha$, we do have $\lim_{n\to\infty} E(\hat{\alpha}) = \alpha$. Thus, to verify consistency, we must still show that $V(\hat{\alpha}) \to 0$ as $n \to \infty$. (See Theorem 14.1.) We must evaluate $E(\hat{\alpha})^2$:

$$E(\hat{\alpha})^2 = \int_0^\alpha (\hat{\alpha})^2 g(\hat{\alpha}) \, d\hat{\alpha} = \int_0^\alpha (\hat{\alpha})^2 \frac{n(\hat{\alpha})^{n-1}}{\alpha^n} \, d\hat{\alpha}$$

$$= \frac{n}{\alpha^n} \frac{(\hat{\alpha})^{n+2}}{n+2} \Big|_0^\alpha = \frac{n}{n+2} \alpha^2.$$

Hence

$$V(\hat{\alpha}) = E(\hat{\alpha})^2 - \left(E(\hat{\alpha})\right)^2 = \frac{n}{n+2} \alpha^2 - \left[\frac{n}{n+1} \alpha\right]^2 = \alpha^2 \left[\frac{n}{(n+2)(n+1)^2}\right].$$

Thus $V(\hat{\alpha}) \to 0$ as $n \to \infty$ and hence consistency is established.

EXAMPLE 14.11. Let us consider one example in which two parameters (both unknown) characterize the distribution. Suppose that X has distribution $N(\mu, \sigma^2)$. Hence the pdf of X is

$$f(x) = \frac{1}{\sqrt{2\pi}\,\sigma} \exp\left(-\frac{1}{2}\left[\frac{x-\mu}{\sigma}\right]^2\right).$$

If $(X_1, \ldots, X_n)$ is a sample of X, its likelihood function is given by

$$L(X_1, \ldots, X_n; \mu, \sigma) = (2\pi\sigma^2)^{-n/2} \exp\left\{-\frac{1}{2}\sum_{i=1}^{n}\left[\frac{X_i - \mu}{\sigma}\right]^2\right\}.$$

Hence

$$\ln L = \left(-\frac{n}{2}\right)\ln(2\pi\sigma^2) - \frac{1}{2}\sum_{i=1}^{n}\left(\frac{X_i - \mu}{\sigma}\right)^2.$$

We must solve simultaneously

$$\frac{\partial \ln L}{\partial \mu} = 0 \quad \text{and} \quad \frac{\partial \ln L}{\partial \sigma} = 0.$$

We have

$$\frac{\partial \ln L}{\partial \mu} = \sum_{i=1}^{n}\frac{(X_i - \mu)}{\sigma^2} = 0,$$

which yields $\hat{\mu} = \overline{X}$, the sample mean. And

$$\frac{\partial \ln L}{\partial \sigma} = -\frac{n}{\sigma} + \sum_{i=1}^{n}\frac{(X_i - \mu)^2}{\sigma^3} = 0,$$

which yields

$$\hat{\sigma}^2 = \frac{1}{n}\sum_{i=1}^{n}(X_i - \mu)^2 = \frac{1}{n}\sum_{i=1}^{n}(X_i - \overline{X})^2.$$

Note that the ML method yields a biased estimate of σ^2, since we have already seen that the unbiased estimate is of the form $1/(n - 1)\sum_{i=1}^{n}(X_i - \overline{X})^2$.

EXAMPLE 14.12. We have previously (Example 14.7) considered the problem of estimating the parameter β in an exponential failure law, by testing n items and noting their times to failure, say $T_1, \ldots, T_n$. Another procedure might be the following. Suppose that we take n items, test them, and after a certain length of time has elapsed, say T_0 hours, we simply count the number of items that have failed, say X. Our sample consists of $X_1, \ldots, X_n$, where $X_i = 1$ if the ith item has failed in the specified period, and 0 otherwise. Thus the likelihood function of the sample is

$$L(X_1, \ldots, X_n; \beta) = p^k(1 - p)^{n-k},$$

where $k = \sum_{i=1}^{n} X_i$ = number of items that have failed and p = Prob (item fails). Now p is a function of the parameter β to be estimated; that is,

$$p = P(T \leq T_0) = 1 - e^{-\beta T_0}.$$

Using the result of Example 14.8, we find that the ML estimate of p is $\hat{p} = k/n$. Applying the invariance property of the ML estimate (noting that p is an increasing function of β), we obtain the ML estimate of β by simply solving the equation

$1 - e^{-\beta T_0} = k/n$. An easy computation yields

$$\hat{\beta} = -\frac{1}{T_0} \ln\left(\frac{n-k}{n}\right),$$

or, for the estimate of $1/\beta$ the mean time to failure,

$$\left(\widehat{\frac{1}{\beta}}\right) = \frac{-T_0}{\ln\left[(n-k)/n\right]}.$$

In all the examples considered above, the ML method yielded equations which were relatively easy to solve. In many problems this is not the case, and often we must resort to (approximate) numerical methods in order to obtain the estimates. The following example illustrates such difficulties.

EXAMPLE 14.13. As already noted, the Gamma distribution has important applications to life testing. Let us suppose, for example, that the time to failure of an electrical generator has life length X whose pdf is given by

$$f(x) = \frac{\lambda^r x^{r-1}}{\Gamma(r)} e^{-\lambda x}, \qquad x \geq 0,$$

where r and λ are two positive parameters we wish to estimate. Suppose that a sample, $X_1, \ldots, X_n$ from X is available. (That is, n generators have been tested and their times to failure recorded.) The likelihood function of the sample is

$$L(X_1, \ldots, X_n; \lambda, r) = \frac{(\lambda)^{nr}(\Pi_{i=1}^n X_i)^{r-1} \exp\left(-\lambda \sum_{i=1}^n X_i\right)}{[\Gamma(r)]^n},$$

$$\ln L = nr \ln \lambda + (r-1) \sum_{i=1}^n \ln X_i - \lambda \sum_{i=1}^n X_i - n \ln \Gamma(r).$$

Thus we must simultaneously solve, $\partial \ln L/\partial \lambda = 0$ and $\partial \ln L/\partial r = 0$. These equations become

$$\frac{\partial \ln L}{\partial \lambda} = \frac{nr}{\lambda} - \sum_{i=1}^n X_i = 0,$$

$$\frac{\partial \ln L}{\partial r} = n \ln \lambda + \sum_{i=1}^n \ln X_i - n\frac{\Gamma'(r)}{\Gamma(r)} = 0.$$

Thus $\partial \ln L/\partial \lambda = 0$ yields directly $\hat{\lambda} = r/\overline{X}$. Hence, after replacing λ by $\hat{\lambda}$, we find that $\partial \ln L/\partial r = 0$ gives

$$\ln r - \frac{\Gamma'(r)}{\Gamma(r)} = \ln \overline{X} - \frac{1}{n} \sum_{i=1}^n \ln X_i.$$

It is evident that we must solve the above equation for r, obtaining $\hat{r}$ and then $\hat{\lambda} = \hat{r}/\overline{X}$. Fortunately, the function $\Gamma'(r)/\Gamma(r)$ has been tabulated. A fairly quick method of obtaining the required solution has been discussed in a paper by D. G. Chapman (*Annals of Mathematical Statistics* **27**, 498–506, 1956). This example illustrates that the solution of the likelihood equations can lead to considerable mathematical difficulties.

As we have suggested earlier, the ML estimates possess an additional property which makes them particularly desirable, especially if the estimates are based on a fairly large sample.

Asymptotic property of maximum likelihood estimates. If $\hat\theta$ is a ML estimate for the parameter θ, based on a random sample $X_1, \ldots, X_n$ from a random variable X, then for n sufficiently large, the random variable $\hat\theta$ has approximately the distribution

$$N\left(\theta, \frac{1}{B}\right), \quad \text{where} \quad B = nE\left[\frac{\partial}{\partial\theta}\ln f(X;\theta)\right]^2; \quad (14.4)$$

here f is the point probability function or pdf of X, depending on whether X is discrete or continuous and where θ is assumed to be a real number.

Notes: (a) The above property is, of course, considerably stronger than the property of consistency which we have mentioned previously. Consistency simply says that if n is sufficiently large, $\hat\theta$ will be "close" to θ. The above actually tells us what the probabilistic behavior of $\hat\theta$ is for large n.
    (b) We shall not prove the above assertion, but simply illustrate its use with an example.

EXAMPLE 14.14. Let us reconsider Example 14.7. We found $\hat\beta = 1/\bar{T}$ to be the ML estimate for β. The pdf of T was given by $f(t;\beta) = \beta e^{-\beta t}$, $t > 0$. The above property tells us that for n sufficiently large, $\hat\beta = 1/\bar{T}$ has the approximate distribution $N(\beta, 1/B)$, where B is given by Eq. (14.4).
    To find B, consider $\ln f(T;\beta) = \ln \beta - \beta T$. Then

$$(\partial/\partial\beta)\ln f(T;\beta) = (1/\beta) - T.$$

Therefore

$$\left[\frac{\partial}{\partial\beta}\ln f(T;\beta)\right]^2 = \frac{1}{\beta^2} - \frac{2T}{\beta} + T^2.$$

Since

$$E(T) = 1/\beta \quad \text{and} \quad E(T^2) = V(T) + [E(T)]^2 = 1/\beta^2 + 1/\beta^2 = 2/\beta^2,$$

we have

$$E\left[\frac{\partial}{\partial\beta}\ln f(T;\beta)\right]^2 = \frac{1}{\beta^2} - \frac{2}{\beta}\frac{1}{\beta} + \frac{2}{\beta^2} = \frac{1}{\beta^2}.$$

Hence we find that for large n, $\hat\beta$ has approximately the distribution $N(\beta, \beta^2/n)$. (This verifies the consistency property of the estimate since $\beta^2/n \to 0$ as $n \to \infty$.)

TABLE 14.4

X (height, m)	Y (temperature, °C)	X (height, m)	Y (temperature, °C)
1142	13	1008	13
1742	7	208	18
280	14	439	14
437	16	1471	14
678	13	482	18
1002	11	673	13
1543	4	407	16
1002	9	1290	7
1103	5	1609	6
475	11	910	9
1049	10	1277	11
566	15	410	14
995	10		

14.5 The Method of Least Squares

EXAMPLE 14.15. We are familiar with the fact that the air temperature decreases with the altitude of the locality. The data given in Table 14.4 and the associated *scatter diagram* (plotted points) (Fig. 14.4) substantiate this. Not only do the plotted points indicate that the temperature Y decreases with the altitude X, but a linear trend is evident.

The observations represent altitude (meters) and temperature (degrees centigrade) in the early morning at a number of reporting stations in Switzerland. The data are from Observatory Basel-St. Margarathen.

What is a reasonable model for the above data? We shall suppose that Y is a random variable whose value depends, among other things, on the value of X. Specifically, we shall suppose that

$$Y = \alpha X + \beta + \epsilon,$$

FIGURE 14.4

where α and β are (unknown) constants, X is the (known) altitude at which Y is measured, and ϵ is a random variable. The analysis of this *linear model* depends on the assumptions we make about the random variable ϵ. (We are essentially saying that the temperature is a random outcome whose value may be decomposed into a strictly random component plus a term which depends on the altitude X in a linear way.) The assumption we shall make about ϵ is the following:

$$E(\epsilon) = 0; \qquad V(\epsilon) = \sigma^2 \quad \text{for } all \ X.$$

That is, the expected value and variance of ϵ do not depend on the value of X. Hence $E(Y) = \alpha X + \beta$ and $V(Y) = \sigma^2$. Note that the model we have stipulated depends on three parameters, α, β, and σ^2. We cannot use the method of maximum likelihood to estimate these parameters unless we make further assumptions about the distribution of ϵ. For nothing has been assumed about the distribution of the random variable ϵ; only an assumption about its expected value and variance has been made. (A modification of this model will be mentioned subsequently.)

Before we discuss how to estimate the pertinent parameters, let us say a brief word about the meaning of a random sample in the present context. Suppose that n values of X are chosen, say $x_1, \ldots, x_n$. (Recall again that X is not a random variable here.) For each x_i, let Y_i be an independent observation from the random variable Y described above. Hence $(x_1, Y_1), \ldots, (x_n, Y_n)$ may be considered as a random sample from the random variable Y for given X values $(x_1, \ldots, x_n)$.

> **Definition.** Suppose that we have $E(Y) = \alpha X + \beta$, where α, β, and X are as described above. Let $(x_1, Y_1), \ldots, (x_n, Y_n)$ be a random sample of Y. The *least-squares estimates* of the parameters α and β are those values of α and of β which minimize
>
> $$\sum_{i=1}^{n} [Y_i - (\alpha x_i + \beta)]^2.$$

Note: The interpretation of the above criterion is quite clear. (See Fig. 14.5.) For each pair (x_i, Y_i) we compute the discrepancy between Y_i, the observed value, and $\alpha x_i + \beta$, the expected value. Since we are only concerned with the magnitude of this discrepancy we square and sum over all the sample points. The line sought is the one for which this sum is smallest.

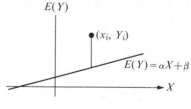

FIGURE 14.5

In order to obtain the required estimates for α and β we proceed as follows. Let $S(\alpha, \beta) = \sum_{i=1}^{n} [Y_i - (\alpha x_i + \beta)]^2$. To minimize $S(\alpha, \beta)$, we must solve the equations

$$\frac{\partial S}{\partial \alpha} = 0 \quad \text{and} \quad \frac{\partial S}{\partial \beta} = 0.$$

Differentiating S with respect to α and β, we obtain

$$\frac{\partial S}{\partial \alpha} = \sum_{i=1}^{n} 2[Y_i - (\alpha x_i + \beta)](-x_i) = -2 \sum_{i=1}^{n} [x_i Y_i - \alpha x_i^2 - \beta x_i],$$

$$\frac{\partial S}{\partial \beta} = \sum_{i=1}^{n} 2[Y_i - (\alpha x_i + \beta)](-1) = -2 \sum_{i=1}^{n} [Y_i - \alpha x_i - \beta].$$

Thus $\partial S/\partial \alpha = 0$ and $\partial S/\partial \beta = 0$ may be written, respectively, as follows:

$$\alpha \sum_{i=1}^{n} x_i^2 + \beta \sum_{i=1}^{n} x_i = \sum_{i=1}^{n} x_i Y_i, \tag{14.5}$$

$$\alpha \sum_{i=1}^{n} x_i + n\beta = \sum_{i=1}^{n} Y_i. \tag{14.6}$$

Thus we have two *linear* equations in the unknowns α and β. The solution may be obtained in the usual way, either by direct elimination or by using determinants. Denoting the solutions by $\hat{\alpha}$ and $\hat{\beta}$, we easily find that

$$\hat{\alpha} = \frac{\sum_{i=1}^{n} Y_i(x_i - \bar{x})}{\sum_{i=1}^{n} (x_i - \bar{x})^2}, \qquad \text{where} \qquad \bar{x} = \frac{1}{n} \sum_{i=1}^{n} x_i, \tag{14.7}$$

$$\hat{\beta} = \bar{Y} - \hat{\alpha}\bar{x}, \qquad \text{where} \qquad \bar{Y} = \frac{1}{n} \sum_{i=1}^{n} Y_i. \tag{14.8}$$

The above solutions are always obtainable and unique provided that

$$\sum_{i=1}^{n} (x_i - \bar{x})^2 \neq 0.$$

However, this condition is satisfied whenever not *all* the x_i's are equal.

The estimate of the parameter σ^2 cannot be obtained by the above methods. Let us simply state that the usual estimate of σ^2, in terms of the least-squares estimates $\hat{\alpha}$ and $\hat{\beta}$, is

$$\hat{\sigma}^2 = \frac{1}{n-2} \sum_{i=1}^{n} [Y_i - (\hat{\alpha}x_i + \hat{\beta})]^2.$$

Notes: (a) $\hat{\alpha}$ is obviously a *linear* function of the sample values $Y_1, \ldots, Y_n$.

(b) $\hat{\beta}$ is also a linear function of $Y_1, \ldots, Y_n$, as the following computation indicates:

$$\hat{\beta} = \bar{Y} - \hat{\alpha}\bar{x}$$

$$= \frac{1}{n} \sum_{i=1}^{n} Y_i - \frac{\bar{x}}{\sum_{i=1}^{n}(x_i - \bar{x})^2} \sum_{i=1}^{n} (x_i - \bar{x}) Y_i$$

$$= \sum_{i=1}^{n} Y_i \left[\frac{1}{n} - \bar{x} \frac{(x_i - \bar{x})}{\sum_{i=1}^{n}(x_i - \bar{x})^2} \right].$$

(c) It is a simple exercise to show that $E(\hat{\alpha}) = \alpha$ and $E(\hat{\beta}) = \beta$. (See Problem 14.34.) That is, both $\hat{\alpha}$ and $\hat{\beta}$ are unbiased estimates.

(d) The variances of $\hat{\alpha}$ and $\hat{\beta}$ are also easily calculated. (See Problem 14.35.) We have

$$V(\hat{\alpha}) = \frac{\sigma^2}{\sum_{i=1}^{n}(x_i - \bar{x})^2}; \qquad V(\hat{\beta}) = \left[\frac{1}{n} + \frac{\bar{x}^2}{\sum_{i=1}^{n}(x_i - \bar{x})^2} \right] \sigma^2. \tag{14.9}$$

(e) The estimates $\hat{\alpha}$ and $\hat{\beta}$ are in fact the best linear unbiased estimates of α and β. That is, among all unbiased, linear estimates, the above have minimum variance. This is a special case of the general *Gauss-Markoff theorem*, which states that under certain conditions least-squares estimates and best linear unbiased estimates are always the same.

(f) The method of least squares may be applied to nonlinear models. For example, if $E(Y) = \alpha X^2 + \beta X + \gamma$, we can estimate α, β, and γ so that

$$\sum_{i=1}^{n} [Y_i - (\alpha x_i^2 + \alpha x_i + \gamma)]^2$$

is minimized.

(g) If we made the additional assumption that the random variable ϵ has distribution $N(0, \sigma^2)$, we can apply the method of maximum likelihood to estimate the parameters α and β. These estimates are the same as the least-squares estimates obtained above. (This is not always true. It is a consequence of the assumption of normality.)

EXAMPLE 14.16. This example is discussed by Y. V. Linnik in *Method of Least Squares and Principles of the Theory of Observations*, Pergamon Press, New York, 1961. The data in this example were obtained by Mendeléjev and reported in *Foundations of Chemistry*. (See Table 14.5.) They concern the solubility of sodium nitrate ($NaNO_3$) in relation to water temperature (°C). At the indicated temperature, the Y parts of $NaNO_3$ dissolve in 100 parts of water. Plotting these data yields the scatter diagram shown in Fig. 14.6.

TABLE 14.5

T	Y	T	Y
0	66.7	29	92.9
4	71.0	36	99.4
10	76.3	51	113.6
15	80.6	68	125.1
21	85.7		

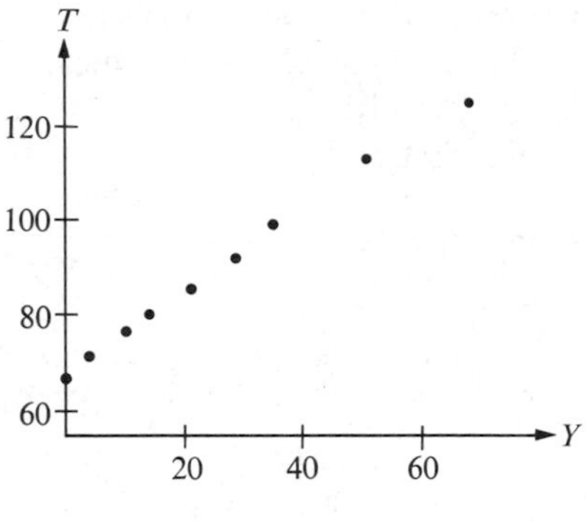

FIGURE 14.6

This suggests a model of the form $E(Y) = bT + a$. Using the method of least squares outlined above, we find that $\hat{b} = 0.87$ and $\hat{a} = 67.5$.

14.6 The Correlation Coefficient

In the previous section we were concerned with pairs of values (X, Y), but, as we repeatedly pointed out, X was not to be considered as a random variable. There are, however, two-dimensional random variables (X, Y) giving rise to a random sample $(X_1, Y_1), \ldots, (X_n, Y_n)$. One of the important parameters associated with a two-dimensional random variable is the correlation coefficient ρ_{xy}.

TABLE 14.6

X (velocity, km/sec)	11.93	11.81	11.48	10.49	10.13	8.87
Y (height, km)	62.56	57.78	53.10	48.61	44.38	40.57

The estimate customarily used for ρ is the *sample correlation coefficient*, defined as follows:

$$r = \frac{\sum_{i=1}^{n} (X_i - \overline{X})(Y_i - \overline{Y})}{\sqrt{\sum_{i=1}^{n} (X_i - \overline{X})^2 \sum_{i=1}^{n} (Y_i - \overline{Y})^2}}.$$

Note that for computational purposes, it is easier to evaluate r as follows:

$$r = \frac{n\sum_{i=1}^{n} X_i Y_i - \sum_{i=1}^{n} X_i \sum_{i=1}^{n} Y_i}{\sqrt{n\sum_{i=1}^{n} X_i^2 - (\sum_{i=1}^{n} X_i)^2} \sqrt{n\sum_{i=1}^{n} Y_i^2 - (\sum_{i=1}^{n} Y_i)^2}}.$$

EXAMPLE 14.17. The data listed in Table 14.6 represent velocity (km/sec) and height (km) of meteor No. 1242 as reported in the "Smithsonian Contributions to Astrophysics" from the *Proceedings of the Symposium on Astronomy and Physics of Meteors.* Cambridge, Mass., Aug. 28–Sept. 1, 1961. A straightforward computation yields $r = 0.94$.

14.7 Confidence Intervals

Thus far we have only been concerned with obtaining a point estimate for an unknown parameter. As we suggested at the beginning of this chapter, there is another approach which often leads to very meaningful results.

Suppose that X has distribution $N(\mu, \sigma^2)$, where σ^2 is assumed to be known, while μ is the unknown parameter. Let $X_1, \ldots, X_n$ be a random sample from X and let $\overline{X}$ be the sample mean.

We know that $\overline{X}$ has distribution $N(\mu, \sigma^2/n)$. Hence $Z = [(\overline{X} - \mu)/\sigma]\sqrt{n}$ has distribution $N(0, 1)$. Note that although Z depends on μ, its probability distribution does not. We can use this fact to our advantage as follows.

Consider

$$2\Phi(z) - 1 = P\left(-z \le \frac{\overline{X} - \mu}{\sigma}\sqrt{n} \le z\right)$$

$$= P\left(-\frac{z\sigma}{\sqrt{n}} - \overline{X} \le -\mu \le +\frac{z\sigma}{\sqrt{n}} - \overline{X}\right)$$

$$= P\left(\overline{X} - \frac{z\sigma}{\sqrt{n}} \le \mu \le \overline{X} + \frac{z\sigma}{\sqrt{n}}\right).$$

This last probability statement must be interpreted *very carefully*. It does *not* mean that the probability of the parameter μ falling into the specified interval

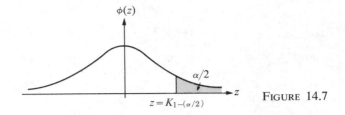

FIGURE 14.7

equals $2\Phi(z) - 1$; μ is a parameter and either is or is not in the above interval. Rather, the above should be interpreted as follows: $2\Phi(z) - 1$ equals the probability that the random interval $(\overline{X} - z\sigma/\sqrt{n}, \overline{X} + z\sigma/\sqrt{n})$ contains μ. Such an interval is called a *confidence interval* for the parameter μ. Since z is at our disposal, we may choose it so that the above probability equals, say $1 - \alpha$. Hence z is defined by the relationship $\Phi(z) = 1 - \alpha/2$. That value of z, denoted by $K_{1-\alpha/2}$, can be obtained from the tables of the normal distribution. (See also Fig. 14.7.) That is, we have $\Phi(K_{1-\alpha/2}) = 1 - \alpha/2$.

To summarize: The interval $(\overline{X} - n^{-1/2}\sigma K_{1-\alpha/2}, \overline{X} + n^{-1/2}\sigma K_{1-\alpha/2})$ is a confidence interval for the parameter μ with *confidence coefficient* $(1 - \alpha)$, or a $(1 - \alpha)$ 100-percent confidence interval.

Suppose that X represents the life length of a piece of equipment. Assume that 100 pieces were tested, yielding an average life length of $\overline{X} = 501.2$ hours. Assume that σ is known to be 4 hours and that we wish to obtain a 95-percent confidence interval for μ. Hence we find the following confidence interval for $\mu = E(X)$:

$$501.2 - \tfrac{4}{10}(1.96), \; 501.2 + \tfrac{4}{10}(1.96), \quad \text{which becomes} \quad (500.4, 502.0).$$

Again a comment is in order. In stating that $(500.4, 502.0)$ is a 95-percent confidence interval for μ, we are not saying that 95 percent of the time the sample mean will lie in *that* interval. The next time we take a random sample, $\overline{X}$ will presumably be different, and hence the endpoints of the confidence interval will be different. We are saying that 95 percent of the time μ will be contained in the interval $(\overline{X} - 1.96\sigma/\sqrt{n}, \overline{X} + 1.96\sigma/\sqrt{n})$. When we make the statement that $500.4 < \mu < 502.0$, we are simply adopting the point of view of believing that something is so which we know to be true most of the time.

Note: The confidence interval constructed is not unique. Just as there may be many (point) estimates for a parameter, so we may be able to construct many confidence intervals. Although we shall not discuss the problem of what we might mean by a "best" confidence interval, let us nevertheless state one obvious fact. If two confidence intervals having the same confidence coefficient are being compared, we would prefer that one which has the smallest expected length.

For the confidence interval considered above, the length L may be written as

$$L = (\overline{X} + n^{-1/2}\sigma K_{1-\alpha/2}) - (\overline{X} - n^{-1/2}\sigma K_{1-\alpha/2}) = 2\sigma n^{-1/2}K_{1-\alpha/2}.$$

Thus L is a constant. Furthermore, solving the preceding equation for n yields

$$n = (2\sigma K_{1-\alpha/2}/L)^2.$$

Hence we can determine n (for given α and σ) so that the confidence interval has a pre-scribed length. In general (as is illustrated in the above example), L will be a decreasing function of n: the smaller we wish to have L the larger we must take n. In the above case we must essentially quadruple n in order to halve L.

14.8 Student's *t*-distribution

The analysis of the above example was very much dependent on the fact that the variance σ^2 was known. How must our procedure be modified if we do not know the value of σ^2?

Suppose that we estimate σ^2 using the unbiased estimate

$$\hat{\sigma}^2 = \frac{1}{n-1} \sum_{i=1}^{n} (X_i - \overline{X})^2.$$

We consider the random variable

$$t = \frac{(\overline{X} - \mu)\sqrt{n}}{\hat{\sigma}}. \tag{14.10}$$

It should be intuitively clear that the probability distribution of the random variable t is considerably more complicated than that of $Z = (\overline{X} - \mu)\sqrt{n}/\sigma$. For in the definition of t *both* the numerator and the denominator are random variables, whereas Z is simply a linear function of $X_1, \ldots, X_n$. To obtain the probability distribution of t we make use of the following facts:

(a) $Z = (\overline{X} - \mu)\sqrt{n}/\sigma$ has distribution $N(0, 1)$.

(b) $V = \sum_{i=1}^{n} (X_i - \overline{X})^2/\sigma^2$ has a chi-square distribution with $(n-1)$ degrees of freedom. (See Theorem 13.4.)

(c) Z and V are independent random variables. (This is not very easy to show, and we shall not verify it here.)

With the aid of the following theorem we can now obtain the pdf of t.

Theorem 14.3. Suppose that the random variables Z and V are independent and have distributions $N(0, 1)$ and χ_k^2, respectively. Define

$$t = \frac{Z}{\sqrt{(V/k)}}.$$

Then the pdf of t is given by

$$h_k(t) = \frac{\Gamma[(k+1)/2]}{\Gamma(k/2)\sqrt{\pi k}} \left(1 + \frac{t^2}{k}\right)^{-(k+1)/2}, \qquad -\infty < t < \infty. \tag{14.11}$$

This distribution is known as *Student's t-distribution* with k degrees of freedom.

Notes: (a) The proof of this theorem is not given here but is suggested in the problem section. (See Problem 14.17.) We have the tools available with which we can find $h_k(t)$ quite easily. We first need to determine the pdf of $\sqrt{V/k}$, which is easily obtained from the known pdf of V. Then we need only apply Theorem 6.5 giving the pdf of the quotient of two independent random variables.

(b) The above theorem may be applied directly to obtain the pdf of $t = (\overline{X} - \mu)\sqrt{n}/\hat{\sigma}$, the random variable considered above. This random variable has Student's t-distribution with $(n - 1)$ degrees of freedom. Observe that although the value of t depends on μ, its distribution does not.

(c) The graph of h_k is symmetric, as indicated in Fig. 14.8. In fact, it resembles the graph of the normal distribution, and the reader can show that

$$\lim_{k \to \infty} h_k(t) = (1/\sqrt{2\pi})e^{-t^2/2}.$$

(d) Because of the importance of this distribution, it has been tabulated. (See Appendix.) For given α, $0.5 < \alpha < 1$, the values of $t_{k,\alpha}$ satisfying the condition

$$\int_{-\infty}^{t_{k,\alpha}} h_k(t)\, dt = \alpha$$

are tabulated. (See Fig. 14.9.) (For values of α satisfying $0 < \alpha < 0.5$, we can use the tabulated values because of the symmetry of the distribution.)

(e) The distribution is named after the English statistician W. S. Gosset, who published his research under the pseudonym "Student."

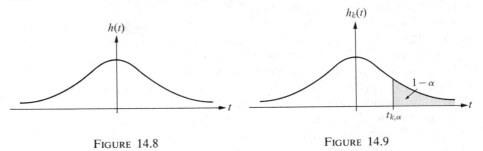

FIGURE 14.8   FIGURE 14.9

We now return to the problem discussed at the beginning of this section. How do we obtain a confidence interval for the mean of a normally distributed random variable if the variance is *unknown*?

In a manner completely analogous to the one used in Section 14.7, we obtain the following confidence interval for μ, with confidence coefficient $(1 - \alpha)$:

$$(\overline{X} - n^{-1/2}\hat{\sigma}t_{n-1,1-\alpha/2},\ \overline{X} + n^{-1/2}\hat{\sigma}t_{n-1,1-\alpha/2}).$$

Thus the above confidence interval has the same structure as the previous one, with the important difference that the known value of σ has been replaced by the estimate $\hat{\sigma}$ and the constant $K_{1-\alpha/2}$, previously obtained from the tables of the normal distribution, has been replaced by $t_{n-1,1-\alpha/2}$, obtained from the tables of the t-distribution.

Note: The length L of the above confidence interval is equal to

$$L = 2n^{-1/2}t_{n-1,1-\alpha/2}\hat{\sigma}.$$

Thus L is not a constant since it depends on $\hat{\sigma}$, which in turn depends on the sample values $(X_1, \ldots, X_n)$.

EXAMPLE 14.18. Ten measurements were made on the resistance of a certain type of wire, yielding the values $X_1, \ldots, X_{10}$. Suppose that $\overline{X} = 10.48$ ohms and $\hat{\sigma} = \sqrt{\frac{1}{9}\sum_{i=1}^{10}(X_i - \overline{X})^2} = 1.36$ ohms. Let us suppose that X has distribution $N(\mu, \sigma^2)$ and that we want to obtain a confidence interval for μ with confidence coefficient 0.90. Hence $\alpha = 0.10$. From the tables of the t-distribution we find that $t_{9,0.95} = 1.83$. Hence the desired confidence interval is

$$\left(10.48 - \frac{1}{\sqrt{10}}(1.36)(1.83),\ 10.48 + \frac{1}{\sqrt{10}}(1.36)(1.83)\right) = (9.69, 11.27).$$

14.9 More on Confidence Intervals

Although we are not attempting to give a general discussion of this topic, we want to continue to consider some important examples.

Sometimes we want to obtain a confidence interval for a particular *function* of an unknown parameter, knowing a confidence interval for the parameter itself. If the function is monotone, this may be accomplished as the following example illustrates.

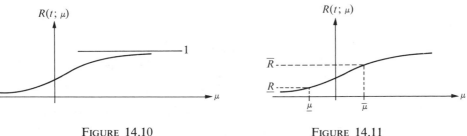

FIGURE 14.10   FIGURE 14.11

EXAMPLE 14.19. Suppose that the life length X of an item has distribution $N(\mu, \sigma^2)$. Assume that σ^2 is known. The reliability of the item for a service time of t hours is given by

$$R(t; \mu) = P(X > t) = 1 - \Phi\left(\frac{t - \mu}{\sigma}\right).$$

Since $\partial R(t; \mu)/\partial \mu > 0$ for all μ, we have that for every fixed t, $R(t; \mu)$ is an increasing function of μ. (See Fig. 14.10.) Thus we may proceed as follows to obtain a confidence interval for $R(t; \mu)$. Let $(\underline{\mu}, \bar{\mu})$ be the confidence interval for μ, obtained in Section 14.7. Let $\underline{R}$ and $\overline{R}$ be the lower and upper endpoints of the desired confidence interval for $R(t; \mu)$, respectively. If we define $\underline{R}$ and $\overline{R}$ by the relationships

$$\underline{R} = 1 - \Phi\left(\frac{t - \mu}{\sigma}\right) \quad \text{and} \quad \overline{R} = 1 - \Phi\left(\frac{t - \bar{\mu}}{\sigma}\right),$$

we find that $P(\underline{R} \le R \le \overline{R}) = P(\mu \le \mu \le \bar{\mu}) = 1 - \alpha$, and hence $(\underline{R}, \overline{R})$ does represent a confidence interval for $R(t; \mu)$ with confidence coefficient $(1 - \alpha)$. (See Fig. 14.11.)

Let us use the sample values obtained in Section 14.7 to illustrate this procedure. Suppose that we want a confidence interval for the reliability of the component when used for $t = 500$ hours. Since we found $\mu = 500.4$ and $\bar{\mu} = 502.0$, we obtain

$$\underline{R} = 1 - \Phi\left(\frac{500 - 500.4}{4}\right) = 0.6554, \quad \overline{R} = 1 - \Phi\left(\frac{500 - 502}{4}\right) = 0.6915.$$

So far we have only been considering *two-sided* confidence intervals. That is, we have obtained two statistics (sometimes called upper and lower confidence bounds), say $L(X_1, \ldots, X_n)$ and $U(X_1, \ldots, X_n)$, such that $P[L \le \theta \le U] = 1 - \alpha$, where θ is the unknown parameter.

Often we are only interested in obtaining *one-sided* confidence intervals of the following forms:

$$P[\theta \le U] = 1 - \alpha \quad \text{or} \quad P[L \le \theta] = 1 - \alpha.$$

Let us illustrate the above with examples.

EXAMPLE 14.20. Suppose that X has distribution $N(\mu, \sigma^2)$ and we want to obtain a one-sided confidence interval for the unknown parameter σ^2. Let $X_1, \ldots, X_n$ be a random sample from X.

From Theorem 13.4 we know that $\sum_{i=1}^{n} (X_i - \overline{X})^2/\sigma^2$ has distribution χ^2_{n-1}. Hence from the tables of the chi-square distribution we can obtain a number $\chi^2_{n-1, 1-\alpha}$ such that

$$P\left[\sum_{i=1}^{n} \frac{(X_i - \overline{X})^2}{\sigma^2} \le \chi^2_{n-1, 1-\alpha}\right] = 1 - \alpha.$$

(See Fig. 14.12.) The above probability may be written as follows:

$$P\left[\sigma^2 \ge \frac{\sum_{i=1}^{n} (X_i - \overline{X})^2}{\chi^2_{n-1, 1-\alpha}}\right] = 1 - \alpha.$$

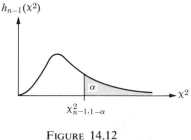

$h_{n-1}(\chi^2)$

$\chi^2_{n-1, 1-\alpha}$

FIGURE 14.12

Hence $\left(\sum_{i=1}^{n} (X_i - \overline{X})^2/\chi^2_{n-1, 1-\alpha}, \infty\right)$ is the required one-sided confidence interval for σ^2 with confidence coefficient $(1 - \alpha)$.

EXAMPLE 14.21. Suppose that X the life length of an electronic device is exponentially distributed with parameter $1/\beta$. Hence $E(X) = \beta$. Let $X_1, \ldots, X_n$ be a sample of X. We have found in Example 14.7 that $\sum_{i=1}^{n} X_i/n$ is the ML estimate of β. From the corollary of Theorem 10.9 we find that $2n\overline{X}/\beta$ has distribution χ_{2n}^2. Hence $P[2n\overline{X}/\beta \geq \chi_{2n,1-\alpha}^2] = \alpha$, where the number $\chi_{2n,1-\alpha}^2$ is obtained from the table of the chi-square distribution.

If we want a (lower) confidence interval for the reliability $R(t; \beta) = P(X > t) = e^{-t/\beta}$, we proceed as follows. Multiply the above inequality by $(-t)$ and rearrange the terms, obtaining

$$P[(-t/\beta) \geq -t\chi_{2n,1-\alpha}^2/\overline{X}2n] = 1 - \alpha.$$

This in turn implies that, since e^x is an increasing function of x,

$$P\left\{ R(t; \beta) = e^{-t/\beta} \geq \exp\left[-\frac{t\chi_{2n,1-\alpha}^2}{\overline{X}2n} \right] \right\} = 1 - \alpha.$$

Hence $(\exp[-t\chi_{2n,1-\alpha}^2/\overline{X}2n], \infty)$ is a one-sided confidence interval for $R(t; \beta)$ with confidence coefficient $(1 - \alpha)$.

As a final illustration of a confidence interval, let us find a confidence interval for the parameter p associated with a *binomially* distributed random variable X. We shall only consider the case where n, the number of repetitions of the experiment giving rise to X, is large enough so that we may use the *normal approximation*.

Let $X/n = h$ represent the relative frequency of an event A in n repetitions of an experiment for which $P(A) = p$. Hence $E(h) = p$ and $V(h) = pq/n$, where $q = 1 - p$.

Using the normal approximation to the binomial distribution, we may write

$$P\left[\frac{|h - p|}{\sqrt{pq/n}} \leq K \right] = P[|h - p| \leq K\sqrt{pq/n}] \simeq \frac{1}{\sqrt{2\pi}} \int_{-K}^{K} e^{-t^2/2} \, dt$$

$$= 2\Phi(K) - 1,$$

where, as always, $\Phi(K) = (1/\sqrt{2\pi})\int_{-\infty}^{K} e^{-t^2/2} \, dt$. Thus, if we set the above probability equal to $(1 - \alpha)$, we may obtain the value of K from the table of the normal distribution. That is, $2\Phi(K) - 1 = 1 - \alpha$ implies $K = K_{1-\alpha/2}$. Since we are interested in obtaining a confidence interval for p, we must rewrite the above inequality $\{|h - p| \leq K\sqrt{pq/n}\}$ as an inequality in p. Now $\{|h - p| \leq K\sqrt{pq/n}\}$ is equivalent to $\{(h - p)^2 \leq K^2(1 - p)p/n\}$. If we consider an (h, p) coordinate system, the above inequality represents the boundary and interior of an ellipse. The shape of the ellipse is determined by K and n: the larger n, the narrower the ellipse.

Consider a point $Q(h, p)$ in the hp-plane. (See Fig. 14.13.) Q will be a "random" point in the sense that its first coordinate h will be determined by the outcome of the experiment. Since Q will lie inside the ellipse if and only if $\{|h - p| \leq K\sqrt{pq/n}\}$, the probability of this occurring will be $2\Phi(K) - 1$. If we want to

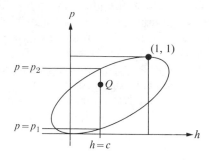

FIGURE 14.13

have this probability equal to $(1 - \alpha)$, we must choose K appropriately, that is $K = K_{1-\alpha/2}$.

Now p is unknown. (That, of course, is our problem.) The line $h = c$ (constant) will cut the ellipse in two places, say $p = p_1$ and $p = p_2$. (It can easily be checked that for given α and h, two distinct values of p will always exist.)

The values p_1 and p_2 may be obtained as the solution of the quadratic equation (in p): $(h - p)^2 = K^2(1 - p)p/n$. The solutions are

$$p_1 = \frac{hn + (K^2/2) - K[h(1 - h)n + (K^2/4)]^{1/2}}{n + K^2},$$

$$p_2 = \frac{hn + (K^2/2) + K[h(1 - h)n + (K^2/4)]^{1/2}}{n + K^2}. \qquad (14.12)$$

Hence $\{|h - p| \leq K\sqrt{pq/n}\}$ is equivalent to $\{p_1 \leq p \leq p_2\}$. Thus if K is chosen so that the former event has probability $(1 - \alpha)$, we have obtained a confidence interval for p, namely (p_1, p_2), with confidence coefficient $(1 - \alpha)$.

To summarize: In order to obtain a confidence interval for $p = P(A)$ with confidence coefficient $(1 - \alpha)$, perform the experiment n times and compute the relative frequency of the event A, say h. Then compute p_1 and p_2 according to Eqs. (14.12), where K is determined from the tables of the normal distribution. Recall that this procedure is valid only if n is sufficiently large to justify the normal approximation.

Note: Recall that in Eq. (14.12), $h = X/n$, where $X =$ number of occurrences of the event A. Thus $X = nh$ and $n - X = n(1 - h)$. If in addition to n, both X and $n - X$ are large, p_1 and p_2 can be approximated, respectively, by

$$p_1 \simeq h - \frac{K}{\sqrt{n}}\sqrt{h(1 - h)}, \qquad p_2 \simeq h + \frac{K}{\sqrt{n}}\sqrt{h(1 - h)}.$$

EXAMPLE 14.22. In a certain production process, 79 items were manufactured during a particular week. Of these, 3 were found to be defective. Thus $h = \frac{3}{79} =$

0.038. Using the above procedure, we obtain $(0.013, 0.106)$ as a confidence interval for $p = P$ (item is defective) with confidence coefficient 0.95.

EXAMPLE 14.23.   A factory has a large stockpile of items, some of which come from a production method now considered inferior while others come from a modern process. Three thousand items are chosen at random from the stockpile. Of these, 1578 are found to have originated from the inferior process. Using the above approximation for p_1 and p_2, the following computation yields a 99 percent confidence interval for $p = $ proportion of items from the inferior process:

$$h = \frac{1578}{3000} = 0.526, \qquad K = 2.576,$$

$$p_1 = 0.526 - \frac{2.576}{\sqrt{3000}} \sqrt{(0.526)(0.474)} = 0.502,$$

$$p_2 = 0.526 + \frac{2.576}{\sqrt{3000}} \sqrt{(0.526)(0.474)} = 0.550.$$

PROBLEMS

14.1.   Suppose that an object is measured independently with two different measuring devices. Let L_1 and L_2 be the measured lengths obtained from the first and second device, respectively.   If both devices are calibrated correctly we may assume that $E(L_1) = E(L_2) = L$, the true length. However, the accuracy of the devices is not necessarily the same. If we measure accuracy in terms of variance, then $V(L_1) \neq V(L_2)$. If we use the linear combination $Z = aL_1 + (1 - a)L_2$ for our estimate of L, we have immediately that $E(Z) = L$. That is, Z is an unbiased estimate of L. For what choice of $a, 0 < a < 1$, is the variance of Z a minimum?

14.2.   Let X be a random variable with expectation μ and variance σ^2. Let $(X_1, \ldots, X_n)$ be a sample of X. There are many other estimates of σ^2 which suggest themselves in addition to the one proposed earlier.   Show that $C\sum_{i=1}^{n-1} (X_{i+1} - X_i)^2$ is an unbiased estimate of σ^2 for an appropriate choice of C.   Find that value of C.

14.3.   Suppose that 200 independent observations $X_1, \ldots, X_{200}$ are obtained from a random variable X.   We are told that $\sum_{i=1}^{200} X_i = 300$ and that $\sum_{i=1}^{200} X_i^2 = 3754$. Using these values, obtain an unbiased estimate for $E(X)$ and $V(X)$.

14.4.   A random variable X has pdf $f(x) = (\beta + 1)x^\beta, 0 < x < 1$.

(a) Obtain the ML estimate of β, based on a sample $X_1, \ldots, X_n$.

(b) Evaluate the estimate if the sample values are 0.3, 0.8, 0.27, 0.35, 0.62, and 0.55.

14.5.   The data in Table 14.4 were obtained on the distribution of depth of sapwood in telephone poles.   (W. A. Shewhart, Economic Control of Quality of Manufactured Products.   Macmillan and Co., New York, 1932, p. 66.)   Assuming that the random variable under consideration has distribution $N(\mu, \sigma^2)$, obtain the ML estimates of μ and σ^2.

TABLE 14.4

Depth of Sapwood (in.)	Frequency	Depth of Sapwood (in.)	Frequency
1.0	2	3.7	123
1.3	29	4.0	82
1.6	62	4.3	48
1.9	106	4.6	27
2.2	153	4.9	14
2.5	186	5.2	5
2.8	193	5.5	1
3.1	188		
3.4	151	Total frequencies: 1370	

14.6. Suppose that T, the time to failure (in hours) of an electronic device, has the following pdf:

$$f(t) = \beta e^{-\beta(t-t_0)}, \qquad t > t_0 > 0,$$
$$= 0, \qquad \text{elsewhere.}$$

(T has an exponential distribution truncated to the left at t_0.) Suppose that n items are tested and the failure times $T_1, \ldots, T_n$ are recorded.

(a) Assuming that t_0 is known, obtain the ML estimate of β.

(b) Assuming that t_0 is unknown but β is known, obtain the ML estimate of t_0.

14.7. Consider the same failure law as described in Problem 14.6. This time N items are tested for T_0 hours ($T_0 > t_0$), and the number of items that fail in that period is recorded, say k. Answer question (a) of Problem 14.6.

14.8. Suppose that X is uniformly distributed over $(-\alpha, \alpha)$. Find the ML estimate of α based on a random sample of size n, $X_1, \ldots, X_n$.

14.9. (a) A procedure is performed until a particular event A occurs for the first time. On each repetition $P(A) = p$. Suppose that n_1 repetitions are required. Then this experiment is repeated, and this time n_2 repetitions are required to yield the event A. If this is done k times, we obtain the sample $n_1, \ldots, n_k$. Based on this sample, obtain the ML estimate of p.

(b) Suppose that k is quite large. Find the approximate value of $E(\hat{p})$ and $V(\hat{p})$, where $\hat{p}$ is the ML estimate obtained in (a).

14.10. A component assumed to have an exponential failure distribution was tested. The following life lengths (in hours) were observed: 108, 212, 174, 130, 198, 169, 252, 168, 143. Using these sample values, obtain the ML estimate for the reliability of the component when used for a period of 150 hours.

14.11. The following data represent the life length of electric lamps (in hours):

1009, 1085, 1123, 1181, 1235, 1249, 1263, 1292, 1327, 1338, 1348,
1352, 1359, 1368, 1379, 1397, 1406, 1425, 1437, 1438, 1441, 1458,
1483, 1488, 1499, 1505, 1509, 1519, 1541, 1543, 1548, 1549, 1610,
1620, 1625, 1638, 1639, 1658, 1673, 1682, 1720, 1729, 1737, 1752,
1757, 1783, 1796, 1809, 1828, 1834, 1871, 1881, 1936, 1949, 2007.

From the above sample values obtain the ML estimate for the reliability of such an electric lamp when used for a 1600-hour operation, assuming that the life length is normally distributed.

14.12. Suppose that two lamps as described in Problem 14.11 are used in (a) a series connection, and (b) a parallel connection. In each case find the ML estimate of the reliability for a 1600-hour operation of the system, based on the sample values given in Problem 14.11.

14.13. Let us assume that α-particles are emitted from a radioactive source according to a Poisson distribution. That is, if X is the number of particles emitted during an interval of t minutes, then $P(X = k) = e^{-\lambda t}(\lambda t)^k/k!$ Instead of recording the actual number of particles emitted, suppose that we note the number of times no particle was emitted. Specifically, suppose that 30 radioactive sources having the same strength are observed for a period of 50 seconds and that in 25 of the cases at least one particle was emitted. Obtain the ML estimate of λ based on this information.

14.14. A random variable X has distribution $N(\mu, 1)$. Twenty observations are taken on X but instead of recording the actual value we only note whether or not X was negative. Suppose that the event $\{X < 0\}$ occurred precisely 14 times. Using this information, obtain the ML estimate of μ.

14.15. Suppose that X has a Gamma distribution; that is, the pdf is given by

$$f(x) = \frac{\lambda(\lambda x)^{r-1}e^{-\lambda x}}{\Gamma(r)}, \qquad x > 0.$$

Suppose that r is known. Let $X_1, \ldots, X_n$ be a sample of X and obtain the ML estimate of λ based on this sample.

14.16. Suppose that X has a Weibull distribution with pdf

$$f(x) = (\lambda\alpha)x^{\alpha-1}e^{-\lambda x^{\alpha}}, \qquad x > 0.$$

Suppose that α is known. Find the ML estimate of λ based on a sample of size n.

14.17. Prove Theorem 14.3. [*Hint:* See Note (a) following this theorem.]

14.18. Compare the value of $P(X \geq 1)$, where X has distribution $N(0, 1)$, with $P(t \geq 1)$, where t has Student's t-distribution with:

(a) 5 df (b) 10 df (c) 15 df (d) 20 df (e) 25 df

14.19. Suppose that X has distribution $N(\mu, \sigma^2)$. A sample of size 30, say $X_1, \ldots, X_{30}$, yields the following values: $\sum_{i=1}^{30} X_i = 700.8$, $\sum_{i=1}^{30} X_i^2 = 16{,}395.8$. Obtain a 95-percent (two-sided) confidence interval for μ.

14.20. Suppose that X has distribution $N(\mu, 4)$. A sample of size 25 yields a sample mean $\bar{X} = 78.3$. Obtain a 99-percent (two-sided) confidence interval for μ.

14.21. Suppose that the life length of a component is normally distributed, say $N(\mu, 9)$. Twenty components are tested and their time to failure $X_1, \ldots, X_{20}$ is recorded. Suppose that $\bar{X} = 100.9$ hours. Obtain a 99-percent (two-sided) confidence interval for the reliability $R(100)$.

14.22. Obtain a one-sided (lower) 99-percent confidence interval for $R(100)$ of Problem 14.21.

14.23. Suppose that X has distribution $N(\mu, \sigma^2)$, where μ and σ^2 are unknown. A sample of size 15 has yielded the values $\sum_{i=1}^{15} X_i = 8.7$ and $\sum_{i=1}^{15} X_i^2 = 27.3$. Obtain a (two-sided) 95-percent confidence interval for σ^2.

14.24. One hundred components were tested, and 93 of these functioned more than 500 hours. Obtain a (two-sided) 95-percent confidence interval for $p = P$ (component functions more than 500 hours). [*Hint:* Use Eq. 14.12.]

14.25. Suppose that X, the length of a bolt, has distribution $N(\mu, 1)$. A large number of bolts are manufactured and subsequently separated into two large stockpiles. Stockpile 1 contains only those bolts for which $X > 5$, while stockpile 2 contains all the remaining bolts. A sample of size n is taken from stockpile 1 and the lengths of the chosen bolts are measured. Thus we obtain a sample $Y_1, \ldots, Y_n$ from the random variable Y which is a normally distributed random variable truncated to the left at 5. Write down the equation to be solved in order to obtain the ML estimate of μ based on the sample $(Y_1, \ldots, Y_n)$, in terms of the tabulated function ϕ and Φ, where $\phi(x) = (1/\sqrt{2\pi})e^{-x^2/2}$ and Φ is the cdf of the $N(0, 1)$ distribution.

14.26. (The F-distribution.) Let X and Y be independent random variables having $\chi_{n_1}^2$ and $\chi_{n_2}^2$ distributions, respectively. Let the random variable F be defined as follows: $F = (X/n_1)/(Y/n_2) = n_2 X/n_1 Y$. (This random variable plays an important role in many statistical applications.) Show that the pdf of F is given by the following expression:

$$h(f) = \frac{\Gamma[(n_1 + n_2)/2]}{\Gamma(n_1/2)\Gamma(n_2/2)} \left(\frac{n_1}{n_2}\right)^{n_1/2} f^{(n_1/2)-1}\left(1 + (n_1/n_2)f\right)^{-(1/2)(n_1+n_2)}, \qquad f > 0.$$

[This is called (Snedecor's) F-distribution with (n_1, n_2) degrees of freedom. Because of its importance, probabilities associated with the random variable F have been tabulated.] [*Hint:* To derive the above pdf, use Theorem 6.5.]

14.27. Sketch the graph of the pdf h as given in Problem 14.26, assuming that $n_1 > n_2 > 2$.

14.28. One reason for the importance of the F-distribution is the following. Suppose that X and Y are independent random variables with distributions $N(\mu_x, \sigma_x^2)$ and $N(\mu_y, \sigma_y^2)$, respectively. Let $X_1, \ldots, X_{n_1}$ and $Y_1, \ldots, Y_{n_2}$ be random samples from X and Y, respectively. Then the statistic $C\sum_{i=1}^{n_1} (X_i - \bar{X})^2/\sum_{i=1}^{n_2} (Y_i - \bar{Y})^2$ has an F-distribution for an appropriate choice of C. Show this and determine C. What are the degrees of freedom associated with this distribution?

14.29. Suppose that the random variable t has Student's t-distribution with 1 degree of freedom. What is the distribution of t^2? Identify it.

14.30. Suppose that X is normally distributed. A random sample of size 4 is obtained and $\bar{X}$ the sample mean is computed. If the sum of the squares of the deviations of these 4 measurements from $\bar{X}$ equals 48, obtain a 95 percent (two-sided) confidence interval for $E(X)$ in terms of $\bar{X}$.

14.31. The following sample of size 5 was obtained from the two-dimensional random variable (X, Y). Using these values, compute the sample correlation coefficient.

x	1	2	3	4	5
y	4	5	3	1	2

14.32. Suppose that $E(Y) = \alpha X + \beta$. A sample of size 50 is available, say (x_i, Y_i), $i = 1, \ldots, 50$ for which $\bar{x} = \bar{Y} = 0, \sum_{i=1}^{50} x_i^2 = 10, \sum_{i=1}^{50} Y_i^2 = 15$ and $\sum_{i=1}^{50} x_i Y_i = 8$.

(a) Determine the least squares estimates of the parameters α and β, say $\hat{\alpha}$ and $\hat{\beta}$.

(b) What is the value of the minimum sum of squares $\sum_{i=1}^{50} [Y_i - (\hat{\alpha} x_i + \hat{\beta})]^2$.

14.33. It might be (erroneously) supposed that an unbiased estimate may always be found for an unknown parameter. That this is not so is illustrated by the following example. Suppose that n repetitions of an experiment are made and a particular event A occurs precisely k times. If there is hypothesized a constant probability $p = P(A)$ that A occurs whenever the experiment is performed, we might be interested in estimating the ratio $r = p/(1 - p)$. To see that no unbiased estimate of $r = p/(1 - p)$ exists [based on the observation of kA's and $(n - k)\bar{A}$'s], suppose that in fact such an estimate does exist. That is, suppose that $\hat{r} = h(k)$ is a statistic for which $E(\hat{r}) = p/(1 - p)$. Specifically, suppose that $n = 2$ and hence $k = 0, 1,$ or 2. Denote the corresponding three values of $\hat{r}$ by a, b, and c. Show that $E(\hat{r}) = p/(1 - p)$ yields a contradiction by noting what happens to the left- and the right-hand side of this equation as $p \to 1$.

14.34. Verify that the least squares estimates $\hat{\alpha}$ and $\hat{\beta}$ as given by Eqs. (14.7) and (14.8) are unbiased.

14.35. Verify the expressions for $V(\hat{\alpha})$ and $V(\hat{\beta})$ as given by Eq. (14.9).

14.36. Suppose that $E(Y) = \alpha X^2 + \beta X + \gamma$, where X is preassigned. Based on a sample (x_i, Y_i), $i = 1, \ldots, n$, determine the least squares estimates of the parameters α, β, and γ.

14.37 With the aid of Table 7, obtain a sample of size 20 from a random variable having $N(2, 4)$ distribution.

(a) Suppose that this sample was obtained from a random variable having distribution $N(\alpha, 4)$. Use the sample values to obtain a 95 percent confidence interval for α.

(b) Same as (a) except that sample is assumed to come from a $N(\alpha, \beta^2)$ distribution with β^2 unknown.

(c) Compare the lengths of the confidence intervals in (a) and (b) and comment.

15

Testing Hypotheses

15.1 Introduction

In this chapter we shall discuss another way to deal with the problem of making a statement about an unknown parameter associated with a probability distribution, based on a random sample. Instead of finding an estimate for the parameter, we shall often find it convenient to hypothesize a value for it and then use the information from the sample to confirm or refute the hypothesized value. The concepts to be discussed in this chapter can be formulated on a very sound theoretical basis. However, we shall not pursue this topic from a formal point of view. Rather, we shall consider a number of procedures which are intuitively very appealing. We shall study some of the properties of the suggested procedures, but shall not attempt to indicate why some of the proposed methods should be preferred to alternative ones. The interested reader can obtain a more theoretical foundation of some of these procedures by pursuing some of the references suggested at the end of the chapter.

Consider the following example.

EXAMPLE 15.1. A manufacturer has been producing shearing pins to be used under certain stress conditions. It has been found that the life length (hours) of these pins is normally distributed, $N(100, 9)$. A new manufacturing scheme has been introduced whose purpose it is to extend the life length of these pins. That is, the hope is that the life length X (corresponding to pins produced by the new process) will have distribution $N(\mu, 9)$, where $\mu > 100$. (We assume that the variance remains the same. This means, essentially, that the variability of the new process is the same as that of the old process.)

Thus the manufacturer and the potential buyer of these pins are interested in testing the following hypothesis:

$$H_0: \mu = 100 \qquad \text{versus} \qquad H_1: \mu > 100.$$

(We are making the tacit assumption that the new process cannot be worse than the old one.) H_0 is called the *null hypothesis*, and H_1 the *alternative hypothesis*.

We are essentially facing a problem similar to the one discussed in Chapter 14. We are studying a random variable and do not know the value of a parameter associated with its distribution. We could resolve this problem as we did previously, by simply estimating μ. However, in many situations we are really interested in making a *specific decision*: Should we accept or reject the hypothesis H_0? Thus we will not retreat to the previous concepts of estimation, but will proceed to develop some concepts which are particularly suitable to resolve the specific problem at hand.

We begin by obtaining a sample of size n from the random variable X. That is, we choose, at random, n pins manufactured by the new process and record how long each of these pins functions, thus obtaining the sample $X_1, \ldots, X_n$. We then compute the arithmetic mean of the numbers, say $\overline{X}$. Since $\overline{X}$ is known to be a "good" estimate of μ, it seems reasonable that we should base our decision to accept or reject H_0 on the value of $\overline{X}$. Since we are concerned about discriminating between $\mu = 100$ and values of μ greater than 100, it seems reasonable that we should reject H_0 if $(\overline{X} - 100)$ is "too large."

Thus we are led (strictly on intuitive grounds) to the following procedure, usually called a *test* of the hypothesis: Reject H_0 if $\overline{X} - 100 > C'$ or, equivalently, if $\overline{X} > C$, (where C is a constant to be determined), and accept it otherwise.

Note that the particular form of the test we are using was in part suggested by the alternative hypothesis H_1. This is a point to which we shall refer again. If in the above situation we had been concerned with testing H_0: $\mu = 100$ versus H_1': $\mu \neq 100$, we would have used the test: Reject H_0 if $|\overline{X} - 100| > C'$. We are now in a position analogous to the one in which we found ourselves when we constructed an estimate $\hat{\theta}$ for the parameter θ. We asked: How "good" is the estimate? What desirable properties does it have? We can ask similar questions about the test we have constructed: How "good" is the test? What properties does it possess? How do we compare it with another test that might be proposed?

In order to answer such questions we must first of all realize that no solution exists, in the usual sense, for the problem we are posing. That is, by simply inspecting some of the pins being manufactured we can never be *sure* that $\mu = 100$. (Again note the analogy to the estimation problem: We do not expect our estimate $\hat{\theta}$ to be equal to θ. We simply hope that it will be "close" to θ.) The same is true here: A test will not always lead to the right decision, but a "good" test should lead to the correct decision "most of the time."

Let us be more precise. There are basically two types of errors we can make. We may reject H_0 when in fact H_0 is true; that is, when the quality of the pins has *not* been improved. This may occur because we happened to choose a few strong pins in our sample which are not typical of the entire production. Or, alternatively, we may accept H_0 when in fact H_0 is false; that is, when the quality of the pins has been improved. Formally we make the following definition.

Definition

> *Type 1 error:* Reject H_0 when H_0 is true.
>
> *Type 2 error:* Accept H_0 when H_0 is false.

It should be clear that we cannot completely avoid making these errors. We shall try to keep the probability of making these errors relatively small.

In order to come to grips with this problem we shall introduce the very important notion of the Operating Characteristic function of the test, say L, which is the following function of the (unknown) parameter μ.

Definition.   The *operating characteristic function* (OC function) of the above test is defined as

$$L(\mu) = P \text{ (accept } H_0 \mid \mu) = P(\overline{X} \le C \mid \mu).$$

That is, $L(\mu)$ is the probability of accepting H_0 considered as a function of μ.

Note: Another function, very closely related to the OC function, is the *power function* defined by

$$H(\mu) = P[\text{reject } H_0 \mid \mu].$$

Hence $H(\mu) = 1 - L(\mu)$.   We shall use the OC function to describe test properties although this could just as easily be done in terms of the power function.

In the specific case being considered, we can obtain the following explicit expression for L: If μ is the true value of $E(X)$, then $\overline{X}$ has distribution $N(\mu, 9/n)$. Hence,

$$L(\mu) = P(\overline{X} \le C \mid \mu) = P\left(\frac{\overline{X} - \mu}{3/\sqrt{n}} \le \frac{C - \mu}{3/\sqrt{n}}\right) = \Phi\left[\frac{(C - \mu)\sqrt{n}}{3}\right],$$

where, as always,

$$\Phi(s) = \frac{1}{\sqrt{2\pi}} \int_{-\infty}^{s} e^{-x^2/2}\, dx.$$

The following properties of $L(\mu)$ are easily verified:

(a) $L(-\infty) = 1$.
(b) $L(+\infty) = 0$.
(c) $dL/d\mu < 0$ for all μ. (Hence L is a strictly decreasing function of μ.)

Thus the graph of the function L has the general appearance of the curve in Fig. 15.1. (The specific shape will, of course, depend on the choice of the constant C and the sample size n.)

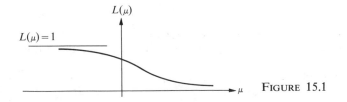

FIGURE 15.1

Consider $1 - L(100)$. This number represents the probability of rejecting H_0 when H_0 is true. That is, $1 - L(100)$ represents the probability of a type-1 error. If n and C are given, then $1 - L(100)$ is completely determined. For example, if we take $n = 50$ and $C = 101$, we obtain

$$1 - L(100) = 1 - \Phi\left[\frac{101 - 100}{3}\sqrt{50}\right]$$

$$= 1 - \Phi(2.37) = 0.009.$$

Thus this particular test would lead us to reject H_0 erroneously about 0.9 percent of the time.

Often we consider the problem from a slightly different point of view. Suppose that the sample size n is given and the probability of a type-1 error is *specified*. That is, $1 - L(100) = \alpha$ or, equivalently, $L(100) = 1 - \alpha$. What should be the value of C?

Specifically, if we take $n = 50$ and choose $\alpha = 0.05$, we obtain C as a solution of the following equation:

$$0.95 = \Phi\left(\frac{C - 100}{3}\sqrt{50}\right).$$

From the table of the normal distribution this yields

$$1.64 = \frac{C - 100}{3}\sqrt{50}.$$

Hence

$$C = 100 + \frac{3(1.64)}{\sqrt{50}} = 100.69.$$

Thus, if we reject the hypothesis whenever the sample mean is greater than 100.69, we are guaranteed a probability of 0.05 that a type-1 error will occur. Since n and C are now known, the OC function is completely specified. Its graph is shown in Fig. 15.2. The value 0.05 is called the *significance level* of the test (or sometimes the *size* of the test). In most problems this value is taken to be less than 0.1.

Note that by specifying α and the sample size n, only the constant C had to be determined in order to specify the test completely. We did this by insisting that the graph of the OC function pass through a specified point, namely $(100, 0.95)$.

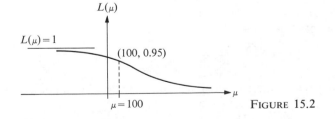

FIGURE 15.2

(It should be clear how the above procedure would be modified had we chosen some value other than 0.05 for the significance level.)

Now tnat the OC function is completely specified we may find the coordinates of any other point. For example, what is the value of $L(102)$?

$$L(102) = \Phi\left(\frac{100.69 - 102}{3}\sqrt{50}\right)$$

$$= \Phi(-3.1) = 0.00097.$$

Thus for the test under consideration, the probability of accepting H_0: $\mu = 100$ when in fact $\mu = 102$ equals 0.00097. Hence the probability of a type-2 error is very small indeed if $\mu = 102$. Since L is a decreasing function of μ, we note that $L(\mu) < 0.00097$ for all $\mu > 102$.

If we want to choose both n and C we must specify two points through which the graph of the OC function should pass. In this way we are able to control not only the probability of type-1 error but also the probability of type-2 error.

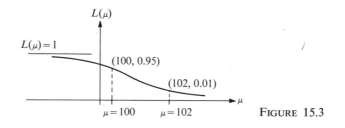

FIGURE 15.3

Suppose that in the example being considered we wish to avoid rejecting H_0 when $\mu \geq 102$. Thus we may set $L(102) = 0.01$, say, and since L is a decreasing function of μ, it follows that $L(\mu) \leq 0.01$ for $\mu > 102$. (See Fig. 15.3.) If we also require a significance level of 0.05, we obtain the following two equations for the determination of n and C:

$$L(100) = 0.95, \qquad L(102) = 0.01.$$

These equations become

$$0.95 = \Phi\left(\frac{C - 100}{3}\sqrt{n}\right), \qquad 0.01 = \Phi\left(\frac{C - 102}{3}\sqrt{n}\right).$$

From the tables of the normal distribution we find that these expressions are equivalent to

$$1.64 = \frac{C - 100}{3}\sqrt{n}, \qquad -2.33 = \frac{C - 102}{3}\sqrt{n}.$$

In order to eliminate n, we divide one equation by the other. Thus

$$(C - 102)(1.64) = (-2.33)(C - 100),$$

from which we obtain

$$C = \frac{(102)(1.64) - (100)(-2.33)}{1.64 - (-2.33)} = 100.8.$$

Once C is known, we may obtain n by squaring either of the above equations: Hence

$$n = \left[\frac{3(1.64)}{C - 100}\right]^2 = 34.6 \simeq 35.$$

15.2 General Formulation: Normal Distribution with Known Variance

We have considered in some detail an example dealing with a hypothesis involving the mean of a normally distributed random variable. While some of the calculations are still fresh in our minds, let us generalize this example as follows.

Suppose that X is a random variable with distribution $N(\mu, \sigma^2)$ where σ^2 is assumed to be known. To test $H_0: \mu = \mu_0$ versus $H_1: \mu > \mu_0$ we propose the following: We obtain a sample of size n, compute the sample mean $\bar{X}$, and reject H_0 if $\bar{X} > C$, where C is a constant to be determined. The OC function of this test is given by

$$L(\mu) = P(\bar{X} \le C) = \Phi\left(\frac{C - \mu}{\sigma}\sqrt{n}\right).$$

The general shape of the OC function is as indicated in Fig. 15.4.

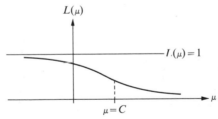

FIGURE 15.4

The general properties of $L(\mu)$ are easily established (see Problem 15.4):

(a) $L(-\infty) = 1$.
(b) $L(+\infty) = 0$.
(c) $L'(\mu) < 0$, and hence L is a strictly decreasing function of μ.   (15.1)
(d) $L''(\mu) = 0$ for $\mu = C$, and hence the graph has an inflection point here.
(e) Increasing n causes the curve to become steeper.

In order to proceed, we must consider two cases.

Case 1. If n is given and we specify the significance level of the test (that is, the probability of type-1 error) at some value α, we may obtain the value of C by solving the following equation:

$$1 - \alpha = \Phi\left(\frac{C - \mu_0}{\sigma}\sqrt{n}\right).$$

Defining K_α to be the relationship $1/\sqrt{2\pi}\int_{-\infty}^{K_\alpha} e^{-t^2/2}\,dt = \alpha$, we may write the above as

$$K_{1-\alpha} = \frac{C - \mu_0}{\sigma}\sqrt{n},$$

where $K_{1-\alpha}$ may be obtained from the table of the normal distribution. Hence we reject H_0 if

$$\overline{X} > \mu_0 + \frac{\sigma}{\sqrt{n}}K_{1-\alpha}.$$

Case 2. If n *and* C are to be determined, we must specify two points on the graph of the OC curve: $1 - L(\mu_0) = \alpha$, the significance level, and $L(\mu_1) = \beta$, the probability of type-2 error for $\mu = \mu_1$. Thus we must solve the following equations for n and C:

$$1 - \alpha = \Phi\left(\frac{C - \mu_0}{\sigma}\sqrt{n}\right); \qquad \beta = \Phi\left(\frac{C - \mu_1}{\sigma}\sqrt{n}\right).$$

These equations may be solved for C and n as indicated previously. We obtain

$$C = \frac{\mu_1 K_{1-\alpha} - \mu_0 K_\beta}{K_{1-\alpha} - K_\beta}, \qquad n = \left[\frac{\sigma K_{1-\alpha}}{C - \mu_0}\right]^2,$$

where $K_{1-\alpha}$ and K_β have been defined above.

In the procedure outlined, we have dealt with the alternative hypothesis $H_1: \mu > 100$ (or, in the general case, $\mu > \mu_0$). In some other context we might wish to consider $H_0: \mu = \mu_0$ versus $H_1': \mu < \mu_0$ or $H_0: \mu = \mu_0$ versus $H_1'': \mu \neq \mu_0$. It should be quite apparent how we modify the above test to such alternative hypothesis. If we consider $H_1': \mu < \mu_0$, we would reject if $\overline{X} < C$ and the OC function would be defined by

$$L(\mu) = P(\overline{X} \geq C) = 1 - \Phi\left(\frac{C - \mu}{\sigma}\sqrt{n}\right).$$

If we consider $H_1'': \mu \neq \mu_0$, we would reject H_0 whenever $|\overline{X} - \mu_0| > C$ and hence the OC function would be defined by

$$L(\mu) = P(|\overline{X} - \mu_0| \leq C)$$
$$= \Phi\left(\frac{C + \mu_0 - \mu}{\sigma}\sqrt{n}\right) - \Phi\left(\frac{-C + \mu_0 - \mu}{\sigma}\sqrt{n}\right).$$

If we choose the *same* significance level for each of the above tests, say α, and plot the graphs of the respective OC functions on the same coordinate system, we obtain the following: (A corresponds to H_1, B to H_1', and D to H_1'').

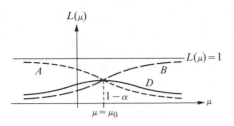

$$\text{FIGURE } 15.5$$

Figure 15.5 gives us a means of comparing the three tests being considered. All the tests have the same significance level. (It should be clearly understood that C is only a generic symbol for a constant and will not be the same in each case. The point is that in each case C has been chosen so that the test will have significance level α.)

If $\mu > \mu_0$, then test A is better than the other two since it will yield a smaller value for the probability of type-2 error. However, if $\mu < \mu_0$, test A is the worst of those being considered while test B is the best. Finally, test D is generally acceptable and yet in any specific case may be improved upon by either test A or test B. Thus we note that it is very important to have a specific alternative hypothesis in mind because the test we choose may depend on it. (We only compared tests having the same significance level; this is not absolutely necessary but the comparisons become somewhat vague if we use tests having different significance levels. Note the similarity to our comparison of certain estimates: we only compared the variances of those estimates which were unbiased.)

In many situations it is fairly clear which alternative hypothesis we should consider. In the above example, for instance, we presumably know that the new manufacturing process would either produce pins of the *same* or *increased* durability and hence we would use test A as suggested. (If the new process were to produce pins of inferior quality, our test would be very poor.) If no such assumption is warranted it would be best to use a test such as D: Reject H_0 if $|\overline{X} - \mu_0| > C$. Tests such as A and B are called *one-tailed* tests while a test such as D is called a *two-tailed* test.

In considering alternative hypotheses, we might find the following analogy useful. Suppose that a person is missing from location M, and we *know* that the individual has gone either to the left or to the right of M, staying on a straight path.

$$\text{M}$$

If 10 persons are available for a search party, how should these persons be dispersed? If nothing whatsoever is known concerning the whereabouts of the individual, it might be reasonable to send a group of 5 persons in each direction, thus dispatching a fairly effective but not too strong search group both to the left and to the right. However, if there is some indication that the person has strayed to the left, then possibly all or most of the available manpower should be sent to the left, making for a very effective search party on the left and for a very inef-fective one on the right. Other considerations might also influence the use of the available resources. For example, suppose that the path to the left leads to flat wooded country while the path to the right leads to the edge of a deep canyon. It is obvious that in this case most of the search would be concentrated on the right because the consequence of being lost on the right are much greater than those on the left.

The analogy should be clear. In testing hypotheses, we too must be concerned about the consequences of our decision to reject H_0 or to accept it. For example, is the error we make in accepting some pins which are actually inferior (believing they are not) as important as not accepting pins which are actually satisfactory (believing they are not)?

Note: In the above formulation we suggested that we often preassign the significance level of the test. This is frequently done. However, there is another approach to the problem which is very common and should be commented on.

Let us return to Example 15.1 where we tested $H_0: \mu = 100$ versus $H_1: \mu > 100$. Suppose that we simply obtain a sample of size 50, compute the sample mean, say $\overline{X}$, and find that $\overline{X} = 100.87$. Should we accept or reject H_0? We may argue as follows: *If* $\mu = 100$, then $\overline{X}$ has distribution $N(100, \frac{9}{50})$. Thus we may compute,

$$P(\overline{X} \geq 100.87) = P\left(\frac{\overline{X} - 100}{3} \sqrt{50} \geq \frac{100.87 - 100}{3} \sqrt{50}\right)$$

$$= 1 - \Phi(2.06) = 0.019699.$$

Since $0.01 < 0.019699 < 0.05$ we shall say that the observed value of $\overline{X}$ is sig-nificant on the 5-percent level but not on the 1-percent level. That is, if we use $\alpha = 0.05$, we would reject H_0, while at the same time if we use $\alpha = 0.01$, we would not reject H_0.

Saying this differently, *if* $\mu = 100$, we obtained a result which should occur only about 1.9 percent of the time. If we feel that in order to accept H_0 a result should have at least probability 0.05 of occurring, then we reject it. If we are satisfied with a probability of 0.01, we accept it.

Note: The above test stipulated that H_0 should be rejected whenever $\overline{X} > C$. Suppose that the sample size $n = 2$. Hence the above criterion becomes $(X_1 + X_2)/2 > C$ or, equivalently, $(X_1 + X_2) > k$. Thus the set of possible sample values (x_1, x_2) has been divided into two regions: $R = \{(x_1, x_2) | x_1 + x_2 > k\}$ and $\overline{R}$.

The specific region R depends of course on the value of k, which in turn depends on the significance level of the test. R, the region of rejection, is sometimes called the *critical region* of the test. (See Fig. 15.6.)

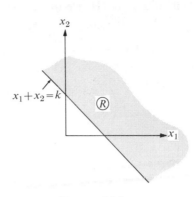

FIGURE 15.6

Quite generally, a test may be described in terms of its critical region R. That is, we reject H_0 if and only if $(x_1, \ldots, x_n) \in R$.

15.3 Additional Examples

Instead of formulating a general theory for testing hypotheses (which does exist and which is quite extensive) we shall continue to consider some examples. In each case, the test we propose will be intuitively appealing. No attempt will be made to indicate that a particular test is best in some sense.

EXAMPLE 15.2. Two production processes are being compared. The output of process A may be characterized as a random variable X with distribution $N(\mu_x, \sigma_x^2)$ while the output of process B may be characterized as a random variable Y with distribution $N(\mu_y, \sigma_y^2)$. We shall suppose that the variability inherent in each of the processes, as measured by the variance, is *known*. We want to test the hypothesis $H_0: \mu_x = \mu_y$ versus the alternative hypothesis $H_1: \mu_x - \mu_y > 0$.
　　We obtain a sample of size n from X, say $X_1, \ldots, X_n$, and a sample of size m from Y, say $Y_1, \ldots, Y_m$. We compute the respective sample means $\bar{X}$ and $\bar{Y}$ and propose the following test for testing the above hypothesis.
　　Reject H_0 if $\bar{X} - \bar{Y} > C$ where C is a constant chosen so that the test has a specified significance level equal to α.
　　The random variable $Z = [(\bar{X} - \bar{Y}) - (\mu_x - \mu_y)]/\sqrt{\sigma_x^2/n + \sigma_y^2/m}$ has distribution $N(0, 1)$. Defining $\mu = \mu_x - \mu_y$, we may express the OC function of the above test as a function of μ as follows:

$$L(\mu) = P(\bar{X} - \bar{Y} \le C \mid \mu) = P\left(Z \le \frac{C - \mu}{\sqrt{\sigma_x^2/n + \sigma_y^2/m}}\right)$$

$$= \Phi\left(\frac{C - \mu}{\sqrt{\sigma_x^2/n + \sigma_y^2/m}}\right).$$

Now $\mu_x = \mu_y$ is equivalent to $\mu = 0$. Hence to determine C we must solve the equation

$$L(0) = 1 - \alpha$$

or

$$1 - \alpha = \Phi\left(\frac{C}{\sqrt{\sigma_x^2/n + \sigma_y^2/m}}\right).$$

Therefore

$$K_{1-\alpha} = \frac{C}{\sqrt{\sigma_x^2/n + \sigma_y^2/m}}$$

where K_α is defined, as before, by the relation $\alpha = (1/\sqrt{2\pi}) \int_{-\infty}^{K_\alpha} e^{-t^2/2} \, dt$. Thus

$$C = K_{1-\alpha}\sigma\sqrt{\sigma_x^2/n + \sigma_y^2/m}.$$

(We will not pursue the question of determining n and m optimally. A discussion of this problem is found in Derman and Klein, *Probability and Statistical Inference for Engineers*. Oxford University Press, New York, 1959.)

EXAMPLE 15.3.   A manufacturer supplies fuses approximately 90 percent of which function properly. A new process is initiated whose purpose it is to increase the proportion of properly functioning fuses. Thus we wish to test the hypothesis $H_0: p = 0.90$ versus $H_1: p > 0.90$, where p is the proportion of properly functioning fuses. (That is, we are testing the hypothesis that no improvement has taken place versus the hypothesis that the new process is superior.) We obtain a sample of 50 fuses manufactured by the new process and we count X, the number of properly functioning fuses. We propose the following test:

Reject H_0 whenever $X > 48$ and accept otherwise.

Assuming that the random variable X has a binomial distribution with parameter p (which is a realistic assumption if the sample is taken from a very large lot), we obtain the following expression for the OC function.

$$L(p) = P(X \le 48) = 1 - P(X \ge 49)$$

$$= 1 - \sum_{k=49}^{50} \binom{50}{k} p^k(1 - p)^{50-k} = 1 - p^{49}(50 - 49p)$$

after some algebraic simplification. Hence we have the following.

(a) $L(0) = 1.$ (c) $L'(p) < 0$ for all p, $0 < p < 1$.
(b) $L(1) = 0.$ (d) $L''(p) = 0$ if $p = 48/49$.

[Properties (c) and (d) are easily verified by straightforward differentiation.] Thus the graph of the above function L has the shape of the curve shown in Fig. 15.7. The significance level α of this test is obtained by evaluating $1 - L(0.9)$. We obtain

$$\begin{aligned} \alpha &= 1 - L(0.9) \\ &= (0.9)^{49}[50 - 44.1] \\ &= 0.034. \end{aligned}$$

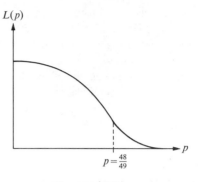

$L(p)$

$p = \frac{48}{49}$

FIGURE 15.7

Notes: (a) The above example may be generalized as follows. Suppose that X is a binomially distributed random variable based on n repetitions of an experiment, with parameter p. To test the hypothesis $H_0: p = p_0$ versus $H_1: p > p_0$, we propose the following test.

Reject H_0 whenever $X > C$, where C is a constant to be determined. (Hence accept H_0 whenever $X \le C$.)

The OC function of this test will be of the form

$$L(p) = \sum_{k=0}^{C} \binom{n}{k} p^k (1 - p)^{n-k}. \tag{15.2}$$

The following properties of L are easily verified. (See Problem 15.5.)

(1) $L(0) = 1$; $L(1) = 0$.
(2) $L'(p) < 0$ for all p, $0 < p < 1$. (Therefore L is strictly decreasing.)
(3) $L''(p) = 0$ if $p = C/(n - 1)$. [Hence L has a point of inflection at $C/(n - 1)$.]

(b) We have said that in some cases we may approximate the binomial distribution with the Poisson distribution. That is, for n large and p small, $P(X = k) \simeq e^{-np}(np)^k/k!$ Using this form of $P(X = k)$, we find that the OC function for the test proposed above becomes

$$R(p) = \sum_{k=0}^{C} \frac{e^{-np}(np)^k}{k!}. \tag{15.3}$$

The following properties of R are also easily verified. (See Problem 15.6.)

(4) $R(0) = 1$; $R(1) = 0$.
(5) $R'(p) < 0$ for all p, $0 < p < 1$.
(6) $R''(p) = 0$ if $p = C/n$.
(7) $R(C/n)$ is a function only of C and not of n.

Let us reconsider the problem of testing $H_0: \mu = \mu_0$ versus $H_1: \mu > \mu_0$, where X has distribution $N(\mu, \sigma^2)$. Previously we assumed that σ^2 is known. Let us now remove this restriction.

Our previous test rejected H_0 whenever $(\overline{X} - \mu_0)\sqrt{n}/\sigma > C$; C was determined by using the fact that $(\overline{X} - \mu_0)\sqrt{n}/\sigma$ has distribution $N(0, 1)$ if $\mu = \mu_0$.

Just as we did when constructing a confidence interval for μ when σ^2 was un-
known, let us estimate σ^2 by $\hat{\sigma}^2 = [1/(n-1)]\sum_{i=1}^{n}(X_i - \overline{X})^2$. Now let us use a
test analogous to the one proposed above: Reject H_0 whenever $(\overline{X} - \mu)\sqrt{n}/\hat{\sigma} > C$.
To determine C we use the fact that $(\overline{X} - \mu_0)\sqrt{n}/\hat{\sigma}$ has Student's t-distribution
with $(n-1)$ degrees of freedom if $\mu = \mu_0$. (See Theorem 14.3.)

Let α, the significance level, be preassigned. Then $\alpha = P[(\overline{X} - \mu_0)\sqrt{n}/\hat{\sigma} > C]$
implies that $C = t_{n-1,1-\alpha}$, obtainable from the table of Student's t-distribution.
(See Fig. 15.8.) Reject H_0 whenever $\overline{X} > \hat{\sigma}t_{n-1,1-\alpha}n^{-1/2} + \mu_0$, thus becomes
our test.

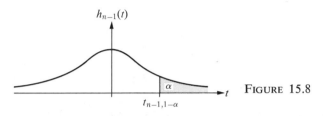

$$h_{n-1}(t)$$

$$t_{n-1,1-\alpha}$$

FIGURE 15.8

EXAMPLE 15.4. Suppose that X, the annual rainfall at a certain locality, is normally distributed with $E(X) = 30.0$ inches. (This value has been established from
a long history of weather data.) In recent years, certain climatological changes
seem to be evident affecting, among other things, the annual precipitation. It is
hypothesized that in fact the annual rainfall has increased. Specifically we want
to test $H_0: \mu = 30.0$ versus $H_1: \mu > 30.0$. The variance is assumed to be unknown since the suggested climatological changes may also affect the variability
of the rainfall, and hence past records on the variance are not meaningful.

Let us assume that the past 8 years have yielded the following annual precipitation (inches):

$$34.1, 33.7, 27.4, 31.1, 30.9, 35.2, 28.4, 32.1.$$

Straightforward calculations yield $\overline{X} = 31.6$ and $\hat{\sigma}^2 = 7.5$. From the tables of
the t-distribution we find that

$$t_{7,0.95} = 1.89.$$

Thus

$$\hat{\sigma}t_{7,0.95}/\sqrt{8} + 30.0 = 31.8 > 31.6.$$

Hence we do not reject H_0 on the 0.05 significance level.

15.4 Goodness of Fit Tests

In most of our discussions, we assumed that the random variable under consideration has a specific distribution. In the previous chapter and in earlier sections of this one, we have learned how to resolve the problem of having an unknown
parameter associated with a probability distribution. However, it may happen

that we are not even sure about the general form of the underlying distribution. Let us consider a few examples.

EXAMPLE 15.5. Merchant vessels of a certain type were exposed to risk of accident through heavy weather, ice, fire, grounding, breakdown of machinery, etc., for a period of 400 days. The number of accidents to each vessel, say X, may be considered as a random variable. The following data were reported.

Number of accidents (X):	0	1	2	3	4	5	6
Number of vessels with X accidents:	1448	805	206	34	4	2	1

Do the above data justify the assumption that X has a Poisson distribution?

EXAMPLE 15.6. Suppose that 20 samples of a particular type of wire are obtained and the resistances (ohms) are measured. The following values are obtained:

$$9.8, \ 14.5, \ 13.7, \ 7.6, \ 10.5, \ 9.3, \ 11.1, \ 10.1, \ 12.7, \ 9.9,$$
$$10.4, \ 8.3, \ 11.5, \ 10.0, \ 9.1, \ 13.8, \ 12.9, \ 10.6, \ 8.9, \ 9.5.$$

If R is the random variable from which the above sample is obtained, are we justified in supposing that R is normally distributed?

EXAMPLE 15.7. Twenty electron tubes were tested and the following life times (hours) were reported:

$$7.2, \ 37.8, \ 49.6, \ 21.4, \ 67.2, \ 41.1, \ 3.8, \ 8.1, \ 23.2, \ 72.1,$$
$$11.4, \ 17.5, \ 29.8, \ 57.8, \ 84.6, \ 12.8, \ 2.9, \ 42.7, \ 7.4, \ 33.4.$$

Are the above data consistent with the hypothesis that T, the random variable being sampled, is exponentially distributed?

The above examples are typical of a large class of problems which arise frequently in applications. A number of statistical techniques are available by which such problems may be analyzed, and we shall consider one of these below.

The problem of testing the hypothesis that a random variable has a certain specified distribution may be considered as a special case of the following general problem.

Consider again the situation giving rise to the multinomial distribution (Section 8.7). An experiment ε is performed n times. Each repetition of ε results in one and only one of the events A_i, $i = 1, 2, \ldots, k$. Suppose that $P(A_i) = p_i$. Let n_i be the number of times A_i occurs among the n repetitions of ε, $n_1 + \cdots + n_k = n$.

We want to test the hypothesis $H_0 : p_i = p_{io}$, $i = 1, \ldots, k$, where p_{io} is a specified value. Karl Pearson (1900) introduced the following "goodness of fit" test for testing the above hypothesis:

$$\text{Reject } H_0 \text{ whenever } D^2 = \sum_{i=1}^{k} \frac{(n_i - np_{io})^2}{np_{io}} > C, \tag{15.4}$$

where C is a constant to be determined.

Notes: (a) Since $E(n_i) = np_{io}$ if $p_i = p_{io}$, the above test criterion has considerable intuitive appeal. For it requires that we reject H_0 whenever the discrepancy between the observed values n_i and the expected values np_{io} is "too large." The above statistic D^2 is sometimes suggestively written as $\sum_{i=1}^{k} (o_i - e_i)^2/e_i$, where o_i and e_i refer to the observed and expected value of n_i, respectively.

(b) It is important to realize that D^2 is a statistic (that is, a function of the observed values $n_1, \ldots, n_k$) and hence a random variable. In fact, D^2 is a discrete random variable assuming a large, finite number of values. The actual distribution of D^2 is very complicated. Fortunately, an approximation is available for the distribution of D^2, valid if n is large, making the procedure suggested above very usable.

Theorem 15.1. If n is sufficiently large, and if $p_i = p_{io}$, the distribution of D^2 has approximately the chi-square distribution with $(k - 1)$ degrees of freedom.

Proof: The following argument is *not* a rigorous proof. It is simply an attempt to make the result plausible.

Consider a special case, namely $k = 2$. Then

$$D^2 = \frac{(n_1 - np_{1o})^2}{np_{1o}} + \frac{(n_2 - np_{2o})^2}{np_{2o}}.$$

Using the fact that $n_1 + n_2 = n$ and that $p_{1o} + p_{2o} = 1$, we may write

$$D^2 = \frac{(n_1 - np_{1o})^2}{np_{1o}} + \frac{(n - n_1 - n(1 - p_{1o}))^2}{np_{2o}}$$

$$= \frac{(n_1 - np_{1o})^2}{np_{1o}} + \frac{(n_1 - np_{1o})^2}{np_{2o}} = (n_1 - np_{1o})^2 \left[\frac{1}{np_{1o}} + \frac{1}{np_{2o}} \right]$$

$$= (n_1 - np_{1o})^2 \left[\frac{np_{2o} + np_{1o}}{n^2 p_{1o} p_{2o}} \right] = \frac{(n_1 - n_1 p_{1o})^2}{np_{1o}(1 - p_{1o})}.$$

Now $n_1 = \sum_{j=1}^{n} Y_{1j}$, where

$$Y_{1j} = 1 \qquad \text{if } A_1 \text{ occurs on the } j\text{th repetition,}$$
$$= 0, \qquad \text{elsewhere.}$$

Thus n_1 may be expressed as the sum of n independent random variables, and according to the Central Limit Theorem, has approximately a normal distribution if n is large. In addition, $E(n_1) = np_{1o}$ and $V(n_1) = np_{1o}(1 - p_{1o})$ if p_{1o} is the true value of p_1. Hence if $p_1 = p_{1o}$, then for large n, the random variable $(n_1 - np_{1o})/\sqrt{np_{1o}(1 - p_{1o})}$ has approximately the distribution $N(0, 1)$.

Thus, according to Theorem 10.8 for large n, the random variable

$$\left[\frac{n_1 - np_{10}}{\sqrt{np_{10}(1 - p_{10})}}\right]^2$$

has approximately the distribution χ_1^2.

We have shown that if n is sufficiently large, D^2 (with $k = 2$) has approximately the distribution χ_1^2. But this is precisely what we claimed in the theorem. The proof for general k follows the same lines: We must show that D^2 may be expressed as the sum of squares of $(k - 1)$ independent random variables each with distribution $N(0, 1)$ if n is large and if $p_i = p_{io}$, and appealing to Theorem 10.8, we find that the above result follows.

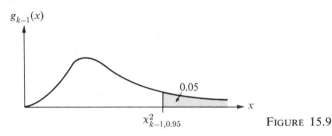

FIGURE 15.9

We can use the result just established to answer the question of "how large" D^2 should be in order that we reject the hypothesis H_0: $p_i = p_{io}$.

Suppose that we want to achieve a probability of type-1 error (that is, significance level) equal to 0.05. This means that we expect to reject H_0 about 5 percent of the time, when in fact H_0 is true. Thus we choose C to satisfy

$$P(D^2 > C \mid p_i = p_{io}) = 0.05.$$

Since D^2 has distribution χ_{k-1}^2 if $p_i = p_{io}$, we may obtain the value of C from the tables of the chi-square distribution; that is, $C = \chi_{k-1,0.95}^2$; here $\chi_{k-1,0.95}^2$ is defined by the relation

$$\int_{\chi_{k-1,0.95}^2}^{\infty} g_{k-1}(x)\, dx = 0.05,$$

where $g_{k-1}(x)$ is the pdf of a random variable with χ_{k-1}^2 distribution. (See Fig. 15.9.)

Let us use the ideas developed above to answer some of the questions raised at the beginning of this section: How can we find a test to decide whether to accept or reject the hypothesis that a particular sample was taken from a random variable with a specified distribution?

At this point we must make a distinction between two types of problems. We might simply hypothesize that the random variable being sampled has *some* normal distribution without specifying the parameters involved. Or we might be more specific and hypothesize that the random variable under consideration has a normal distribution with specified mean and variance. The two problems can be treated in similar ways, but the latter (when we specify the hypothesized distribution *completely*) is somewhat simpler and we shall consider it first.

Case 1. Testing for a completely specified distribution.

EXAMPLE 15.8. Suppose we believe that the life length T of light bulbs is exponentially distributed with parameter $\beta = 0.005$. (That is, the expected time to failure is 200 hours.) We obtain a sample of 150 bulbs, test them, and record their burning time, say $T_1, \ldots, T_{150}$. Consider the following four mutually exclusive events:

$$A_1: 0 \leq T < 100; \qquad A_2: 100 \leq T < 200;$$
$$A_3: 200 \leq T < 300; \qquad A_4: T \geq 300.$$

Suppose that we record n_i, the number of times (among the 150 failure times) the event A_i occurred, and find the following: $n_1 = 47$, $n_2 = 40$, $n_3 = 35$, and $n_4 = 28$. In order to compute the statistic D^2 we must evaluate p_i, $i = 1, 2, 3, 4$. Now

$$p_1 = P(T \leq 100) = 1 - e^{-0.005(100)} = 1 - e^{-0.5} = 0.39,$$
$$p_2 = P(100 \leq T < 200) = 1 - e^{-0.005(200)} - 0.39 = 0.24,$$
$$p_3 = P(200 \leq T < 300) = 1 - e^{-0.005(300)} - (1 - e^{-0.005(200)}) = 0.15,$$
$$p_4 = P(T > 300) = e^{-0.005(300)} = 0.22.$$

We can now compute

$$D^2 = \sum_{i=1}^{4} \frac{(n_i - np_{io})^2}{np_{io}}$$

$$= \frac{(47 - 58.5)^2}{58.5} + \frac{(40 - 36)^2}{36} + \frac{(35 - 22.5)^2}{22.5} + \frac{(28 - 33)^2}{33} = 11.56.$$

From the tables of the chi-square distribution, we find that $P(D^2 > 11.56) < 0.01$, where D^2 has approximately the chi-square distribution with $4 - 1 = 3$ degrees of freedom. Hence we would reject (on the 1 percent level) the hypothesis that the data represent a sample from an exponential distribution with parameter $\beta = 0.005$.

The above example illustrates the general procedure which we use to test the hypothesis that $X_1, \ldots, X_n$ represents a sample from a random variable with a completely specified distribution:

(a) Divide the real line into k mutually exclusive intervals, $A_1, \ldots, A_k$.
(b) Let n_i be the number of sample values falling into A_i, $i = 1, 2, \ldots, k$.
(c) Let $p_{io} = P(A_i)$. These values can be computed since the hypothesis completely specifies the distribution.

(d) Compute D^2 and reject the hypothesis if $D^2 > C$, where C is obtained from the tables of the chi-square distribution. If a significance level of α is desired, $C = \chi^2_{k-1,\,1-\alpha}$.

Note: We shall not discuss the question of how the above intervals A_i should be chosen or how many should be chosen. Let us only state the following rule: If $np_{io} < 5$ for any A_i combine the data with A_{i+1} or A_{i-1}. That is, we do not want to subdivide the sample space of the random variable into such fine parts that the expected number of occurrences in any particular subdivision is less than 5. [A readable discussion of this problem may be found in a paper by W. G. Cochran, entitled "The χ^2-Test of Goodness of Fit," appearing in *Ann. Math. Stat.* **23**, 315–345 (1952).]

Case 2. Testing for a distribution if parameters must be estimated.

In many problems we have only reason to suppose that the random variable being sampled has a distribution of a certain *type* without being able to specify the parameters. For example, we know that certain assumptions we have made can lead to a Poisson distribution, to an exponential distribution, etc. In order to apply the technique suggested in the previous section we must *know* the parameter values of the distribution.

If we do not know these values, the obvious approach is to first estimate the unknown parameters, and then use these *estimates* in order to evaluate the probabilities p_i. Two questions arise.

(a) How should the parameters be estimated?

(b) If $\hat{p}_i$ (that is, the estimated parameter) is used in place of p_{io} in the expression for D^2, how does this affect the distribution of D^2? (The fact that it will affect the distribution should be clear if we realize that originally the p_{io}'s were constants, while now the $\hat{p}_i$'s are themselves random variables, thus giving the random variable D^2 a much more complicated structure.)

We shall state (without proof) some answers to the above questions.

(a) The estimates usually used for the parameters are those obtained from the method of maximum likelihood.

(b) If the number of parameters to be estimated equals $r < k$, then for large n, the random variable D^2 has again a chi-square distribution, this time with $n - k - r$ degrees of freedom.

Note: This last fact is quite remarkable. It means that D^2 has the same underlying χ^2-distribution as before, the only difference being that one degree of freedom is lost for every parameter that must be estimated.

EXAMPLE 15.9. Consider the data for ash content in coal as given in Example 14.6. Suppose that we wish to test the hypothesis that these data were obtained from a normally distributed random variable. We must first estimate the pertinent parameters μ and σ^2. We previously obtained the M.L. estimates $\hat{\mu} = 17.0$

and $\hat{\sigma}^2 = 7.1$. Let us subdivide the possible values of X into the following five categories:

$$A_1: X < 12; \quad A_2: 12 \leq X < 15; \quad A_3: 15 \leq X < 18;$$
$$A_4: 18 \leq X < 21; \quad A_5: X \geq 21.$$

Let n_i be the number of times A_i occurs. We find that

$$n_1 = 7; \quad n_2 = 49; \quad n_3 = 109; \quad n_4 = 67; \quad n_5 = 18.$$

Next we must compute $p_i = P(A_i)$, using the estimated values $\hat{\mu}$ and $\hat{\sigma}^2$ obtained above. We have

$$p_1 = P(X < 12) = P\left(\frac{X - 17}{2.7} < \frac{12 - 17}{2.7}\right) = \Phi(-1.85) = 0.03,$$
$$p_2 = P(12 \leq X < 15) = \Phi(-0.74) - \Phi(-1.85) = 0.20,$$
$$p_3 = P(15 \leq X < 18) = \Phi(0.37) - \Phi(-0.74) = 0.41,$$
$$p_4 = P(18 \leq X < 21) = \Phi(1.48) - \Phi(0.37) = 0.29,$$
$$\hat{p}_5 = P(X \geq 21) = 1 - \Phi(1.48) = 0.07.$$

Now we can evaluate

$$D^2 = \sum_{i=1}^{5} \frac{(n_i - 250\hat{p}_i)^2}{250\hat{p}_i}$$

$$= \frac{(7 - 7.5)^2}{7.5} + \frac{(49 - 50)^2}{50} + \frac{(109 - 102.5)^2}{102.5}$$

$$+ \frac{(67 - 72.5)^2}{72.5} + \frac{(18 - 17.5)^2}{17.5}$$

$$= 0.82.$$

Since D^2 has $5 - 1 - 2 = 2$ degrees of freedom, we find from the tables of the chi-square distribution that $P(D^2 \geq 0.82) \simeq 0.65$, and hence we would accept the hypothesis of normality.

EXAMPLE 15.10. Let us consider the data presented in Example 15.5. Does the random variable $X =$ number of accidents during the specified 400 day period have a Poisson distribution?

Let us first estimate the parameter λ of the distribution. In Chapter 14 we found that the ML estimate of λ is given by the sample mean. Thus

$$\hat{\lambda} = \frac{0(1448) + 1(805) + 2(206) + 3(34) + 4(4) + 5(2) + 6(1)}{1448 + 805 + 206 + 34 + 4 + 2 + 1}$$

$$= \frac{1351}{2500}$$

$$= 0.54.$$

Let $A_1: X = 0$; $A_2: X = 1$; $A_3: X = 2$; $A_4: X = 3$; $A_5: X \geq 4$. Then $n_1 = 1448$; $n_2 = 805$; $n_3 = 206$; $n_4 = 34$; $n_5 = 7$, and we obtain

$$\hat{p}_1 = P(X = 0) = e^{-(0.54)} = 0.58 \qquad \text{and} \qquad n\hat{p}_1 = 1450$$

$$\hat{p}_2 = P(X = 1) = e^{-0.54}(0.543) = 0.31 \qquad \text{and} \qquad n\hat{p}_2 = 775$$

$$\hat{p}_3 = P(X = 2) = e^{-0.54} \frac{(0.543)^2}{2} = 0.08 \qquad \text{and} \qquad n\hat{p}_3 = 200$$

$$\hat{p}_4 = P(X = 3) = e^{-0.54} \frac{(0.543)^3}{6} = 0.01 \qquad \text{and} \qquad n\hat{p}_4 = 25$$

$$\hat{p}_5 = P(X \geq 4) = 1 - P(X < 4) = 0.02 \qquad \text{and} \qquad n\hat{p}_5 = 50$$

We can now evaluate

$$D^2 = \sum_{i=1}^{5} \frac{(n_i - n\hat{p}_i)^2}{n\hat{p}_i} = \frac{(1448 - 1450)^2}{1450} + \frac{(805 - 775)^2}{775} + \frac{(206 - 200)^2}{200}$$

$$+ \frac{(34 - 25)^2}{25} + \frac{(7 - 50)^2}{50} = 42.2.$$

Since there are five categories and we estimated one parameter, the random variable D^2 has the approximate distribution χ_3^2. We find from the tables of the chi-square distribution that $P(D^2 \geq 42.2) \sim 0$ and hence we would reject the hypothesis.

PROBLEMS

15.1. Suppose that X has distribution $N(\mu, \sigma^2)$ with σ^2 known. To test $H_0: \mu = \mu_0$ versus $H_1: \mu < \mu_0$ the following procedure is proposed: Obtain a sample of size n and reject H_0 whenever the sample mean $\overline{X} < C$, where C is a constant to be determined.

(a) Obtain an expression for the OC function $L(\mu)$ in terms of the tabulated normal distribution.

(b) If the significance level of the test is to be $\alpha = 0.01$, obtain an expression for C.

(c) Suppose that $\sigma^2 = 4$ and assume that we are testing $H_0: \mu = 30$ versus $H_1: \mu < 30$. Determine the sample size n and the constant C in order to satisfy the conditions $L(30) = 0.98$, and $L(27) = 0.01$.

(d) Suppose that the following sample values are obtained from X:

27.1, 29.3, 31.5, 33.0, 30.1, 30.9, 28.4, 32.4, 31.6, 28.9, 27.3, 29.1.

Would you reject H_0 (versus H_1) as stated in (c) on the 5-percent significance level?

15.2. Consider the situation described in Problem 15.1 except that the alternative hypothesis is of the form $H_1: \mu \neq \mu_0$. Hence we reject H_0 whenever $|\overline{X} - \mu_0| > C$. Answer questions (a) and (b) above.

15.3. Suppose that X has a Poisson distribution with parameter λ. To test $H_0: \lambda = \lambda_0$ versus $H_1: \lambda > \lambda_0$, the following test is proposed. Obtain a sample of size n, compute the sample mean $\overline{X}$, and reject H_0 whenever $\overline{X} > C$, where C is a constant to be determined.

(a) Obtain an expression for the OC function of the above test, say $L(\lambda)$. [*Hint:* Use the reproductive property of the Poisson distribution.]

(b) Graph the OC function.

(c) Suppose that we are testing $H_0: \lambda = 0.2$ versus $H_1: \lambda > 0.2$. A sample of size $n = 10$ is obtained, and we reject H_0 if $\overline{X} > 0.25$. What is the significance level of this test?

15.4.   Establish Properties of Eq. (15.1) for the OC function $L(\mu) = \Phi[(C - \mu)\sqrt{n}/\sigma]$.

15.5.   Verify the properties for the OC function $L(p)$ as defined in Eq. (15.2).

15.6.   Verify the properties for the OC function $R(p)$ as defined in Eq. (15.3).

15.7.   It is known that a large shipment of volt meters contains a certain proportion of defectives, say p. To test $H_0: p = 0.2$ versus $H_1: p > 0.2$, the following procedure is used. A sample of size 5 is obtained and X, the number of defective voltmeters, is counted. If $X \leq 1$, accept H_0; if $X > 4$, reject H_0; and if $\overline{X} = 2$, 3 or 4 take a second sample of size 5. Let Y be the number of defective meters in the second sample. Reject H_0 if $Y \geq 2$ and accept otherwise. (Assume that the lot being sampled is sufficiently large so that X and Y may be supposed to be independent binomially distributed random variables.)

(a) Obtain an expression for $L(p)$, the OC function of the above test, and sketch its graph.

(b) Find the size of the above test.

(c) What is the probability of type-2 error if $p = 0.5$?

15.8.   If $n = 4$ and $k = 3$, how many possible values may the random variable D^2, as defined in Eq. (15.4), assume?

15.9.   (a) Compute the expected value of the random variable D^2 as defined in Eq. (15.4).

(b) How does this value compare with the (asymptotic) expected value of D^2 obtained from the chi-square distribution which may be used to approximate the distribution of D^2 when n is large?

15.10.   Three kinds of lubricants are being prepared by a new process. Each lubricant is tested on a number of machines, and the result is then classified as acceptable or non-acceptable. The data in Table 15.1 represent the outcome of such an experiment. Test the hypothesis that the probability p of a lubricant resulting in an acceptable outcome is the *same* for all three lubricants. [*Hint:* First estimate p from the sample.]

TABLE 15.1

	Lubricant 1	Lubricant 2	Lubricant 3
Acceptable	144	152	140
Nonacceptable	56	48	60
Total	200	200	200

15.11. In using various failure laws we have found that the exponential distribution plays a particularly important role, and hence it is important to be able to decide whether a particular sample of failure times comes from an underlying exponential distribution. Suppose that 335 light bulbs have been tested and the following summary of T their life length (in hours) is available:

Life length (in hours)	$0 \leq T < 100$	$100 \leq T < 200$	$200 \leq T < 300$	$300 \leq T < 400$	$T \geq 400$
Number of bulbs	82	71	68	62	52

From the actual failure times which were reported, it was found that T, the sample mean, was equal to 123.5 hours. Using this information, test the hypothesis that T, the time to failure is exponentially distributed.

15.12. Suppose that the random variable X has the following pdf:

$$f(x) = \frac{\alpha \cos \alpha x}{\sin (\pi/2)\alpha}, \qquad 0 < x < \pi/2, \qquad \text{where } 0 < \alpha < 1.$$

To test $H_0: \alpha = \alpha_0$ the following test is proposed. Obtain one observation from X, say X_1, and reject H_0 if $X_1 > 1$.

(a) Obtain an expression for the OC function of this test, say $L(\alpha)$, and graph it.
(b) What is the size of this test if $\alpha_0 = \pi/4$?

15.13. In a network of 165 cells, the number of graphite grains in each cell was counted. The data in Table 15.2 were obtained. Test the hypothesis that the number of grains in each cell is a random variable with a Poisson distribution. [*Hint:* Pool the observations ≤ 2 and also those ≥ 10.]

TABLE 15.2

Number of graphite grains in a cell	Observed	Number of graphite grains in a cell	Observed
0	1	7	17
1	1	8	22
2	5	9	21
3	7	10	4
4	20	11	2
5	34	12	1
6	30		

References

The following list of references, although by no means exhaustive, should give the interested reader the opportunity of finding a great deal of supplementary and complementary reading on many of the subjects which are examined in the text. Besides finding a number of topics discussed which are not included in this presentation, the reader will find certain subjects considered either in greater detail or from a somewhat different point of view.

In addition to noting the textbooks listed below, the reader should also be aware of certain professional journals in which important contributions are made. Many of these journals are, of course, written for persons who have considerably greater background in and understanding of this subject area than may be gained in one semester's study. However, there are several journals which occasionally contain very lucid articles which are within the grasp of a student who has mastered the subject of this text. Among these are the *Journal of the American Statistical Association* and *Technometrics*. The latter is in fact subtitled "A Journal of Statistics for the Physical, Chemical and Engineering Sciences", and hence should be of particular interest to the student for whom this text is chiefly intended.

Many of the listed books contain discussions of most of the topics covered in this text. However, a few of the references are somewhat more specialized and are particularly pertinent for a few chapters only. In those cases the specific chapters are listed after the references.

BAZOVSKY, I., *Reliability Theory and Practice*, Prentice-Hall, Inc., Englewood Cliffs, New Jersey, 1961 (11).

BERMAN, SIMEON M., *The Elements of Probability*, Addison-Wesley Publishing Co., Inc., Reading, Mass., 1969.

BOWKER, A. H. and G. J. LIEBERMAN, *Engineering Statistics*, Prentice-Hall, Inc., Englewood Cliffs, New Jersey, 1959.

DERMAN, C. and M. KLEIN, *Probability and Statistical Inference for Engineers*, Oxford University Press, New York, 1959.

EHRENFELD, S., and S. B. LITTAUER, *Introduction to Statistical Method*, McGraw-Hill Book Company, New York, 1964.

FELLER, W., *An Introduction to Probability Theory and Its Applications*, Vol. 1, 3rd Ed., John Wiley and Sons, Inc., New York, 1968 (1, 2, 3).

FREUND, J. E., *Mathematical Statistics*, Prentice-Hall, Inc., Englewood Cliffs, New Jersey, 1962.

FRY, T. C., *Probability and Its Engineering Uses*, 2nd Ed., D. Van Nostrand, Princeton, New Jersey, 1964.

GNEDENKO, B. V., *The Theory of Probability* (Translated from the Russian), Chelsea Publishing Company, Inc., New York, 1962.

GUTTMAN, I. and S. S. WILKS, *Introductory Engineering Statistics*, John Wiley and Sons, Inc., New York, 1965.

LINDGREN, B. W., *Statistical Theory*, 2nd Ed., The Macmillan Company, New York, 1968.

LINDGREN, B. W. and G. W. MCELRATH, *Introduction to Probability and Statistics*, 3rd Ed., The Macmillan Company, New York, 1969 (13, 14, 15).

LLOYD, D. K. and M. LIPOW, *Reliability: Management, Methods, and Mathematics*, Prentice-Hall, Inc., Englewood Cliffs, New Jersey, 1962 (11).

McCORD, J. R., III and R. M. MORONEY, JR., *Introduction to Probability Theory*, The Macmillan Company, New York, 1964.

MILLER, I. and J. E. FREUND, *Probability and Statistics for Engineers*, Prentice-Hall, Inc., Englewood Cliffs, New Jersey, 1965.

MOOD, A. M. and F. A. GRAYBILL, *Introduction to the Theory of Statistics*, 2nd Ed., McGraw-Hill Book Company, New York, 1963.

PARZEN, E., *Modern Probability Theory and Its Applications*, John Wiley and Sons, Inc., New York, 1960.

WADSWORTH, B. P. and J. G. BRYAN, *Introduction to Probability and Random Variables*, McGraw-Hill Book Company, Inc., New York, 1960.

Appendix

TABLE 1.   Values of the Standard Normal Distribution Function*

$$\Phi(z) = \int_{-\infty}^{z} \frac{1}{\sqrt{2\pi}} e^{-u^2/2} \, du = P(Z \le z)$$

z	0	1	2	3	4	5	6	7	8	9
−3.0	0.0013	0.0010	0.0007	0.0005	0.0003	0.0002	0.0002	0.0001	0.0001	0.0000
−2.9	0.0019	0.0018	0.0017	0.0017	0.0016	0.0016	0.0015	0.0015	0.0014	0.0014
−2.8	0.0026	0.0025	0.0024	0.0023	0.0023	0.0022	0.0021	0.0021	0.0020	0.0019
−2.7	0.0035	0.0034	0.0033	0.0032	0.0031	0.0030	0.0029	0.0028	0.0027	0.0026
−2.6	0.0047	0.0045	0.0044	0.0043	0.0041	0.0040	0.0039	0.0038	0.0037	0.0036
−2.5	0.0062	0.0060	0.0059	0.0057	0.0055	0.0054	0.0052	0.0051	0.0049	0.0048
−2.4	0.0082	0.0080	0.0078	0.0075	0.0073	0.0071	0.0069	0.0068	0.0066	0.0064
−2.3	0.0107	0.0104	0.0102	0.0099	0.0096	0.0094	0.0091	0.0089	0.0087	0.0084
−2.2	0.0139	0.0136	0.0132	0.0129	0.0126	0.0122	0.0119	0.0116	0.0113	0.0110
−2.1	0.0179	0.0174	0.0170	0.0166	0.0162	0.0158	0.0154	0.0150	0.0146	0.0143
−2.0	0.0228	0.0222	0.0217	0.0212	0.0207	0.0202	0.0197	0.0192	0.0188	0.0183
−1.9	0.0287	0.0281	0.0274	0.0268	0.0262	0.0256	0.0250	0.0244	0.0238	0.0233
−1.8	0.0359	0.0352	0.0344	0.0336	0.0329	0.0322	0.0314	0.0307	0.0300	0.0294
−1.7	0.0446	0.0436	0.0427	0.0418	0.0409	0.0401	0.0392	0.0384	0.0375	0.0367
−1.6	0.0548	0.0537	0.0526	0.0516	0.0505	0.0495	0.0485	0.0475	0.0465	0.0455
−1.5	0.0668	0.0655	0.0643	0.0630	0.0618	0.0606	0.0594	0.0582	0.0570	0.0559
−1.4	0.0808	0.0793	0.0778	0.0764	0.0749	0.0735	0.0722	0.0708	0.0694	0.0681
−1.3	0.0968	0.0951	0.0934	0.0918	0.0901	0.0885	0.0869	0.0853	0.0838	0.0823
−1.2	0.1151	0.1131	0.1112	0.1093	0.1075	0.1056	0.1038	0.1020	0.1003	0.0985
−1.1	0.1357	0.1335	0.1314	0.1292	0.1271	0.1251	0.1230	0.1210	0.1190	0.1170
−1.0	0.1587	0.1562	0.1539	0.1515	0.1492	0.1469	0.1446	0.1423	0.1401	0.1379
−0.9	0.1841	0.1814	0.1788	0.1762	0.1736	0.1711	0.1685	0.1660	0.1635	0.1611
−0.8	0.2119	0.2090	0.2061	0.2033	0.2005	0.1977	0.1949	0.1922	0.1894	0.1867
−0.7	0.2420	0.2389	0.2358	0.2327	0.2297	0.2266	0.2236	0.2206	0.2177	0.2148
−0.6	0.2743	0.2709	0.2676	0.2643	0.2611	0.2578	0.2546	0.2514	0.2483	0.2451
−0.5	0.3085	0.3050	0.3015	0.2981	0.2946	0.2912	0.2877	0.2843	0.2810	0.2776
−0.4	0.3446	0.3409	0.3372	0.3336	0.3300	0.3264	0.3228	0.3192	0.3156	0.3121
−0.3	0.3821	0.3783	0.3745	0.3707	0.3669	0.3632	0.3594	0.3557	0.3520	0.3483
−0.2	0.4207	0.4168	0.4129	0.4090	0.4052	0.4013	0.3974	0.3936	0.3897	0.3859
−0.1	0.4602	0.4562	0.4522	0.4483	0.4443	0.4404	0.4364	0.4325	0.4286	0.4247
−0.0	0.5000	0.4960	0.4920	0.4880	0.4840	0.4801	0.4761	0.4721	0.4681	0.4641

*B. W. Lindgren, *Statistical Theory*, The Macmillan Company, 1960.

TABLE 1 (*Continued*)

$$\Phi(z) = \int_{-\infty}^{z} \frac{1}{\sqrt{2\pi}} e^{-u^2/2} \, du = P(Z \le z)$$

z	0	1	2	3	4	5	6	7	8	9
0.0	0.5000	0.5040	0.5080	0.5120	0.5160	0.5199	0.5239	0.5279	0.5319	0.5359
0.1	0.5398	0.5438	0.5478	0.5517	0.5557	0.5596	0.5636	0.5675	0.5714	0.5753
0.2	0.5793	0.5832	0.5871	0.5910	0.5948	0.5987	0.6026	0.6064	0.6103	0.6141
0.3	0.6179	0.6217	0.6255	0.6293	0.6331	0.6368	0.6406	0.6443	0.6480	0.6517
0.4	0.6554	0.6591	0.6628	0.6664	0.6700	0.6736	0.6772	0.6808	0.6844	0.6879
0.5	0.6915	0.6950	0.6985	0.7019	0.7054	0.7088	0.7123	0.7157	0.7190	0.7224
0.6	0.7257	0.7291	0.7324	0.7357	0.7389	0.7422	0.7454	0.7486	0.7517	0.7549
0.7	0.7580	0.7611	0.7642	0.7673	0.7703	0.7734	0.7764	0.7794	0.7823	0.7852
0.8	0.7881	0.7910	0.7939	0.7967	0.7995	0.8023	0.8051	0.8078	0.8106	0.8133
0.9	0.8159	0.8186	0.8212	0.8238	0.8264	0.8289	0.8315	0.8340	0.8365	0.8389
1.0	0.8413	0.8438	0.8461	0.8485	0.8508	0.8531	0.8554	0.8577	0.8599	0.8621
1.1	0.8643	0.8665	0.8686	0.8708	0.8729	0.8749	0.8770	0.8790	0.8810	0.8830
1.2	0.8849	0.8869	0.8888	0.8907	0.8925	0.8944	0.8962	0.8980	0.8997	0.9015
1.3	0.9032	0.9049	0.9066	0.9082	0.9099	0.9115	0.9131	0.9147	0.9162	0.9177
1.4	0.9192	0.9207	0.9222	0.9236	0.9251	0.9265	0.9278	0.9292	0.9306	0.9319
1.5	0.9332	0.9345	0.9357	0.9370	0.9382	0.9394	0.9406	0.9418	0.9430	0.9441
1.6	0.9452	0.9463	0.9474	0.9484	0.9495	0.9505	0.9515	0.9525	0.9535	0.9545
1.7	0.9554	0.9564	0.9573	0.9582	0.9591	0.9599	0.9608	0.9616	0.9625	0.9633
1.8	0.9641	0.9648	0.9656	0.9664	0.9671	0.9678	0.9686	0.9693	0.9700	0.9706
1.9	0.9713	0.9719	0.9726	0.9732	0.9738	0.9744	0.9750	0.9756	0.9762	0.9767
2.0	0.9772	0.9778	0.9783	0.9788	0.9793	0.9798	0.9803	0.9808	0.9812	0.9817
2.1	0.9821	0.9826	0.9830	0.9834	0.9838	0.9842	0.9846	0.9850	0.9854	0.9857
2.2	0.9861	0.9864	0.9868	0.9871	0.9874	0.9878	0.9881	0.9884	0.9887	0.9890
2.3	0.9893	0.9896	0.9898	0.9901	0.9904	0.9906	0.9909	0.9911	0.9913	0.9916
2.4	0.9918	0.9920	0.9922	0.9925	0.9927	0.9929	0.9931	0.9932	0.9934	0.9936
2.5	0.9938	0.9940	0.9941	0.9943	0.9945	0.9946	0.9948	0.9949	0.9951	0.9952
2.6	0.9953	0.9955	0.9956	0.9957	0.9959	0.9960	0.9961	0.9962	0.9963	0.9964
2.7	0.9965	0.9966	0.9967	0.9968	0.9969	0.9970	0.9971	0.9972	0.9973	0.9974
2.8	0.9974	0.9975	0.9976	0.9977	0.9977	0.9978	0.9979	0.9979	0.9980	0.9981
2.9	0.9981	0.9982	0.9982	0.9983	0.9984	0.9984	0.9985	0.9985	0.9986	0.9986
3.0	0.9987	0.9990	0.9993	0.9995	0.9997	0.9998	0.9998	0.9999	0.9999	1.0000

TABLE 2.   Binomial Distribution Function

$$1 - F(x - 1) = \sum_{r=x}^{r=n} \binom{n}{r} p^r q^{n-r}$$

$n = 10$ $x = 10$	$n = 10$ $x = 9$	$n = 10$ $x = 8$	$n = 10$ $x = 7$	p
0.0000000	0.0000000	0.0000000	0.0000000	0.01
.0000000	.0000000	.0000000	.0000000	.02
.0000000	.0000000	.0000000	.0000000	.03·
.0000000	.0000000	.0000000	.0000000	.04
.0000000	.0000000	.0000000	.0000001	.05
.0000000	.0000000	.0000000	.0000003	.06
.0000000	.0000000	.0000000	.0000008	.07
.0000000	.0000000	.0000001	.0000020	.08
.0000000	.0000000	.0000002	.0000045	.09
.0000000	.0000000	.0000004	.0000091	.10
.0000000	.0000000	.0000008	.0000173	.11
.0000000	.0000000	.0000015	.0000308	.12
.0000000	.0000001	.0000029	.0000525	.13
.0000000	.0000002	.0000051	.0000856	.14
.0000000	.0000003	.0000087	.0001346	.15
.0000000	.0000006	.0000142	.0002051	.16
.0000000	.0000010	.0000226	.0003042	.17
.0000000	.0000017	.0000350	.0004401	.18
.0000001	.0000027	.0000528	.0006229	.19
.0000001	.0000042	.0000779	.0008644	.20
.0000002	.0000064	.0001127	.0011783	.21
.0000003	.0000097	.0001599	.0015804	.22
.0000004	.0000143	.0002232	.0020885	.23
.0000006	.0000207	.0003068	.0027228	.24
.0000010	.0000296	.0004158	.0035057	.25
.0000014	.0000416	.0005362	.0044618	.26
.0000021	.0000577	.0007350	.0056181	.27
.0000030	.0000791	.0009605	.0070039	.28
.0000042	.0001072	.0012420	.0086507	.29
.0000059	.0001437	.0015904	.0105921	.30
.0000082	.0001906	.0020179	.0128637	.31
.0000113	.0002505	.0025384	.0155029	.32
.0000153	.0003263	.0031673	.0185489	.33
.0000206	.0004214	.0039219	.0220422	.34
.0000276	.0005399	.0048213	.0260243	.35
.0000366	.0006865	.0058864	.0305376	.36
.0000481	.0008668	.0071403	.0356252	.37
.0000628	.0010871	.0086079	.0413301	.38
.0000814	.0013546	.0103163	.0476949	.39
.0001049	.0016777	.0122946	.0547619	.40
.0001342	.0020658	.0145738	.0625719	.41
.0001708	.0025295	.0171871	.0711643	.42
.0002161	.0030809	.0201696	.0805763	.43
.0002720	.0037335	.0235583	.0908427	.44
.0003405	.0045022	.0273918	.1019949	.45
.0004242	.0054040	.0317105	.1140612	.46
.0005260	.0064574	.0365560	.1270655	.47
.0006493	.0076828	.0419713	.1410272	.48
.0007979	.0091028	.0480003	.1559607	.49
.0009766	.0107422	.0546875	.1718750	.50

TABLE 2 (*Continued*)

$$1 - F(x - 1) = \sum_{r=x}^{r=n} \binom{n}{r} p^r q^{n-r}$$

$n = 10$ $x = 6$	$n = 10$ $x = 5$	$n = 10$ $x = 4$	$n = 10$ $x = 3$	$n = 10$ $x = 2$	$n = 10$ $x = 1$	p
0.0000000	0.0000000	0.0000020	0.0001138	0.0042662	0.0956179	0.01
.0000000	.0000007	.0000305	.0008639	.0161776	.1829272	.02
.0000001	.0000054	.0001471	.0027650	.0345066	.2625759	.03
.0000007	.0000218	.0004426	.0062137	.0581538	.3351674	.04
.0000028	.0000637	.0010285	.0115036	.0861384	.4012631	.05
.0000079	.0001517	.0020293	.0188378	.1175880	.4613849	.06
.0000193	.0003139	.0035761	.0283421	.1517299	.5160177	.07
.0000415	.0005857	.0058013	.0400754	.1878825	.5656115	.08
.0000810	.0010096	.0088338	.0540400	.2254471	.6105839	.09
.0001469	.0016349	.0127952	.0701908	.2639011	.6513216	.10
.0002507	.0025170	.0177972	.0884435	.3027908	.6881828	.11
.0004069	.0037161	.0239388	.1086818	.3417250	.7214990	.12
.0006332	.0052967	.0313048	.1307642	.3803692	.7515766	.13
.0009505	.0073263	.0399642	.1545298	.4184400	.7786984	.14
.0013832	.0098741	.0499698	.1798035	.4557002	.8031256	.15
.0019593	.0130101	.0613577	.2064005	.4919536	.8250988	.16
.0027098	.0168038	.0741472	.2341305	.5270412	.8448396	.17
.0036694	.0213229	.0883411	.2628010	.5608368	.8625520	.18
.0048757	.0266325	.1039261	.2922204	.5932435	.8784233	.19
.0063694	.0327935	.1208739	.3222005	.6241904	.8926258	.20
.0081935	.0398624	.1391418	.3525586	.6536289	.9053172	.21
.0103936	.0478897	.1586739	.3831197	.6815306	.9166422	.22
.0130167	.0569196	.1794024	.4137173	.7078843	.9267332	.23
.0161116	.0669890	.2012487	.4441949	.7326936	.9357111	.24
.0197277	.0781269	.2241249	.4744072	.7559748	.9436865	.25
.0239148	.0903542	.2479349	.5042200	.7777550	.9507601	.26
.0287224	.1036831	.2725761	.5335112	.7980705	.9570237	.27
.0341994	.1181171	.2979405	.5621710	.8169646	.9625609	.28
.0403932	.1336503	.3239164	.5901015	.8344869	.9674476	.29
.0473490	.1502683	.3503893	.6172172	.8506917	.9717525	.30
.0551097	.1679475	.3772433	.6434445	.8656366	.9755381	.31
.0637149	.1866554	.4043626	.6687212	.8793821	.9788608	.32
.0732005	.2063514	.4316320	.6929966	.8919901	.9817716	.33
.0835979	.2269866	.4589388	.7162304	.9035235	.9843166	.34
.0949341	.2485045	.4861730	.7383926	.9140456	.9865373	.35
.1072304	.2708415	.5132284	.7594627	.9236190	.9884708	.36
.1205026	.2939277	.5400038	.7794292	.9323056	.9901507	.37
.1347603	.3176870	.5664030	.7982887	.9401661	.9916070	.38
.1500068	.3420385	.5923361	.8160453	.9472594	.9928666	.39
.1662386	.3668967	.6177194	.8327102	.9536426	.9939534	.40
.1834452	.3921728	.6424762	.8483007	.9593705	.9948888	.41
.2016092	.4177749	.6665372	.8628393	.9644958	.9956920	.42
.2207058	.4436094	.6898401	.8763538	.9690684	.9963797	.43
.2407033	.4695813	.7123307	.8888757	.9731358	.9969669	.44
.2615627	.4955954	.7339621	.9004403	.9767429	.9974670	.45
.2832382	.5215571	.7546952	.9110859	.9799319	.9978917	.46
.3056772	.5473730	.7744985	.9208530	.9827422	.9982511	.47
.3288205	.5729517	.7933480	.9297839	.9852109	.9985544	.48
.3526028	.5982047	.8112268	.9379222	.9873722	.9988096	.49
.3769531	.6230469	.8281250	.9453125	.9892578	.9990234	.50

TABLE 3.   Poisson Distribution Function*

$$1 - F(x - 1) = \sum_{r=x}^{r=\infty} \frac{e^{-a}a^r}{r!}$$

x	$a = 0.2$	$a = 0.3$	$a = 0.4$	$a = 0.5$	$a = 0.6$
0	1.0000000	1.0000000	1.0000000	1.0000000	1.0000000
1	.1812692	.2591818	.3296800	.393469	.451188
2	.0175231	.0369363	.0615519	.090204	.121901
3	.0011485	.0035995	.0079263	.014388	.023115
4	.0000568	.0002658	.0007763	.001752	.003358
5	.0000023	.0000158	.0000612	.000172	.000394
6	.0000001	.0000008	.0000040	.000014	.000039
7			.0000002	.000001	.000003

x	$a = 0.7$	$a = 0.8$	$a = 0.9$	$a = 1.0$	$a = 1.2$
0	1.0000000	1.0000000	1.0000000	1.0000000	1.0000000
1	.503415	.550671	.593430	.632121	.698806
2	.155805	.191208	.227518	.264241	.337373
3	.034142	.047423	.062857	.080301	.120513
4	.005753	.009080	.013459	.018988	.033769
5	.000786	.001411	.002344	.003660	.007746
6	.000090	.000184	.000343	.000594	.001500
7	.000009	.000021	.000043	.000083	.000251
8	.000001	.000002	.000005	.000010	.000037
9				.000001	.000005
10					.000001

x	$a = 1.4$	$a = 1.6$	$a = 1.8$	$a = 1.9$	$a = 2.0$
0	1.000000	1.000000	1.000000	1.000000	1.000000
1	.753403	.798103	.834701	.850431	.864665
2	.408167	.475069	.537163	.566251	.593994
3	.166502	.216642	.269379	.296280	.323324
4	.053725	.078813	.108708	.125298	.142877
5	.014253	.023682	.036407	.044081	.052653
6	.003201	.006040	.010378	.013219	.016564
7	.000622	.001336	.002569	.003446	.004534
8	.000107	.00260	.000562	.000793	.001097
9	.000016	.000045	.000110	.000163	.000237
10	.000002	.000007	.000019	.000030	.000046
11		.000001	.000003	.000005	.000008

* E. C. Molina, *Poisson's Exponential Binomial Limit*, D. Van Nostrand, Inc., 1947.

TABLE 3 (*Continued*)

$$1 - F(x - 1) = \sum_{r=x}^{r=\infty} \frac{e^{-a}a^r}{r!}$$

x	$a = 2.5$	$a = 3.0$	$a = 3.5$	$a = 4.0$	$a = 4.5$	$a = 5.0$
0	1.000000	1.000000	1.000000	1.000000	1.000000	1.000000
1	.917915	.950213	.969803	.981684	.988891	.993262
2	.712703	.800852	.864112	.908422	.938901	.959572
3	.456187	.576810	.679153	.761897	.826422	.875348
4	.242424	.352768	.463367	·.566530	.657704	.734974
5	.108822	.184737	.274555	.371163	.467896	.559507
6	.042021	.083918	.142386	.214870	.297070	.384039
7	.014187	.033509	.065288	.110674	.168949	.237817
8	.004247	.011905	.026739	.051134	.086586	.133372
9	.001140	.003803	.009874	.021363	.040257	.068094
10	.000277	.001102	.003315	.008132	.017093	.031828
11	.000062	.000292	.001019	.002840	.006669	.013695
12	.000013	.000071	.000289	.000915	.002404	.005453
13	.000002	.000016	.000076	.000274	.000805	.002019
14		.000003	.000019	.000076	.000252	.000698
15		.000001	.000004	.000020	.000074	.000226
16			.000001	.000005	.000020	.000069
17				.000001	.000005	.000020
18					.000001	.000005
19						.000001

TABLE 4.   Critical Values for Student's t-Distribution*

Pr{Student's $t \leq$ tabled value} $= \gamma$

f	0.75	0.90	0.95	0.975	0.99	0.995
1	1.0000	3.0777	6.3138	12.7062	31.8207	63.6574
2	0.8165	1.8856	2.9200	4.3027	6.9646	9.9248
3	0.7649	1.6377	2.3534	3.1824	4.5407	5.8409
4	0.7407	1.5332	2.1318	2.7764	3.7469	4.6041
5	0.7267	1.4759	2.0150	2.5706	3.3649	4.0322
6	0.7176	1.4398	1.9432	2.4469	3.1427	3.7074
7	0.7111	1.4149	1.8946	2.3646	2.9980	3.4995
8	0.7064	1.3968	1.8595	2.3060	2.8965	3.3554
9	0.7027	1.3830	1.8331	2.2622	2.8214	3.2498
10	0.6998	1.3722	1.8125	2.2281	2.7638	3.1693
11	0.6974	1.3634	1.7959	2.2010	2.7181	3.1058
12	0.6955	1.3562	1.7823	2.1788	2.6810	3.0545
13	0.6938	1.3502	1.7709	2.1604	2.6503	3.0123
14	0.6924	1.3450	1.7613	2.1448	2.6245	2.9768
15	0.6912	1.3406	1.7531	2.1315	2.6025	2.9467
16	0.6901	1.3368	1.7459	2.1199	2.5835	2.9208
17	0.6892	1.3334	1.7396	2.1098	2.5669	2.8982
18	0.6884	1.3304	1.7341	2.1009	2.5524	2.8784
19	0.6876	1.3277	1.7291	2.0930	2.5395	2.8609
20	0.6870	1.3253	1.7247	2.0860	2.5280	2.8453
21	0.6864	1.3232	1.7207	2.0796	2.5177	2.8314
22	0.6858	1.3212	1.7171	2.0739	2.5083	2.8188
23	0.6853	1.3195	1.7139	2.0687	2.4999	2.8073
24	0.6848	1.3178	1.7109	2.0639	2.4922	2.7969
25	0.6844	1.3163	1.7081	2.0595	2.4851	2.7874
26	0.6840	1.3150	1.7056	2.0555	2.4786	2.7787
27	0.6837	1.3137	1.7033	2.0518	2.4727	2.7707
28	0.6834	1.3125	1.7011	2.0484	2.4671	2.7633
29	0.6830	1.3114	1.6991	2.0452	2.4620	2.7564
30	0.6828	1.3104	1.6973	2.0423	2.4573	2.7500
31	0.6825	1.3095	1.6955	2.0395	2.4528	2.7440
32	0.6822	1.3086	1.6939	2.0369	2.4487	2.7385
33	0.6820	1.3077	1.6924	2.0345	2.4448	2.7333
34	0.6818	1.3070	1.6909	2.0322	2.4411	2.7284
35	0.6816	1.3062	1.6896	2.0301	2.4377	2.7238
36	0.6814	1.3055	1.6883	2.0281	2.4345	2.7195
37	0.6812	1.3049	1.6871	2.0262	2.4314	2.7154
38	0.6810	1.3042	1.6860	2.0244	2.4286	2.7116
39	0.6808	1.3036	1.6849	2.0227	2.4258	2.7079
40	0.6807	1.3031	1.6839	2.0211	2.4233	2.7045
41	0.6805	1.3025	1.6829	2.0195	2.4208	2.7012
42	0.6804	1.3020	1.6820	2.0181	2.4185	2.6981
43	0.6802	1.3016	1.6811	2.0167	2.4163	2.6951
44	0.6801	1.3011	1.6802	2.0154	2.4141	2.6923
45	0.6800	1.3006	1.6794	2.0141	2.4121	2.6896

*D. B. Owen, *Handbook of Statistical Tables*, Addison-Wesley Publishing Co., 1962. (Courtesy Atomic Energy Commission, Washington, D.C.)

TABLE 4 (*Continued*)

Pr{Student's $t \leq$ tabled value} $= \gamma$

f	0.75	0.90	0.95	0.975	0.99	0.995
46	0.6799	1.3002	1.6787	2.0129	2.4102	2.6870
47	0.6797	1.2998	1.6779	2.0117	2.4083	2.6846
48	0.6796	1.2994	1.6772	2.0106	2.4066	2.6822
49	0.6795	1.2991	1.6766	2.0096	2.4049	2.6800
50	0.6794	1.2987	1.6759	2.0086	2.4033	2.6778
51	0.6793	1.2984	1.6753	2.0076	2.4017	2.6757
52	0.6792	1.2980	1.6747	2.0066	2.4002	2.6737
53	0.6791	1.2977	1.6741	2.0057	2.3988	2.6718
54	0.6791	1.2974	1.6736	2.0049	2.3974	2.6700
55	0.6790	1.2971	1.6730	2.0040	2.3961	2.6682
56	0.6789	1.2969	1.6725	2.0032	2.3948	2.6665
57	0.6788	1.2966	1.6720	2.0025	2.3936	2.6649
58	0.6787	1.2963	1.6716	2.0017	2.3924	2.6633
59	0.6787	1.2961	1.6711	2.0010	2.3912	2.6618
60	0.6786	1.2958	1.6706	2.0003	2.3901	2.6603
61	0.6785	1.2956	1.6702	1.9996	2.3890	2.6589
62	0.6785	1.2954	1.6698	1.9990	2.3880	2.6575
63	0.6784	1.2951	1.6694	1.9983	2.3870	2.6561
64	0.6783	1.2949	1.6690	1.9977	2.3860	2.6549
65	0.6783	1.2947	1.6686	1.9971	2.3851	2.6536
66	0.6782	1.2945	1.6683	1.9966	2.3842	2.6524
67	0.6782	1.2943	1.6679	1.9960	2.3833	2.6512
68	0.6781	1.2941	1.6676	1.9955	2.3824	2.6501
69	0.6781	1.2939	1.6672	1.9949	2.3816	2.6490
70	0.6780	1.2938	1.6669	1.9944	2.3808	2.6479
71	0.6780	1.2936	1.6666	1.9939	2.3800	2.6469
72	0.6779	1.2934	1.6663	1.9935	2.3793	2.6459
73	0.6779	1.2933	1.6660	1.9930	2.3785	2.6449
74	0.6778	1.2931	1.6657	1.9925	2.3778	2.6439
75	0.6778	1.2929	1.6654	1.9921	2.3771	2.6430
76	0.6777	1.2928	1.6652	1.9917	2.3764	2.6421
77	0.6777	1.2926	1.6649	1.9913	2.3758	2.6412
78	0.6776	1.2925	1.6646	1.9908	2.3751	2.6403
79	0.6776	1.2924	1.6644	1.9905	2.3745	2.6395
80	0.6776	1.2922	1.6641	1.9901	2.3739	2.6387
81	0.6775	1.2921	1.6639	1.9897	2.3733	2.6379
82	0.6775	1.2920	1.6636	1.9893	2.3727	2.6371
83	0.6775	1.2918	1.6634	1.9890	2.3721	2.6364
84	0.6774	1.2917	1.6632	1.9886	2.3716	2.6356
85	0.6774	1.2916	1.6630	1.9883	2.3710	2.6349
86	0.6774	1.2915	1.6628	1.9879	2.3705	2.6342
87	0.6773	1.2914	1.6626	1.9876	2.3700	2.6335
88	0.6773	1.2912	1.6624	1.9873	2.3695	2.6329
89	0.6773	1.2911	1.6622	1.9870	2.3690	2.6322
90	0.6772	1.2910	1.6620	1.9867	2.3685	2.6316

TABLE 5. Critical Values for the Chi-square Distribution*

Pr$\{\chi^2$ r.v. with f degrees of freedom $\leq$ tabled value$\} = \gamma$

f	0.005	0.01	0.025	0.05	0.10	0.25
1	-	-	0.001	0.004	0.016	0.102
2	0.010	0.020	0.051	0.103	0.211	0.575
3	0.072	0.115	0.216	0.352	0.584	1.213
4	0.207	0.297	0.484	0.711	1.064	1.923
5	0.412	0.554	0.831	1.145	1.610	2.675
6	0.676	0.872	1.237	1.635	2.204	3.455
7	0.989	1.239	1.690	2.167	2.833	4.255
8	1.344	1.646	2.180	2.733	3.490	5.071
9	1.735	2.088	2.700	3.325	4.168	5.899
10	2.156	2.558	3.247	3.940	4.865	6.737
11	2.603	3.053	3.816	4.575	5.578	7.584
12	3.074	3.571	4.404	5.226	6.304	8.438
13	3.565	4.107	5.009	5.892	7.042	9.299
14	4.075	4.660	5.629	6.571	7.790	10.165
15	4.601	5.229	6.262	7.261	8.547	11.037
16	5.142	5.812	6.908	7.962	9.312	11.912
17	5.697	6.408	7.564	8.672	10.085	12.792
18	6.265	7.015	8.231	9.390	10.865	13.675
19	6.844	7.633	8.907	10.117	11.651	14.562
20	7.434	8.260	9.591	10.851	12.443	15.452
21	8.034	8.897	10.283	11.591	13.240	16.344
22	8.643	9.542	10.982	12.338	14.042	17.240
23	9.260	10.196	11.689	13.091	14.848	18.137
24	9.886	10.856	12.401	13.848	15.659	19.037
25	10.520	11.524	13.120	14.611	16.473	19.939
26	11.160	12.198	13.844	15.379	17.292	20.843
27	11.808	12.879	14.573	16.151	18.114	21.749
28	12.461	13.565	15.308	16.928	18.939	22.657
29	13.121	14.257	16.047	17.708	19.768	23.567
30	13.787	14.954	16.791	18.493	20.599	24.478
31	14.458	15.655	17.539	19.281	21.434	25.390
32	15.134	16.362	18.291	20.072	22.271	26.304
33	15.815	17.074	19.047	20.867	23.110	27.219
34	16.501	17.789	19.806	21.664	23.952	28.136
35	17.192	18.509	20.569	22.465	24.797	29.054
36	17.887	19.233	21.336	23.269	25.643	29.973
37	18.586	19.960	22.106	24.075	26.492	30.893
38	19.289	20.691	22.878	24.884	27.343	31.815
39	19.996	21.426	23.654	25.695	28.196	32.737
40	20.707	22.164	24.433	26.509	29.051	33.660
41	21.421	22.906	25.215	27.326	29.907	34.585
42	22.138	23.650	25.999	28.144	30.765	35.510
43	22.859	24.398	26.785	28.965	31.625	36.436
44	23.584	25.148	27.575	29.787	32.487	37.363
45	24.311	25.901	28.366	30.612	33.350	38.291

*D. B. Owen, *Handbook of Statistical Tables*, Addison-Wesley Publishing Co., 1962. (Courtesy Atomic Energy Commission, Washington, D.C.)

TABLE 5 (*Continued*)

Pr $\{\chi^2$ r.v. with f degrees of freedom $\leq$ tabled value$\} = \gamma$

f	0.75	0.90	0.95	0.975	0.99	0.995
1	1.323	2.706	3.841	5.024	6.635	7.879
2	2.773	4.605	5.991	7.378	9.210	10.597
3	4.108	6.251	7.815	9.348	11.345	12.838
4	5.385	7.779	9.488	11.143	13.277	14.860
5	6.626	9.236	11.071	12.833	15.086	16.750
6	7.841	10.645	12.592	14.449	16.812	18.548
7	9.037	12.017	14.067	16.013	18.475	20.278
8	10.219	13.362	15.507	17.535	20.090	21.955
9	11.389	14.684	16.919	19.023	21.666	23.589
10	12.549	15.987	18.307	20.483	23.209	25.188
11	13.701	17.275	19.675	21.920	24.725	26.757
12	14.845	18.549	21.026	23.337	26.217	28.299
13	15.984	19.812	22.362	24.736	27.688	29.819
14	17.117	21.064	23.685	26.119	29.141	31.319
15	18.245	22.307	24.996	27.488	30.578	32.801
16	19.369	23.542	26.296	28.845	32.000	34.267
17	20.489	24.769	27.587	30.191	33.409	35.718
18	21.605	25.989	28.869	31.526	34.805	37.156
19	22.718	27.204	30.144	32.852	36.191	38.582
20	23.828	28.412	31.410	34.170	37.566	39.997
21	24.935	29.615	32.671	35.479	38.932	41.401
22	26.039	30.813	33.924	36.781	40.289	42.796
23	27.141	32.007	35.172	38.076	41.638	44.181
24	28.241	33.196	36.415	39.364	42.980	45.559
25	29.339	34.382	37.652	40.646	44.314	46.928
26	30.435	35.563	38.885	41.923	45.642	48.290
27	31.528	36.741	40.113	43.194	46.963	49.645
28	32.620	37.916	41.337	44.461	48.278	50.993
29	33.711	39.087	42.557	45.722	49.588	52.336
30	34.800	40.256	43.773	46.979	50.892	53.672
31	35.887	41.422	44.985	48.232	52.191	55.003
32	36.973	42.585	46.194	49.480	53.486	56.328
33	38.058	43.745	47.400	50.725	54.776	57.648
34	39.141	44.903	48.602	51.966	56.061	58.964
35	40.223	46.059	49.802	53.203	57.342	60.275
36	41.304	47.212	50.998	54.437	58.619	61.581
37	42.383	48.363	52.192	55.668	59.892	62.883
38	43.462	49.513	53.384	56.896	61.162	64.181
39	44.539	50.660	54.572	58.120	62.428	65.476
40	45.616	51.805	55.758	59.342	63.691	66.766
41	46.692	52.949	56.942	60.561	64.950	68.053
42	47.766	54.090	58.124	61.777	66.206	69.336
43	48.840	55.230	59.304	62.990	67.459	70.616
44	49.913	56.369	60.481	64.201	68.710	71.893
45	50.985	57.505	61.656	65.410	69.957	73.166

TABLE 6.   Random Digits*

07018	31172	12572	23968	55216	85366	56223	09300	94564	18172
52444	65625	97918	46794	62370	59344	20149	17596	51669	47429
72161	57299	87521	44351	99981	55008	93371	60620	66662	27036
17918	75071	91057	46829	47992	26797	64423	42379	91676	75127
13623	76165	43195	50205	75736	77473	07268	31330	07337	55901
27426	97534	89707	97453	90836	78967	00704	85734	21776	85764
96039	21338	88169	69530	53300	29895	71507	28517	77761	17244
68282	98888	25545	69406	29470	46476	54562	79373	72993	98998
54262	21477	33097	48125	92982	98382	11265	25366	06636	25349
66290	27544	72780	91384	47296	54892	59168	83951	91075	04724
53348	39044	04072	62210	01209	43999	54952	68699	31912	09317
34482	42758	40128	48436	30254	50029	19016	56837	05206	33851
99268	98715	07545	27317	52459	75366	43688	27460	65145	65429
95342	97178	10401	31615	95784	77026	33087	65961	10056	72834
38556	60373	77935	64608	28949	94764	45312	71171	15400	72182
39159	04795	51163	84475	60722	35268	05044	56420	39214	89822
41786	18169	96649	92406	42773	23672	37333	85734	99886	81200
95627	30768	30607	89023	60730	31519	53462	90489	81693	17849
98738	15548	42263	79489	85118	97073	01574	57310	59375	54417
75214	61575	27805	21930	94726	39454	19616	72239	93791	22610
73904	89123	19271	15792	72675	62175	48746	56084	54029	22296
33329	08896	94662	05781	59187	53284	28024	45421	37956	14252
66364	94799	62211	37539	80172	43269	91133	05562	82385	91760
68349	16984	86532	96186	53893	48268	82821	19526	63257	14288
19193	99621	66899	12351	72438	99839	24228	32079	53517	18558
49017	23489	19172	80439	76263	98918	59330	20121	89779	58862
76941	77008	27646	82072	28048	41589	70883	72035	81800	50296
55430	25875	26446	25738	32962	24266	26814	01194	48587	93319
33023	26895	65304	34978	43053	28951	22676	05303	39725	60054
87337	74487	83196	61939	05045	20405	69324	80823	20905	68727
81773	36773	21247	54735	68996	16937	18134	51873	10973	77090
74279	85087	94186	67793	18178	82224	17069	87880	54945	73489
34968	76028	54285	90845	35464	68076	15868	70063	26794	81386
99696	78454	21700	12301	88832	96796	59341	16136	01803	17537
55282	61051	97260	89829	69121	86547	62195	72492	33536	60137
31337	83886	72886	42598	05464	88071	92209	50728	67442	47529
94128	97990	58609	20002	76530	81981	30999	50147	93941	80754
06511	48241	49521	64568	69459	95079	42588	98590	12829	64366
69981	03469	56128	80405	97485	88251	76708	09558	86759	15065
23701	56612	86307	02364	88677	17192	23082	00728	78660	74196
09237	24607	12817	98120	30937	70666	76059	44446	94188	14060
11007	45461	24725	02877	74667	18427	45658	40044	59484	59966
60622	78444	39582	91930	97948	13221	99234	99629	22430	49247
79973	43668	19599	30021	68572	31816	63033	14597	28953	21162
71080	71367	23485	82364	30321	42982	74427	25625	74309	15855
09923	26729	74573	16583	37689	06703	21846	78329	98578	25447
63094	72826	65558	22616	33472	67515	75585	90005	19747	08865
19806	42212	41268	84923	21002	30588	40676	94961	31154	83133
17295	74244	43088	27056	86338	47331	09737	83735	84058	12382
59338	27190	99302	84020	15425	14748	42380	99376	30496	84523

* The Rand Corporation, *A Million Random Digits with 100,000 Normal Deviates*, The Free Press, 1955.

TABLE 6 (*Continued*)

96124	73355	01925	17210	81719	74603	30305	29383	69753	61156
31283	54371	20985	00299	71681	22496	71241	35347	37285	02028
49988	48558	20397	60384	24574	14852	26414	10767	60334	36911
82790	45529	48792	31384	55649	08779	94194	62843	11182	49766
51473	13821	75776	24401	00445	61570	80687	39454	07628	94806
07785	02854	91971	63537	84671	03517	28914	48762	76952	96837
16624	68335	46052	07442	41667	62897	40326	75187	36639	21396
28718	92405	07123	22008	83082	28526	49117	96627	38470	78905
33373	90330	67545	74667	20398	58239	22772	34500	34392	92989
36535	48606	11139	82646	18600	53898	70267	74970	35100	01291
47408	62155	47467	14813	56684	56681	31779	30441	19883	17044
56129	36513	11202	82142	13717	49900	35367	43255	06993	17418
35459	10460	33925	75946	26708	63004	89286	24880	38838	76022
61955	55992	36520	08005	48783	08773	45424	44359	25248	75881
85374	69791	18857	92948	90933	90290	97232	61348	22204	43440
15556	39555	09325	16717	74724	79343	26313	39585	56285	22525
75454	90681	73339	08810	89716	99234	36613	43440	60269	90899
27582	90856	04254	23715	00086	12164	16943	62099	32132	93031
89658	47708	01691	22284	50446	05451	68947	34932	81628	22716
57194	77203	26072	92538	85097	58178	46391	58980	12207	94901
64219	53416	03811	11439	80876	38314	77078	85171	06316	29523
53166	78592	80640	58248	68818	78915	57288	85310	43287	89223
58112	88451	22892	29765	20908	49267	18968	39165	03332	94932
14548	36314	05831	01921	97159	55540	00867	84294	54653	81281
21251	15618	40764	99303	38995	97879	98178	03701	70069	80463
30953	63369	05445	20240	35362	82072	29280	72468	94845	97004
12764	79194	36992	74905	85867	18672	28716	17995	63510	67901
72393	71563	42596	87316	80039	75647	66121	17083	07327	39209
11031	40757	10904	22385	39813	63111	33237	95008	09057	50820
91948	69586	45045	67557	86629	67943	23405	86552	17393	24221
18537	07384	13059	47389	97265	11379	24426	09528	36035	02501
66885	11985	38553	97029	88433	78988	88864	03876	48791	72613
96177	71237	08744	38483	16602	94343	18593	84747	57469	08334
37321	96867	64979	89159	33269	06367	09234	77201	92195	89547
77905	69703	77702	90176	04883	84487	88688	09360	42803	88379
53814	14560	43698	86631	87561	90731	59632	52672	24519	10966
16963	37320	40740	79330	04318	56078	23196	49668	80418	73842
87558	58885	65475	25295	59946	47877	81764	85986	61687	04373
84269	55068	10532	43324	39407	65004	35041	20714	20880	19385
94907	08019	05159	64613	26962	30688	51677	05111	51215	53285
45735	14319	78439	18033	72250	87674	67405	94163	16622	54994
11755	40589	83489	95820	70913	87328	04636	42466	68427	79135
51242	05075	80028	35144	70599	92270	62912	08859	87405	08266
00281	25893	94848	74342	45848	10404	28635	92136	42852	40812
12233	65661	10625	93343	21834	95563	15070	99901	09382	01498
88817	57827	02940	66788	76246	85094	44885	72542	31695	83843
75548	53699	90888	94921	04949	80725	72120	80838	38409	72270
42860	40656	33282	45677	05003	46597	67666	70858	41314	71100
71208	72822	17662	50330	32576	95030	87874	25965	05261	95727
44319	22313	89649	47415	21065	42846	78055	64776	64993	48051

TABLE 7.   Random Normal Deviates

	00	01	02	03	04	05	06	07	08	09
00	.31	-.51	-1.45	-.35	.18	.09	.00	.11	-1.91	-1.07
01	.90	-.36	.33	-.28	.30	-2.62	-1.43	-1.79	-.99	-.35
02	.22	.58	.87	-.02	.04	.12	-.17	.78	-1.31	.95
03	-1.00	.53	-1.90	-.77	.67	.56	-.94	.16	2.22	-.08
04	-.12	-.43	.69	.75	-.32	-.71	-1.13	-.79	-.26	-.86
05	.01	.37	-.36	.68	.44	.43	1.18	-.68	-.13	-.41
06	.16	-.83	-1.88	.89	-.39	.93	-.76	-.12	.66	2.06
07	1.31	-.82	-.36	.36	.24	-.95	.41	-.77	.78	-.27
08	-.38	-.26	-1.73	.06	-.14	1.59	.96	-1.39	.51	-.05
09	.38	.42	-1.39	-.22	-.28	-.03	2.48	1.11	-1.10	.40
10	1.07	2.26	-1.68	-.04	.19	1.38	-1.53	-1.41	.09	-1.91
11	-1.65	-1.29	-1.03	.06	2.18	-.55	-.34	-1.07	.80	1.77
12	1.02	-.67	-1.11	.08	-1.92	-.97	-.70	-.04	-.72	-.47
13	.06	1.43	-.46	-.62	-.11	.36	.64	-.27	.72	.68
14	.47	-1.84	.69	-1.07	.83	-.25	-.91	-1.94	.96	.75
15	.10	1.00	-.54	.61	-1.04	-.33	.94	.56	.62	.07
16	-.71	.04	.63	-.26	-1.35	-1.20	1.52	.63	-1.29	1.16
17	-.94	-.94	.56	-.09	.63	-.36	.20	-.60	-.29	.94
18	.29	.62	-1.09	1.84	-.11	.19	-.45	.23	-.63	-.06
19	.57	.54	-.21	.09	-.57	-.10	-1.25	-.26	.88	-.26
20	.24	.19	-.67	3.04	1.26	-1.21	.52	-.05	.76	-.09
21	-1.47	1.20	.70	-1.80	-1.07	.29	1.18	.34	-.74	1.75
22	-.01	.49	1.16	.17	-.48	.81	1.40	-.17	.57	.64
23	-.63	-.26	.55	-.21	-.07	-.37	.47	-1.69	.05	-.96
24	.85	-.65	-.94	.12	-1.67	.28	-.42	.14	-1.15	-.41
25	1.07	-.36	1.10	.83	.37	-.20	-.75	-.50	.18	1.31
26	1.18	-2.09	-.61	.44	.40	.42	-.61	-2.55	-.09	-1.33
27	.47	.88	.71	.31	.41	-1.96	.34	-.17	1.73	-.33
28	.26	.90	.11	.28	.76	-.12	-1.01	1.29	-.71	2.15
29	.39	-.88	-.15	-.38	.55	-.41	-.02	-.74	-.48	.46
30	-1.01	-.89	-1.23	.07	-.07	-.08	-.08	-1.95	-.34	-.29
31	1.36	.18	.85	.55	.00	-.43	.27	-.39	.25	.69
32	1.02	-2.49	1.79	.04	-.03	.85	-.29	-.77	.28	-.33
33	-.53	-1.13	.75	-.39	.43	.10	-2.17	.37	-1.85	.96
34	.76	1.21	-.68	.26	.93	.99	1.12	-1.72	-.04	-.73
35	.07	-.23	-.88	-.23	.68	.24	1.38	-2.10	-.79	-.27
36	.27	.61	.43	-.38	.68	-.72	.90	-.14	-1.61	-.88
37	.93	.72	-.45	2.80	-.12	.74	-1.47	.39	-.61	-2.77
38	1.03	-.43	.95	-1.49	-.63	.22	.79	-2.80	-.41	.61
39	-.32	1.41	-.23	-.36	.60	-.59	.36	.63	.73	.81
40	1.41	.64	.06	.25	-1.75	.39	1.84	1.23	-1.27	-.75
41	.25	-.70	.33	.12	.04	1.03	-.64	.08	1.63	.34
42	-1.15	.57	.34	-.32	2.31	.74	.85	-1.25	-.17	.14
43	.72	.01	.50	-1.42	.26	-.74	-.55	1.86	-.17	-.10
44	-.92	.15	-.66	.83	.50	.24	-.40	1.90	.35	.69
45	-.42	.62	.24	.55	-.06	.14	-1.09	-1.53	.30	-1.56
46	-.54	1.21	-.53	.29	1.04	-.32	-1.20	.01	.05	.20
47	-.13	-.70	.07	.69	.88	1.18	.61	-.46	-1.54	.50
48	-.29	.36	1.44	-.44	.53	-.14	.66	.00	.33	-.36
49	1.90	-1.21	-1.87	-.27	-1.86	-.49	.25	.25	.14	1.73

TABLE 7   (*Continued*)

	10	11	12	13	14	15	16	17	18	19
00	-.73	.25	-2.08	.17	-1.04	-.23	.74	.23	.70	-.79
01	-.87	-.74	1.44	-.79	-.76	-.42	1.93	.88	.80	-.53
02	1.18	.05	.10	-.15	.05	1.06	.82	.90	-1.38	.51
03	-2.09	1.13	-.50	.37	-.18	-.16	-1.85	-.90	1.32	-.83
04	-.32	1.06	1.14	-.23	.49	1.10	-.27	-.64	.47	-.05
05	.90	-.86	.63	-1.62	-.52	-1.55	.78	-.54	-.29	.19
06	-.16	-.22	-.17	-.81	.49	.96	.53	1.73	.14	1.21
07	.15	-1.12	.80	-.30	-.77	-.91	.00	.94	-1.16	.44
08	-1.87	.72	-1.17	-.36	-1.42	-.46	-.58	.03	2.08	1.11
09	.87	.95	.05	.46	-.01	.85	1.19	-1.61	-.10	-.87
10	.52	.12	-1.04	-.56	-.91	-.13	.17	1.17	-1.24	-.84
11	-1.39	-1.18	1.67	2.88	-2.06	.10	.05	-.55	.74	.33
12	-.94	-.46	-.85	-.29	.54	.71	.90	-.42	-1.30	.50
13	-.51	.04	-.44	-1.87	-1.06	1.18	-.39	.22	-.55	-.54
14	-1.50	-.21	-.89	.43	-1.81	-.07	-.66	-.02	1.77	-1.54
15	-.48	1.54	1.88	.66	-.62	.28	-.34	2.42	-1.65	2.06
16	.89	-.23	.57	.23	1.81	1.02	.33	1.23	1.31	.06
17	.38	1.52	-1.32	2.13	-.14	.28	-.46	.25	.65	1.18
18	-.53	.37	.19	-2.41	.16	.36	-.15	.14	-.15	-.73
19	.15	.62	-1.29	1.84	.80	-.65	1.72	-1.77	.07	.46
20	-.81	-.22	1.16	1.09	-.73	-.15	.87	-.88	.92	-.04
21	-1.61	2.51	-2.17	.49	-1.24	1.16	.97	.15	.37	.18
22	.26	-.48	-.43	-2.08	.75	1.59	1.78	-.55	.85	-1.87
23	-.32	.75	-.35	2.10	-.70	1.29	.94	.20	-1.16	.89
24	-1.00	1.37	.68	.00	1.87	-.14	.77	-.12	.89	-.73
25	.66	.04	-1.73	.25	.26	1.46	-.77	-1.67	.18	-.92
26	-.20	-1.53	.59	-.15	-.15	-.11	.68	-.14	-.42	-1.51
27	1.01	-.44	-.20	-2.05	-.27	-.50	-.27	-.45	.83	.49
28	-1.81	.45	.27	.67	-.74	-.17	-1.11	.13	-1.18	-1.41
29	-.40	1.34	1.50	.57	-1.78	.08	.95	.69	.38	.71
30	-.01	.15	-1.83	1.18	.11	.62	1.86	.42	.03	-.14
31	-.23	-.19	-1.08	.44	-.41	-1.32	.14	.65	-.76	.76
32	-1.27	.13	-.17	-.74	-.44	1.67	-.07	-.99	.51	.76
33	-1.72	1.70	-.61	.18	.48	-.26	-.12	-2.83	2.35	1.25
34	.78	1.55	-.19	.43	-1.53	-.76	.83	-.46	.48	-.43
35	1.86	1.12	-2.09	1.82	-.71	-1.76	-.20	-.38	.82	-1.08
36	-.50	-.93	-.68	-1.62	-.88	.05	-.27	.23	-.58	-.24
37	1.02	-.81	-.62	1.46	-.31	-.37	.08	.59	-.27	.37
38	-1.57	.10	.11	-1.48	1.02	2.35	.27	-1.22	-1.26	2.22
39	2.27	-.61	.61	-.28	-.39	-.45	-.89	1.43	-1.03	-.01
40	-2.17	-.69	1.33	-.26	.15	-.10	-.78	.64	-.70	.14
41	.05	-1.71	.21	.55	-.60	-.74	-.90	2.52	-.07	-1.11
42	-.38	1.75	.93	-1.36	-.60	-1.76	-1.10	.42	1.44	-.58
43	.40	-1.50	.24	-.66	.83	.37	-.35	.16	.96	.79
44	.39	.66	.19	-2.08	.32	-.42	-.53	.92	.69	-.03
45	-.12	1.18	-.08	.30	-.21	.45	-1.84	.26	.90	.85
46	1.20	-.91	-1.08	-.99	1.76	-.80	.51	.25	-.11	-.58
47	-1.04	1.28	2.50	1.56	-.95	-1.02	.45	-1.90	-.02	-.73
48	-.32	.56	-1.03	.11	-.72	.53	-.27	-.17	1.40	1.61
49	1.08	.56	.34	-.28	-.37	.46	.03	-1.13	.34	-1.08

Answers to Selected Problems

CHAPTER 1

1.1. (a) $\{5\}$ (b) $\{1, 3, 4, 5, 6, 7, 8, 9, 10\}$ (c) $\{2, 3, 4, 5\}$
(d) $\{1, 5, 6, 7, 8, 9, 10\}$ (e) $\{1, 2, 5, 6, 7, 8, 9, 10\}$

1.2. (a) $\{x \mid 0 \le x < \frac{1}{4}\} \cup \{x \mid \frac{3}{2} \le x \le 2\}$
(b) $\{x \mid 0 \le x < \frac{1}{4}\} \cup \{x \mid \frac{1}{2} < x \le 1\} \cup \{x \mid \frac{3}{2} \le x \le 2\}$
(c) $\{x \mid 0 \le x \le \frac{1}{2}\} \cup \{x \mid 1 < x \le 2\}$
(d) $\{x \mid \frac{1}{4} \le x \le \frac{1}{2}\} \cup \{x \mid 1 < x < \frac{3}{2}\}$

1.3. (a) True   (b) True   (c) False   (d) False   (e) True

1.4. (a) $A = \{(0, 0), (1, 0), (2, 0), (0, 1), (1, 1), (2, 1), (0, 2), (1, 2)\}$
(b) $B = \{(0, 0), (1, 0), (2, 0), (3, 0), (4, 0), (5, 0), (6, 0), (1, 1), (2, 1), (3, 1),$
$(4,1), (5, 1), (6, 1), (2, 2), (3, 2), (4, 2), (5, 2), (6, 2), (2, 3), (3, 3), (4, 3), (5, 3),$
$(6, 3), (2, 4), (3, 4), (4, 4), (5, 4), (6, 4), (3, 5), (4, 5), (5, 5), (6, 5), (3, 6),$
$(4, 6), (5, 6), (6, 6)\}$

1.6. $\{DD, NDD, DNDD, DNDN, DNND, DNNN, NDND, NDNN, NNDD,$
$NNDN, NNND, NNNN\}$

1.10. (a) $\{(x, y) \mid 0 \le x < y \le 24\}$

1.11. (a) $A \cup B \cup C$ (b) $[A \cap \overline{B} \cap \overline{C}] \cup [\overline{A} \cap B \cap \overline{C}] \cup [\overline{A} \cap \overline{B} \cap C]$
(d) $\overline{A \cap B \cap C}$

1.15. $\frac{1}{13}, \frac{4}{13}, \frac{8}{13}$ **1.16.** (a) $1 - z$ (b) $y - z$ **1.17.** $\frac{5}{8}$

CHAPTER 2

2.1. (a) $\frac{13}{18}$ (b) $\frac{1}{6}$ **2.2.** (a) $\frac{1}{12}$ (b) $\frac{1}{20}$ **2.3.** (a) $\frac{2}{3}$ (b) $\frac{5}{8}$

2.4. (a) $\dfrac{\binom{400}{90}\binom{1100}{110}}{\binom{1500}{200}}$ (b) $1 - \left[\dfrac{\binom{400}{0}\binom{1100}{200} + \binom{400}{1}\binom{1100}{199}}{\binom{1500}{200}}\right]$

2.5. $\frac{4}{45}$ **2.6.** (a) $\frac{5}{8}$ (b) $\frac{7}{8}$ (c) $\frac{3}{4}$

2.7. (a) $\frac{3}{8}$ (b) $\frac{1}{120}$ (c) $\frac{7}{8}$ (d) $\frac{5}{8}$ (e) $\frac{1}{2}$ (f) $\frac{91}{120}$ (g) $\frac{1}{8}$

2.8. 120   **2.9.** 720

2.10. 455   **2.11.** (a) 120   (b) 2970

2.12. (a) 4^8 (b) $4 \cdot 3^7$ (c) 70   (d) 336

2.13. $(N-1)!/(N-n)!N^{n-1}$

2.14. (a) 360   (b) 1296   **2.15.** $a+b$

2.16. (a) $2/n$ (b) $2(n-1)/n^2$ **2.18.** 0.24

2.20. 120

2.21. $\dfrac{\dbinom{r}{r-1}\dbinom{n-r}{k-r}}{\dbinom{n}{k-1}} \cdot \dfrac{1}{n-k+1}$

2.22. $\dfrac{10!}{10^r(10-r)!}$

CHAPTER 3

3.1. $\left(\dfrac{x}{x+y}\right)\left(\dfrac{z+1}{z+v+1}\right)+\left(\dfrac{y}{x+y}\right)\left(\dfrac{z}{z+v+1}\right)$

3.2. (a) $\frac{1}{6}$ (b) $\frac{1}{3}$ (c) $\frac{1}{2}$ **3.3.** $\frac{5}{9}$

3.4. (a) $\frac{2}{105}$ (b) $\frac{2}{5}$ **3.6.** (a) $\frac{33}{95}$ (b) $\frac{14}{95}$ (c) $\frac{48}{95}$

3.7. $\frac{2}{3}$ **3.9.** 0.362, 0.406, 0.232

3.12. (a) $\frac{1}{4}$ (b) $\frac{3}{4}$ **3.13.** $1-(1-p)^n$

3.15. $\frac{5}{16}$ **3.17.** (a) 0.995   (b) 0.145

3.20. (a) $2p^2+2p^3-5p^4+2p^5$ (b) $p+3p^2-4p^3-p^4+3p^5-p^6$

3.23. $\frac{3}{16}$ **3.25.** (a) 0.50   (b) 0.05

3.34. $\beta_n = \frac{1}{2}+(2p-1)^n(\beta-\frac{1}{2})$ **3.35.** (a) 0.65   (b) 0.22   (c) 8/35

3.37. $p_n = \dfrac{\alpha}{\alpha+\beta}+\dfrac{\beta}{(\alpha+\beta)(\alpha+\beta+1)^{n-1}}$

3.39. $(n-1)p^2/(1-2p+np^2)$

CHAPTER 4

4.1. $P(X=0)=\frac{1}{64}$, $P(X=1)=\frac{9}{64}$, $P(X=2)=\frac{27}{64}$, $P(X=3)=\frac{27}{64}$

4.2. (a) $\dbinom{4}{x}\left(\dfrac{1}{5}\right)^x\left(\dfrac{4}{5}\right)^{4-x}$ (b) $\dbinom{5}{x}\dbinom{20}{4-x}\Big/\dbinom{25}{4}$

4.3. (a) $\frac{1}{3}$ (b) $\frac{1}{16}$ (c) $\frac{1}{7}$ **4.9.** $\frac{27}{64}$

4.10. $a=e^{-b}$

4.11. $P(X>b\mid X<b/2)=-7b^3/(b^3+8)$

4.13. (a) $a=(2/b)-b$ **4.14.** (a) $F(t)=5t^4-4t^5$

4.15. (a) $a=\frac{1}{2}$ (c) $\frac{3}{8}$ **4.16.** (b) $F(x)=3x^2-2x^3$

4.17. (a) $f(x)=\frac{1}{5}, 0<x<5$ (b) $f(x)=(1/\pi)(x-x^2)^{-1/2}, 0<x<1$
     (c) $f(x)=3e^{3x}, x<0$

4.20. (a) $\alpha=3$ (c) $\alpha=\frac{5}{4}$

4.23. (b) $P(X = k) = \binom{10}{k}(0.09)^k(0.91)^{10-k}$

4.25. (a) $\frac{1}{4}$ **4.28.** $\frac{3}{5}$

4.29. (a) $k = \beta$ (b) $r = 1$ **4.30.** $-\frac{1}{3} \le x \le \frac{1}{4}$

CHAPTER 5

5.1. $g(y) = \frac{1}{2}(4 - y)^{-1/2}, 3 < y < 4$

5.2. (a) $g(y) = \frac{1}{6}, 7 < y < 13$ (b) $h(z) = 1/2z, e < z < e^3$

5.3. (a) $g(y) = \frac{1}{3}y^{-2/3}e^{-y^{1/3}}, y > 0$

5.6. (a) $g(y) = (1/\pi)(1 - y^2)^{-1/2}, -1 < y < 1$
     (b) $h(z) = (2/\pi)(1 - z^2)^{-1/2}, 0 < z < 1$ (c) $f(w) = 1, 0 < w < 1$

5.7. (a) $g(v) = (3/2\pi)[(3v/4\pi)^{-1/3} - 1], 0 < v < 4\pi/3$
     (b) $h(s) = (3/4\pi)[1 - (s/4\pi)^{1/2}], 0 < s < 4\pi$

5.8. $g(p) = \frac{1}{8}(2/p)^{1/2}, 162 < p < 242$

5.10. (a) $g(0) = 1; g(y) = 0, y \ne 0$
     (b) $g(0) = a/k; g(y) = (x_0 - a)/ky_0, 0 < y < y_0[(k - a)/(x_0 - a)]$
     (c) $g(0) = a/k; g(y) = (x_0 - a)/ky_0, 0 < y < y_0; g(y_0) = 1 - x_0/k$

5.13. 0.71

CHAPTER 6

6.2. (a) $k = \frac{1}{8}$ (b) $h(x) = x^3/4, 0 < x < 2$
     (c) $g(y) = \begin{cases} \frac{1}{3} - y/4 + y^3/48, 0 \le y \le 2 \\ \frac{1}{3} - y/4 + (5/48)y^3, -2 \le y \le 0 \end{cases}$

6.3. (a) $\frac{5}{6}$ (b) $\frac{7}{24}$ (c) $\frac{5}{32}$ **6.5.** $k = 1/(1 - e^{-1})^2$

6.6. (a) $k = \frac{1}{2}$ (b) $h(x) = 1 - |x|, -1 < x < 1$
     (c) $g(y) = 1 - |y|, -1 < y < 1$

6.8. $h(z) = 1/2z^2, z \ge 1$
           $= 1/2, 0 < z < 1$

6.11. $g(h) = (1600 - 9h^2)/80h^2, 8 < h < \frac{40}{3}$
           $= \frac{1}{5}, \frac{20}{3} < h \le 8$
           $= (5h^2 - 80)/16h^2, 4 \le h \le \frac{20}{3}$

6.12. $h(i) = e^{-(2/i)^{1/2}}[-(2/i) - 2(2/i)^{1/2} - 2]$
                        $+ e^{-(1/i)^{1/2}}[(1/i) + 2(1/i)^{1/2} + 2], i > 0$

6.13. $h(w) = .6 + 6w - 12w^{1/2}, 0 < w < 1$

6.14. (a) $g(x) = e^{-x}, x > 0$ (b) $h(y) = ye^{-y}, y > 0$

CHAPTER 7

7.3. 3.4   **7.4.** $\frac{1}{3}(2C_3 + C_2 - 3C_1)$

7.6. \$0.03   **7.8.** $7\frac{2}{15}$

7.9. \$50   **7.10.** (a) $C = \frac{1}{6}$ (b) $E(D) = \frac{19}{9}$

7.12. (b) $E(Z) = \frac{8}{3}$ **7.13.** (b) $E(W) = \frac{4}{3}$

7.14. 154

7.15. $E(Y) = 10$, $V(Y) = 3$, $E(Z) = (e/2)(e^2 - 1)$, $V(Z) = (e^2/2)(e^2 - 1)$

7.18. $E(Y) = 0$, $V(Y) = \frac{1}{2}$, $E(Z) = 2/\pi$, $V(Z) = (\pi^2 - 8)/2\pi^2$, $E(W) = \frac{1}{2}$, $V(W) = \frac{1}{12}$

7.20. Case 2: $E(Y) = (y_0/2k)(x_0 - a)$, $V(Y) = \dfrac{(x_0 - a)y_0^2}{k}\left[\dfrac{1}{3} - \dfrac{x_0 - a}{4k}\right]$

7.24. $a = \frac{1}{5}$, $b = 2$ **7.25.** (a) $E(V) = \frac{3}{2}$ (b) $E(P) = \frac{4}{3}$

7.26. $V(X) = \frac{4}{3}a^2$ **7.27.** $E(S) = \frac{65}{6}$

7.30. (a) $g(x) = x/2$, $0 < x < 2$; $h(y) = 1/2 - y/8$, $0 < y < 4$
(b) $V(X) = \frac{2}{9}$

7.31. $E(Z) \simeq \mu_x/\mu_y + 2(\mu_x/\mu_y^3)\sigma_y^2$; $V(Z) \simeq (1/\mu_y^2)\sigma_x^2 + (\mu_x^2/\mu_y^4)\sigma_y^2$

7.48. $P(X_i = 1) = p(r/n) + q[(n - r)/n]$

7.32. $E(Z) \simeq \frac{29}{27}$; $V(Z) \simeq \frac{3}{27}$ **7.35.** $\frac{1}{4}$; 0

7.46. $\frac{48}{65}$ **7.48.** $P(X_i = 1) = p(r/n) + q[(n - r)/n]$

CHAPTER 8

8.1. 0.219

8.3. (a) 0.145   (b) 4   (c) 2   (d) 1 or 2   (e) 1.785   (f) 0.215

8.4. 0.3758 (binomial), 0.4060 (Poisson)

8.5. 0.067   **8.6.** $P(X = 0) = 0.264$

8.7. $E(P) = \$32.64$ **8.9.** (b) 0.027

8.10. 0.215   **8.12.** (a) $(0.735)^7$ (b) $1 - (0.265)^7$

8.16. (a) $(1 - p_1)(1 - p_2)(1 - p_3)(1 - p_4)$ (c) 0.0964

8.17. (a) 0.064   **8.20.** $(2 - \ln 3)/3$ **8.24.** \$19,125

CHAPTER 9

9.1. (a) 0.2266   (b) 0.2902   **9.2.** 0.3085

9.3. 0.21   **9.5.** (a) D_2 (b) D_2

9.6. $E(Y) = (2/\pi)^{1/2}$, $V(Y) = (\pi - 2)/\pi$

9.10. $C = 0.433\sigma + \mu$ **9.11.** 0.5090

9.12. $E(L) = \$0.528$

9.13. (a) $\frac{1}{4}$; 0.0456   (b) $\frac{1}{4}$; 0.069   (c) $\frac{1}{4}$; 0.049

9.15. \$23.40   **9.17.** (a) \$0.077

9.24. 0.10, 0.80

9.25. $E(Y) \simeq \ln \mu - (1/2\mu^2)\sigma^2$; $V(Y) \simeq (1/\mu^2)\sigma^2$

9.28. (b) $E(X) = np[(1 - p^{n-1})/(1 - p^n)]$

9.32. 0.15

CHAPTER 10

10.1. (a) $M_X(t) = (2/t^2)[e^t(t-1)+1]$ (b) $E(X) = \frac{2}{3}, V(X) = \frac{1}{18}$

10.3. (a) $M_X(t) = \lambda e^{ta}/(\lambda - t)$ (b) $E(X) = (a\lambda + 1)/\lambda, V(X) = 1/\lambda^2$

10.4. (a) $M_X(t) = \frac{1}{6}(e^t + e^{2t} + e^{3t} + e^{4t} + e^{5t} + e^{6t})$

10.6. (a) $(1 - t^2)^{-1}$ **10.8.** (b) $E(X) = 3.2$

10.9. (a) 0.8686   **10.12.** 0.30

10.13. 0.75   **10.14.** 0.579

10.18. $\frac{1}{3}$ **10.19.** $(e^{3t} - 1)/3te^t$

CHAPTER 11

11.1. 83.55 hours, 81.77 hours, 78.35 hours

11.2. 48.6 hours

11.3. $f(t) = C_0 \exp[-C_0 t], 0 \le t \le t_0$
$= C_1 \exp[-C_0 t_0 + C_1(t_0 - t_1)], t > t_0$

11.4. (a) $f(t) = Ce^{-C(t-A)}, t \ge A$ **11.5.** (b) $R(t) = [A/(A+t)]^{r+1}$

11.7. (a) Approximately $(0.5)^6$ **11.9.** 0.007

11.10. $R(t) = 2e^{-0.06t} - e^{-0.09t}$ **11.11.** (a) 0.014

11.12. (a) $m = \ln(\sqrt{2})$ (b) $m = 0.01$ (c) $+\infty$

11.13. $R(t) = \exp(-C_0 t), 0 < t < t_0$
$= \exp[t(C_1 t_0 - C_0) - (C_1/2)(t^2 + t_0^2)], t > t_0$

11.14. (a) $R(t) = e^{-\beta_1 t} + e^{-\beta_2 t} + e^{-\beta_3 t} - e^{-(\beta_1+\beta_2)t} - e^{-(\beta_1+\beta_3)t}$
$- e^{-(\beta_2+\beta_3)t} + e^{-(\beta_1+\beta_2+\beta_3)t}$

11.15. (a) $R_S = 1 - (1 - R^n)^k$ **11.16.** (a) $R_S = [1 - (1 - R)^k]^n$

11.18. (a) 0.999926   (b) 0.99   (c) 0.68

11.19. (a) $R_S = [1 - (1 - R_A)(1 - R_B)(1 - R_C)][1 - (1 - R_{A'})(1 - R_{B'})]$
(b) $R_S = 1 - R_C(1 - R_{A'})(1 - R_{B'})$
$- (1 - R_C)(1 - R_A R_{A'})(1 - R_B R_{B'})$

11.22. $M_X(t) = -2\lambda[1/(t - \lambda) - 1/(t - 2\lambda)]$

CHAPTER 12

12.1. (a) $n = 392$ (b) $n = 5000$ **12.2.** (a) 0.083

12.3. (a) 0.9662   (b) $n = 24$ **12.4.** 0.1814

12.5. (a) 0.1802   (b) $n = 374$ **12.6.** (a) 0.1112   (b) 0.1915

12.7. $g(r) = (15,000)^{-1}(r^3 - 60r^2 + 600r), 0 \le r \le 10$
$= (15,000)^{-1}(-r^3 + 60r^2 - 1200r + 8000), 10 \le r \le 20$

12.9. (a) $g(s) = 50[1 - e^{-0.2(s+0.01)}]$, if $-0.01 \le s \le 0.01$
$= 50[e^{-0.2(s-0.01)} - e^{-0.2(s+0.01)}]$, if $s > 0.01$

12.12. $f(s) = \frac{1}{2}\left[\Phi\left(\frac{s - 99}{2}\right) - \Phi\left(\frac{s - 101}{2}\right)\right]$

12.13. (a) $\dfrac{p_1 p_2 (1 - p_1)^{k-1}}{p_2 - p_1}\left[1 - \left(\dfrac{1 - p_2}{1 - p_1}\right)^{k-1}\right]$ (b) 0.055

CHAPTER 13

13.3. $P(M = m) = [1 - (1 - p)^m]^n - [1 - (1 - p)^{m-1}]^n$

13.4. (a) 0.13   (b) 0.58   **13.5.** (a) 0.018   (b) 0.77

13.6. 0.89   **13.8.** (a) $(1 - 2t)^{-1}$

CHAPTER 14

14.1. $\dfrac{V(L_2)}{V(L_1) + V(L_2)}$ **14.2.** $C = 1/2(n - 1)$

14.3. 16.1   **14.4.** $-1 - n/\sum_{i=1}^{n} \ln X_i$

14.6. (a) $1/(\overline{T} - t_0)$ **14.7.** (a) $\dfrac{1}{T_0 - t_0} \ln\left(\dfrac{n}{n - k}\right)$

14.9. (a) $k/\sum_{i=1}^{k} n_i$ **14.13.** 0.034

14.14. -0.52 **14.15.** $r/\overline{X}$

14.16. $n/\sum_{i=1}^{n} X_i^\alpha$

14.29. F distribution with $(1, 1)$ degrees of freedom

14.31. $(-4/5)$ **14.32.** 8.6

CHAPTER 15

15.1. (a) $1 - \Phi\left(\dfrac{C - \mu}{\sigma} \sqrt{n}\right)$ (b) $C = -2.33 \dfrac{\sigma}{\sqrt{n}} + \mu_0$

15.2. (a) $\Phi\left[\dfrac{C + \mu_0 - \mu}{\sigma} \sqrt{n}\right] - \Phi\left[\dfrac{-C + \mu_0 - \mu}{\sigma} \sqrt{n}\right]$

(b) $C = 2.575\sigma n^{-1/2}$

15.3. (a) $\displaystyle\sum_{k=0}^{[nC]} \dfrac{e^{-n\lambda}(n\lambda)^k}{k!}$,   where $[nC]$ = greatest integer $\leq nC$.   (c) 0.3233

15.7. (a) $(1 - p)^4[5p + (1 + 25p^5)(1 - p) + 55p^4(1 - p)^2$

15.8. 15 $+ 60p^3(1 - p)^3 + 10p^2(1 - p)^3]$

15.9. (a) $(k - 1)$ (b) Same   **15.12.** (a) $\sin \alpha/\sin\left(\dfrac{\pi}{2}\alpha\right)$

Index